AF847852

Environmental Science and Engineering
Subseries: Environmental Science

Series Editors: R. Allan • U. Förstner • W. Salomons

Peter Knoepfel

in cooperation with: Christoph Bättig, Wyn Grant, Corinne Larrue, Anthony Perl, Stéphane Nahrath, Kathrin Peter, Franziska Teuscher, Frédéric Varone, Helmut Weidner, Willi Zimmermann and many others

Environmental Policy Analyses

Learning from the Past for the Future - 25 Years of Research

With 41 Figures

Springer

AUTHOR:

PROF. PETER KNOEPFEL
IDHEAP
RTE DE LA MALADIÈRE 21
CH-1022 CHAVANNES
SWITZERLAND
E-mail: Peter.Knoepfel@idheap.unill.ch

ISSN 1863-5520
ISBN 10 3-540-73148-2 **Springer Berlin Heidelberg New York**
ISBN 13 978-3-540-73148-1 **Springer Berlin Heidelberg New York**

Library of Congress Control Number: 2007929275

This work is subject to copyright. All rights are reserved, whether the whole or part of the material is concerned, specifically the rights of translation, reprinting, reuse of illustrations, recitation, broadcasting, reproduction on microfilm or in any other way, and storage in data banks. Duplication of this publication or parts thereof is permitted only under the provisions of the German Copyright Law of September 9, 1965, in its current version, and permission for use must always be obtained from Springer-Verlag. Violations are liable to prosecution under the German Copyright Law.

Springer is a part of Springer Science+Business Media
springeronline.com
© Springer-Verlag Berlin Heidelberg 2007

The use of general descriptive names, registered names, trademarks, etc. in this publication does not imply, even in the absence of a specific statement, that such names are exempt from the relevant protective laws and regulations and therefore free for general use.

Cover design: E. Kirchner, Heidelberg
Production: A. Oelschläger
Typesetting: Camera-ready by the Author

Printed on acid-free paper 30/2132/AO 543210

Preface

Having had the opportunity of receiving considerable public money (mainly from the Swiss National Science Foundation) for numerous environmental policy studies in Switzerland and abroad during the last 25 years, I have decided to put the results – published mainly in German and French but also in English in many scientific journals, monographs and book sections – in a single book on behalf of my students, younger scholars interested in policy analysis dating from the period of the first environmental policies, and my friends teaching this subject in Switzerland and abroad.

The criteria used to select the fifteen contributions were their impact on the scientific debate at the time of their publication, their validity for present conceptualizations of policy analysis in the field of the environment, and their contribution to the development of policy analysis as a whole[1]. Moreover, the selected contributions mainly focus on empirical applications in the field of environmental policies or the interface between environmental and non-environmental policies. Some have led to concrete policy recommendations of which most have been partly implemented since, at least within my own country. This research also sometimes addressed regional political administrative arrangements leading to administrative reforms and new strategic directions of implementing agencies.

Last but not least, the common denominator of the four sections is continuity and change. Although the book deals with the past, the selected works should contribute to present and future conceptualizations of transformed environmental policies to be analyzed within the larger context of new concepts in sustainable regulation of natural resource use. The concepts for such enlarged analysis, which are presently under way not only in our institute but also in many other academic bodies, will constitute the core of the next book on actual changes in policy analysis having taken

---

[1] Refer to the original manual written in French (Knoepfel, P., Larrue, C., Varone, F. 2001. *Analyse et pilotage des politiques publiques*. Bâle: Helbing & Lichtenhahn (série Analyse des politiques publiques / Politikanalyse no 2); 2nd edition: Knoepfel, P., Larrue, C., Varone, F. 2006. *Analyse et pilotage des politiques publiques*. Zurich/Coire: Rüegger Verlag (série Analyse des politiques publiques / Politikanalyse no 2 – 2ème édition). This book will also appear in a Spanish (co-author: Joan Subirats), a British (co-author Michael Hill), a German (co-authors Adrian Vatter, Christoph Knill), and a Mexican version (co-authors Miriam Hinojosa and Roberto Garza Leonard) in 2007 and 2008.

place in the first decades of this third millennium in the field of natural resources.

Scientific research is a collective work in which many people normally are involved. This book is a good example because not only are half of the contributions co-authored but all of them stem from fruitful exchanges between various research teams formed by the anonymous author "many others". These include former assistants working at my chair[2], colleagues of the international comparative research network of the SO_2-study[3], colleagues of European study groups[4], and again many others.

18th October 2006
Peter Knoepfel

---

[2] Such as Corinne Larrue (1983–1986), Rita Imhof (1990–1992), Martin Benninghoff (1993–1998), Ingrid-Kissling-Näf (1990–1995), Frédéric Varone (1997–1999), Jérôme Savary (since 2001-2007), Serge Terribilini (1995–1998), Sonja Wälti (1995–1997), Corine Mauch (2000–2002), Emmanuel Reynard (1999–2002), Kurt Bisang (1999-2002), Stéphane Nahrath (since 1999).

[3] Meinof Dierkes, Helmut Weidner, Kenneth Hanf, Volker Prittwitz, Helmut Schreiber, Robert Klammer (Germany), Michael Hill, Particia Garrard (Great Britain), Corinne Larrue, Richard Darbéra, Henrique Magalhas (France), Bruno Dente, Rodolfo Lewanski (Italy), Theo van der Tak, Theo Toonen (Netherlands).

[4] Environmental policy instruments (European Science Foundation – 1990–92): Bruno Dente, Angela Liberatore (Italy), Helmut Weidner (Germany), Corinne Larrue (France), Joan Subirats (Spain), Hans Bressers, Peter Jan Klok (Netherlands), James Cameron (Great Britain); members of the concerted action "Ecological state" – 4th framework programme of the European Union (1996–1998): Corinne Larrue (France), Joan Subirats (Spain), Hans Bressers (Netherlands), Lennart J. Lundqvist (Sweden) or the European project Euwareness (2000–02), 5th framework programme: Hans Bressers, Stefan Kuks (Netherlands), David Aubin, Frédéric Varone (Belgium), Corinne Larrue, Jean-Marc Dziedzizki (France), Meritxell Costejà, Nuria Font, Joan Subirats (Spain), Bruno Dente, Alessandria Goria, Nicola Lugaresi (Italy), Ingrid Kissling-Näf, Corine Mauch, Emmanuel Reynard, Adèle Thorens (Switzerland)

Table of contents

Section I:
Classics – still valid?

When comparing the first version of the manual contribution *Environmental Policy (1992)* with the second one (2002), I became aware of the most important changes that arose in the period covered, between 1982 and 2002. They mainly consisted of a globalization and a transformation of the scope of environmental policies. These policies increasingly played a crucial role in fighting serious survival problems of humanity that did not yet exist on the political agendas of individual countries or international organizations during the 1980s (e.g., climate change). Furthermore, the new interpretation of environmental policies to include comprehensive resource management policies[5] considerably enlarged their scope. Both of these changes, which evolved relatively rapidly after the Earth Summit of Rio de Janeiro (1992), compelled social scientists dealing with the environment to revise pre-existing policy concepts. Nevertheless, the contributions in this section should convince readers that these early concepts, which still heavily shape regional and national classical environmental policies, must be considered as a solid basis for such new approaches.

This is especially true for the analysis of the main actors intervening in both environmental policies and resource policies (although the latter focus more on actors using natural resources for non protective purposes), for environmental monitoring as well as for political administrative management and implementation structures. In *Approaches to an Effective Framework for Environmental Management (1993)* we recommend, amongst other implementation principles, the use of regional action plans that structure individual implementation activities which, considering natural resource management today, have turned out to be an indispensable intermediary phase in the policy cycle (observed not only in environmental policies but in most implementation processes of public policies under financial pressure). As a teacher of policy analysis not limited to environmental policies, this article gave me the opportunity to extrapolate analytical concepts from environmental to other public policies and thus consider the former as an excellent framework for practical policy analysis as a whole. This is particularly true for the "iron triangle" which turned out to be the general basic actors' structure of every public policy.

This holds even truer for *Formulation and Implementation of Air Quality Control Programmes: Patterns of Interest Consideration (1982)*, which Helmut Weidner and I wrote in the early 1980s. The purpose of the oldest article of this volume was initially quite pragmatic. Within our big international comparative study on the implementation of SO_2 air quality policies, we had to develop a common framework for a comparative reading of very

[5] Cf. e.g. chapter 15 of this book.

different legal systems in approximately a dozen European countries dealing with this first generation of air quality policies. Many countries were lacking either ambient air quality standards or emission standards for individual plants, and their political administrative programmes did not contain any regulations on how to measure the actual concentration of this pollutant in the air (leaving this nasty work to technicians) nor clear-cut rules on which administrative branches using which financial and personnel resources should implement these policies at the regional level. Most programmes were somehow lacking some elements. In order to identify them, it became necessary to construct an ideal concept for a comprehensive reading of such programmes.

I remember the highly critical comments of our colleagues within the German Research Foundation's study group which was dedicated to the study of policy implementation processes under the amiable direction of Renate Mayntz and Fritz Scharpf in the late 1970s, when we presented for the first time this "core-shell-concept". Using diagrams in scientific articles was quite unusual at that time and the inclusion of institutional elements ("outer shells" such as programming elements of institutional administrative frameworks that govern implementation as well as resources and procedures) into programme analyses was far from being accepted. Nevertheless, this concept became quoted and taught by many colleagues towards the end of the 1980s and in the 1990s, and it probably became one of the most important analytical elements of our current policy analysis approach for practitioners and theoretical scholars. Indeed, the distinction between substantive and institutional elements of public policies should not be limited to political administrative programmes as one core product of the policy cycle but extended to each of the other five products of the cycle (problem definition, political administrative arrangement for implementation, action plans, outputs and evaluative statements[6]). Evidence of the important interactions between substantive and institutional elements of such programmes may have first become apparent in the experiences the European Union and federal states with their purely substantive policy programmes which led to extremely varied policy outputs and outcomes on the level of measurable air pollution loads amongst the various implementation areas.

This insight into regional differences in policy implementation processes within a single country, be it federal or centralised, represents the basic message of *Explaining Differences in the Performance of Clean Air Policies: An International and Interregional Comparative Study (1986)*,

[6] Cf. Knoepfel, P., Larrue, C., Varone, F., 2006, Analyse et pilotage des politiques publiques, 2ème éd, Zürich/Chur: Rüegger, p. 124ss.

which Helmut Weidner and I wrote on the basis of the same SO_2 air quality control policy study in the middle of the 1980s. The article is not at all revolutionary for this period; it simply reflects the increasing concern of the scientific community regarding newly detected implementation deficits and shortcomings that had been described in the American literature since the end of the 1960s. Viewed from today's perspective, this last chapter of the section might still be valid because the typology is built on the two dimensions of outcomes and administrative costs. This approach has become quite usual within the debate on new public management concepts which in many cases, however, do not study real world outcomes of policies but limit themselves to simple efficiency oriented comparisons between administrative costs and output production. The article was one of the first environmental policy implementation studies that considered both regional variations across countries and related outcome data. Remember that many implementation studies conducted in the 1980s and even today still do not consider regional data and qualitative explanatory variables such as varying administrative structures, varying problem pressures, and actor constellations.

1
Environmental Policy (2002)

Peter Knoepfel[1]

1.1 Major environmental problems in the early 21st century

The highly industrialized western countries, which have been actively pursuing environmental policy since the 1960s with an emphasis on air and water pollution control, have achieved considerable reductions in sulphur dioxide and dust emissions. Thanks to large-scale public investment, adequate facilities for waste water treatment have also become common. On the other hand, emissions of other equally relevant pollutants, such as nitrogen oxides and heavy metals, have not been reduced to the same extent; in the case of hydrocarbons a considerable need for reduction remains. In spite of the climate protection policies initiated in the mid-1990s, carbon dioxide emissions are still increasing, in the northern hemisphere in particular. These emissions are mainly caused by fossil fuel consumption in all kinds of combustion processes and considered as a prime cause of the green-house effect (main source: traffic). In spite of major investments in domestic and industrial water treatment plants, the success achieved in general water-pollution control can only be described as marginal. And in spite of increasing efforts in the area of agro-environmental policy, there has been no significant reduction in the contamination of ground and drinking water and in the eutrophication of stagnant surface waters arising from widespread fertilizer use. Other issues of concern in this context include the rapid extinction of plant and animal species throughout the world and the ongoing and virtually unhindered irreversible destruction of natural areas.

The situation is even worse in the industrialized countries of Eastern Europe, Asia and South America. There, concentrations of sulphur dioxide, nitrogent oxides and dust in the ambient air have reached levels that pose an unequivocal threat to human health. It is still quite common for sewage

---

[1] in: M. Hawkesworth, M. Kogan (eds): Encyclopaedia of government and politics, vol. 2, London (Routledge), 2003. Copyright: We thank the editor for authorizing the publication of this article with some minor linguistic adaptations proposed by Susan Cox.

and industrial effluents to be discharged into surface water with practically no prior purification treatment. Towards the end of the 1990s, there was a slight improvement in the situation in those eastern European countries that have applied to join the EU and are starting to implement the EU *acquis* (legislation). Prospective development in the Third World, be it endogenous or exogenous in origin, prompts prognoses of a dramatic deterioration in already imbalanced ecosystems. Forecasts in the early 21st century also predict an exponential growth in ecological hazards of global proportions, in particular disastrous changes in climatic conditions. Ever-widening gaps in the social distribution of environmental quality can be observed both between the northern and southern hemispheres and between the countries within them, and the ecological question has emerged as a key issue in a veritable "war against poverty" (2nd Earth Summit in Johannesburg, September 2002). Ecosystems that are already suffering from the effects of excessive pollution are being exposed to increasing impacts while efforts are made to improve conditions in regions where environmental quality is comparatively stable.

This alarming situation is largely the outcome of the anthropogenic emission of pollutants. The most important groups of polluters are:

1. basic and heavy industry (producing, for example, dust, sulphur dioxide, hydrocarbons, carbon dioxide, nitrogen oxides, toxic substances in solid and liquid waste);
2. transport (producing, for example, nitrogen oxides, lead, noise, waste);
3. agriculture (causing, for example, erosion, damage to flora and fauna and contamination of soil and ground water through the excessive use of fertilizers);
4. public infrastructure works and construction (causing, for example, noise, surface sealing of the soil, destruction of landscape).

The rapid change in institutional factors associated with the ongoing processes of globalization represents a further serious threat to environmental quality. This world-wide movement may indeed facilitate transfers of environmentally sound (abatement) technologies from north to south/east. However, it also results in the introduction of capitalist market mechanisms to the field of goods and services provided by all kinds of natural resources which run the risk of becoming privatized and placed in the hands of large international companies. These processes may lead to the serious and uncontrolled over-exploitation of the resources, as the relevant governments are politically unable to intervene. Thus, environmental degradation becomes a question of the scope of property rights to natural resources, institutional guarantees of good domestic governance and the sufficiently stringent regulation of the increasingly liberalized world markets.

1.2 Environmental policy: a general definition

The term "environmental policy" includes all government measures aimed at: 1) assessing the state of environmental pollution; 2) evaluating this pollution in relation to the threat it poses to either human welfare (anthropocentric) or ecosystems (ecocentric); and 3) controlling polluting activities by means of regulations, economic incentives and/or training, moral persuasion, information campaigns and collaborative contractual arrangements with selected target groups. These measures are shaped by a set of policy goals that are more or less quantifiable, depending on the existence of precise emission standards or environmental quality standards ("immission" standards). Environmental policy goals can involve the reduction of emissions or local immissions, the freezing of current emission or immission levels or the deceleration of predicted growth rates. In principle, pollution control can be practised either by means of basic preventive measures (changes in the emission source structure) or by limiting certain activities that give rise to emissions (control of immission levels). The latter strategy currently prevails in most industrialized countries.

As a rule, formal environmental policy is the administrative responsibility of special departments or environment ministries and it normally accounts for only a small part of administrative action relating to the environment. Many of the activities of other departments or ministries, such as energy, agriculture, spatial planning, industry, transport and foreign trade, often have far greater relevance with regard to the state of the environment than formal environmental policy. For example, decisions concerning land-use that are taken in the context of spatial planning policy establish basic determinants for the structure of anthropogenic emission sources in local ecosystems. Due to the ecological indifference that prevails in many countries, however, the above-mentioned policy areas are undoubtedly responsible for the emergence and growth of different kinds of environmental damage as opposed to its limitation.

1.3 A brief history of environmental policy

The term "environmental policy" was almost unheard of before the early 1960s. President Eisenhower's "fifteen goals for America" of 1960, which proclaimed the key policy objectives for the decade to come, made no mention of environmental problems. Yet only five years later, President Johnson's "Great Society" programme stated environmental protection as an important issue, and by 1969 the US Congress had enacted one of the first pieces of modern environmental legislation in the world, the National Environmental Protection Act. The government of the Federal Republic of Germany adopted its first environmental programme in 1971. As was the case in many other countries, this programme was prepared in anticipation of the landmark UN conference, "Man and the Biosphere", which was held in Stockholm in 1972. The conference concluded with a declaration, which established "the environment" as an issue of profound political importance – probably for the first time, although concerned scientists (for example, the Club of Rome) had already made this claim in the course of the 1960s. Prior to this historic date, environment-related policy was merely concerned with more or less technical measures involving public health concerns (such as water quality control and clean-air measures in notorious smog areas), consumer protection and protection of fair competition (such as legislation pertaining to food and toxic substances) and safety problems (such as safety at work and the prevention of boiler explosions). In addition to this, various measures were implemented in the area of nature and heritage conservation (such as the designation of nature reserves and the protection of natural and historical monuments). The latter actually emerged as early as the 1930s and were often more focused on scientific than environmental protection concerns.

Although governments pursued individual and diverging paths in their attempts to establish systematic environmental quality control as initiated by the Stockholm conference, by the late 1980s the status of national environmental legislation of OECD countries had more or less reached a common level. However, due to the differences in geographic, topographic and demographic conditions, as well as the differences in environmental conditions (such as high levels of pollution in urban conglomerations) and the structure of national industry, the speed at which individual countries introduced environmental regulations varied considerably. Differences in the perception of environmental problems, political judgement and political strength also influenced the pace of development.

Accordingly, in most western European countries, the first targets of environmental regulation included public-health oriented water protection and waste disposal and, in particular, waste incineration. In the United States, with its vast land resources, the standard practice of waste dumping (landfills) was maintained for a long time. However, population density on the other side of the Atlantic is much higher and space for landfills became increasingly scarce. Thus, waste incineration became more common and this prompted calls for corresponding measures to control the resulting air pollution. In reaction to the smog disasters of the 1950s, the United Kingdom (1956), France (1961) and Italy (1966) adopted air-pollution abatement measures at quite an early stage, while other countries, such as West

Germany, did not introduce systematic clean-air measures until the 1970s. Centralized environmental legislation also lagged behind local and regional anti-pollution measures (for example, in Switzerland). In east European countries serious attempts to create and implement environmental policy comparable to western standards only commenced in 1989, shortly before the political upheavals in those countries. These attempts were mostly last-ditch measures hastily taken to appease environmentalist protests, which could no longer be disregarded. During the 1980s, in coordination with the EU, most western European countries created comprehensive legal frameworks in the fields of chemicals, toxic wastes, industrial hazards and soil protection. During the 1990s, nature, biodiversity and landscape were the focus of increased regulation and towards the end of the century, these sectors were joined by regulations on genetically modified organisms.

By the late 1970s, a number of organizations established in the wake of the Stockholm conference were dealing with environmental policy at international level: for example, the United Nations Environmental Programme (UNEP), the Organization for Economic Co-operation and Development (OECD) and the Economic Commission for Europe (ECE). In the meantime, several important international conventions were ratified by a number of countries. These included the Geneva Convention on long-range transboundary air pollution (1979, in force since 1983); the ECE's Helsinki Protocol on the reduction of sulphur-dioxide emissions or their transboundary fluxes by at least 30 per cent (1985, in force since 1987); the ECE Protocol on the reduction of nitrogen oxides or their transboundary fluxes (1987, in force since 1990); the Vienna Convention on the protection of the ozone layer (1985) and the Montreal Protocol (1987), both of which attempted to eliminate the use of chlorinated fluorocarbons (CFCs) used as propelling agents in aerosols, cooling agents, cleaning agents, blowing agents etc.; the Basle Convention on the control of transboundary transport of hazardous waste and its disposal (1989); and the International Convention on the abatement of the greenhouse effect caused by carbon-dioxide emissions from combustion (1990).

Most of the international agreements established over the past ten years concern climate change (Kyoto Protocol 1997 (2002); further control of long range transboundary air pollution: emission of volatile organic compounds or their transboundary fluxes (Geneva 1991); heavy metals (Aarhus 1998); persistent organic pollutants (Aarhus 1998); fauna, flora and habitats (Bern 1982); biological diversity (Rio de Janeiro 1992). Furthermore, a number of regional conventions (UN, ECE) and protocols concerning transboundary watercourses and lakes (Helsinki 1992; London 1999), environmental impact assessments in transboundary contexts

(Espoo, 1991); industrial accidents (Helsinki 1992) and the protection of the Alps (Salzburg 1992, with nine protocols) have also been adopted. The most wide-ranging international agreement of the 1990s is without doubt the Rio Declaration on Environment and Development, which was adopted at the first Earth Summit, and the related Agenda 21 plan of action, according to which the signature states committed themselves to the development of national strategies for sustainable development, including ecological, societal and economic improvements of their policies to be evaluated at the second Earth Summit in Johannesburg in September 2002.

1.4 Environmental politics

The traditional political parties in most western industrialized nations discovered the environment as an issue long after the launch of the first environment-related measures, i.e. in the late 1970s and even into the 1980s when various incidences of serious environmental damage became manifest (for example, *Waldsterben*, endangered maritime ecosystems, deterioration of soil quality and the biological death of water bodies). There is no doubt that the environment first found expression as a political issue not in the context of the established political parties, but in the context of the grassroots movements of the mid-1970s which were motivated by ecological disputes concerning infrastructure projects, such as nuclear power plants, international airports, highways, etc. Grassroots movements were, and are, characteristic of the post-materialist or "eco-humanistic" values which increasingly gained ground and, in overtaking the issue of the environment in the narrower sense, led to the development of alternative concepts of human needs. Typical examples of these grassroots movements include the peace movement, the anti-nuclear movement, the women's liberation movement and movements promoting ecologically sound urban renewal.

During the late 1980s these opposition movements, which were dominated by the well-educated new urban middle classes, evolved into new political parties ("green parties") which have altered and will continue to alter the party-political spectrum of most western European countries (as evidenced, for example, in Germany, Sweden, Belgium, Ireland, Austria, France and Switzerland). These new parties were initially characterized by a loose and strongly decentralized structure with major emphasis being placed on local and regional activities and an ideology that originally often ran contrary to traditional left-wing/right-wing classifications. Most of these parties have, however, swung somewhat to the left since the mid-

1990s and they have become increasingly concerned about other policy areas aside from the environment, such as energy, transport and foreign and defence policy.

The new political groups and parties established in Eastern Europe in the wake of the revolutionary events of 1989 have also included environmental policy as a key element of their programmes. This comes as no great surprise as the dramatic deterioration of the environment was one of the factors that prompted the decline of the former Communist regimes.

1.5 Environmental policies

Long before the public and traditional political parties recognized the importance of environmental issues, many central and regional administrations were already involved in different ways in the struggle against pollution and engaging in close co-operation with scientists and the trades and industries concerned. The first-generation pollution-abatement measures focused on regions with particularly high pollution levels (immissions), initially so as to reduce health hazards and later to restrict the damage to plants, animals and ecosystems to an acceptable level. This "acceptable level" of pollution was defined in legislation on environmental quality standards. Two major strategies were adopted to bring about compliance with these standards: 1) measures were introduced to reduce the emission of pollutants by households, industry and traffic; and 2) measures were also introduced to dilute or bring about the dilution or long-range distribution of pollutants in water or air, thereby reducing local concentration levels. The best-known example of the latter strategy is the notorious high-stack policy, which was pursued mainly in the United Kingdom, northern France and the Ruhr area in West Germany. The consequences of the high-stack policy which facilitated long-range transboundary fluxes of air pollution, mainly to Scandinavia, were disastrous. Scandinavian lakes and soil were seriously affected by acidification. This partly explains Scandinavia's leading role in the promotion of international conventions on the reduction of sulphur-dioxide and, more recently, nitrogen-oxide emissions. Further examples of such immission-oriented strategies, which can mostly be found in the area of air and water pollution control, include smog regulations and the designation of smog alarm zones (for example, the French *zones d'alertes*) which affect a curb on activities causing emissions during unfavourable weather conditions, but allow the polluting activities to be resumed once conditions have improved.

All of these immission-oriented regulations, which tend to be confined to a few highly polluted areas, have one major disadvantage: they only reduce the total emission of pollutants marginally. For that reason they have been gradually replaced by prevention-oriented concepts which prescribe the use of state-of-the-art pollution abatement technology in old and new facilities, irrespective of local immissions. If such general and nationwide emission cutbacks still fail to meet immission standards, regional and local authorities are then called on to implement stricter emissions limits. Nowadays (2002), most large European, American and Asian cities use complementary smog-alarm schemes to protect their inhabitants against extreme air pollution, most of which is caused by motor vehicles (main pollutants: nitrogen oxide, small particles and terrestrial ozone). This kind of emissions-oriented (preventive) concept with all its consequences was first explicitly formulated in Switzerland's Environmental Protection Act of 1983.

Most countries have also adopted the "polluter-pays" principle by now, thus passing the cost of environmental protection measures on to those who create the need for such measures. Nonetheless, numerous provisions remain in force involving the allocation of all kinds of public funding, mostly in connection with water pollution control, waste disposal, and, in the future, noise abatement. In general, however, it is widely agreed that the consistent internalization of costs in accordance with the "polluter pays principle" enhances environmental protection, although there are some negative implications for income distribution. For example, the passing on of the cost of waste-water purification, waste disposal, air pollution or noise abatement to the polluters by way of cost-equivalent taxes affects lower-income groups harder in relative terms than middle or higher-income groups. Apart from the payment of financial incentives to polluters to minimize harmful activities, these negative repercussions on income distribution are the main reason that public funding is still provided in most areas of environmental policy.

1.6 International comparisons

National variations in the standards applied under environmental legislation, especially emission and immission standards, affect the competitive position of companies operating on international markets. Since many kinds of pollution do not stop at national borders, any country's environmental-policy measures may affect those of its neighbours. It is understandable, therefore, that administrators, politicians and scientists often refer to international comparisons when evaluating environmental protection measures. The following basic findings are confirmed in the extensive literature on the subject.

National environmental quality standards (immission standards) are still heterogeneous. It is an established fact that immission standards are easier to implement at international level than, for example, emission limits. Consequently, the EU set immission standards for several air and water pollutants, despite opposition from a sceptical France, but under pressure from the United Kingdom, which, due to the benefits arising from its geographic location, is strongly in favour of immission standards. According to EU law, Member States may, however, observe previously defined stricter standards.

Environmental quality standards have become an integral part of environmental policy in most western industrialized countries, and increasingly also in some east European and Asian countries. Significant variation exists, however, with respect to the severity, enforcement and effectiveness of these standards. This is primarily due to the fact that, apart from a few exceptions (for example, Austria, the Netherlands, Switzerland and Germany), governments have failed to base environmental policy on a broad ecocentric concept which would extend the scope of protection objectives to ecosystems as such and to all groups that run a higher risk of being harmed by pollution than the average population. Instead, most countries still make do with an anthropocentric concept of environmental protection. Consequently, these countries' environmental quality standards are significantly less stringent. This heterogeneity of standards is furthermore the result of the varying levels of political importance attached to environmental quality standards, the inequalities between the various actors concerned with and affected by environmental protection measures (for example, polluting industries, ecoindustry, environmental groups, consumer organizations etc.) and, not least, the obvious discrepancies in the effectiveness of environmental quality standards themselves. Observation of the differences in the effectiveness of standards often reveals in-built implementation failure due to insufficient monitoring capacities, incomplete information on monitoring results or a lack of legal provisions making non-compliance actionable.

There is even greater cross-national variation with respect to emission and product-related standards with a direct effect on competition. This is already evident in the various lists of substances now under regulation. In negotiations at EU level, for example, the United Kingdom obstinately opposed emission standards and has clung to the traditional case-by-case application of the best practicable means approach as implemented by its Pollution Inspectorate. However, most other European countries, Japan and the United States have enacted an increasing number of pollutant-specific emission standards, for air and water pollutants in particular.

The EU was committed to harmonizing the various emission standards in the interest of fair competition and this strategy was successfully implemented in many environmental policy fields in the course of the 1990s. As things stand, the internationally agreed levels for industrial emission standards still basically reflect the political strength and influence of the different sectors within the industrial system concerned.

Furthermore, international comparison reveals that, in spite of significant differences in the national policy styles underlying programme formulation, the implementation of national environmental programmes by individual regional administrations displays both similarities and differences that cannot be explained on the basis of the noted differences in the programmes themselves (standards etc.). Instead, the relevant distinguishing factors include the different constellations of actors, local environmental quality conditions, the size of the staff, budget and technical equipment at the disposal of regional authorities and also the strategic choices they make. Accordingly, despite the extensive variation between environmental policy programmes, various empirical studies have identified a certain conformity between the impacts of implementation, or at least a distinct similarity concerning those variables guaranteeing a certain quality of implementation. In the United States, for example, the process of programme formulation is highly controversial, while British or Swedish policy processes aim to achieve a mutual consensus and "cosy accommodation" between government agencies and industry with quite different results. On the other hand, however, when measured against the behaviour of polluters, the implementation effects achieved at regional level do not reflect the significant variances that might be expected considering the differences in the underlying programmes. In environmental policy – as in other policy areas – the decisive factor for successful performance is the political will and determination of the responsible authorities to take the necessary measures. Experience has shown that potential impediments at local or regional level cannot easily be overcome by rigorous national programmes. And, *vice versa*, weak national (and international) programmes cannot hold back progressive local or regional authorities in the long run. There is valid scientific evidence indicating that regional authorities faced with pressing problems and strong political mobilization can implement environmental policies more successfully than their colleagues in other regions, in which intervention is successfully blocked and/or the urgency of environmental problems is played down through the withholding of relevant information in spite of strict national programmes and standards.

1.7 Current problems facing environmental policy

One prerequisite for the effective implementation of environmental policy is the systematic monitoring of environmental quality. Most OECD countries and international organizations (OECD, ECE, EU) have made considerable progress in this field and the data produced is increasingly recorded and processed in national information systems, including integrated ecological monitoring systems (a good example of such a system can be found in Sweden) together with information on selected pollution fluxes, flora and fauna, land-use in open and built-up areas and anthropospheric issues (health, economy, politics). Such systems will make it possible to carry out continuous, long-term environmental monitoring, which is necessary not only for the comprehensive evaluation of current environmental policy, but also for the sensitive anticipation of future developments and the corresponding readjustment of environmental policy. Furthermore, environmental monitoring data must be made available to every interested citizen (as is the case in Japan and should be the case in all EU countries in accordance with EU Directive 90/313) and for environmental impact assessments. Citizens and local authorities should be able to participate in environmental monitoring so as to speed up the process between problem perception and remedial action. According to the UN Agenda 21 this data should be complemented with comprehensive data on social and economic sustainability (example: Eurostat, Switzerland).

Today, most western industrialized countries have a more or less comprehensive legal basis at their disposal, with which they can fight environmental pollution, at least with respect to water, air, waste, chemicals and more recently, soils, biodiversity and landscape protection. The environmental-policy action based on this legislation, however, is still largely divided on the basis of the different individual environmental media. Consequently, intervention frequently takes the form of the shifting of problems from one environmental medium to another. Fluxes of harmful substances from production to consumption and the disposal of waste substances in the soil cannot be dealt with appropriately within such fragmented political administrative arrangements. One important task facing environmental policy in the early 21st century is to replace sectoral administrative action with an integrated ecological management of pollution fluxes. Thus, it will be necessary to reform the current administrative structures that have been established on the basis of the outdated logic of segmentation in environmental policy and, in many cases, have been responsible for this problem-shifting. Hierarchically structured administrations and legal dogmas based on single linear causalities need to be replaced with co-operative administrative networks and the legal acceptance of multiple responsibilities (with a corresponding reversal of the burden of proof on the part of the polluter).

Such an ecology-based restructuring of administration and management is particularly necessary at regional level as regional agencies and nongovernmental organizations (environmental protection organizations) have always played the main role in implementing the various environmental laws and regulations. Furthermore, considerably stronger links need to be established between spatial planning and environmental policy, especially in urban and peri-urban areas with heavily polluting individual transport infrastructures. This has been an objective in the United States and United Kingdom for some time now, but very few serious steps have been taken in this direction in most of continental Europe. Spatial planning will have to integrate qualitative land-use concepts and ecological options into its concepts, while environmental policy will have to put a stronger emphasis on land-use issues. All EU member states (EU Directive 85/337) and most other European countries have introduced procedures for environmental impact assessments similar to those provided as early as 1969 under the US National Environmental Protection Act. As yet, however, environmental impact assessment still has to prove its worth as a key instrument in the implementation of land-use-oriented ecological policy.

Most environment agencies in western industrialized countries finally introduced economic instruments in environmental policy in the course of the 1990s. These instruments are still viewed with reluctance in certain quarters which stress *inter alia* the negative social impacts. Practical experience shows, however, that these problems can be overcome. Today, the most widely applied economic control instruments are probably those found in waste management (charges for waste disposal), water policy (water treatment taxes), clean-air policy (limited marketable emission certificates sold so as to accelerate air quality improvement in urban agglomerations) and, more recently, in agro-environmental and landscape policy (direct payments for ecological services provided by farmers).

It is neither realistic nor constitutionally admissible to call for an ecological police state to enforce compliance with environmental regulations. Thus, more or less voluntary co-operation between government agencies, industry, agriculture and households has become inevitable. Such public-private partnerships, which also feature increasingly in other policy fields as the so-called new governance models, emerged in many countries towards the late 1990s. Today, they form part of the actors' landscape in the context of most environment-related policies (e.g. energy-saving, CO_2-reduction, waste, nature conservation). Such partnership arrangements will certainly lead to more effective public action because they include all of the relevant (private, collective and public) actors on a more or less participatory basis. They undoubtedly introduce an additional (secondary) source of legitimacy into policy making and implementation, hence also reinforcing their primary (democratic) legitimation. Nevertheless, such arrangements can also run the risks of a lack of transparency and social bias (mainly favouring middle-class interests to the disadvantage of lower-class citizens). They could also constitute a threat to democracy and the rule of law if they facilitate policy decisions that replace other decisions taken by democratically elected governments. Legitimate public environmental interests can be seriously threatened by unequal partnerships dominated by strong economic interest groups. In order to avoid this risk, governments must take an interest in establishing strong "environmental capacities" incorporating highly skilled actors from politics, environmental organisations and (eco-)business.

The 1980s taught us that undifferentiated and widely dispersed appeals rarely lead to success. Instead, the necessary teaching and learning opportunities should be provided through problem-specific consulting, as practised, for example, in the official agricultural advertising services of the Netherlands, France, Germany and Switzerland. Apart from sufficient funding, several other requirements must be met in order to create effective consulting services which include: co-operation between interconnected administrative units and industrial associations; the immediate transfer of information from ecological research to consulting and practice and *vice versa*; the incorporation of the ecological message in operational consulting (including a corresponding account of ecological costs); and, finally, continuity, credibility and individuality which can be guaranteed not least through a degree of independence of policy and the administration. Consulting services structured in this way will become necessary in the areas of households (waste, energy), transport (promotion of public transport), and, of course, trade and industry (along the lines of the British example). Moreover, if consulting services are to be attractive to their clientele, what they teach should be economically utilizable in a broad sense.

In summary, ecology is set to become a central element of domestic policy in both industrialized and developing countries in the 21st century. Thus, environmental protection will probably relinquish some of its importance as a policy area of its own standing. However, it will instead gain access to many other single-issue policy areas, above all in the context of spatial planning. For economic reasons alone, it is unlikely to prove feasible to go on developing independent policies in the areas of infrastructure, energy, agriculture and industry and later make them ecologically compatible at great expense and effort. On the contrary, the products of these policies will only find demand on the "voter and consumer markets" as goods with a "practical value" if they are conceived and developed in relation to ecological considerations throughout the entire process. For this reason, environmental policy in industrialized western and eastern nations must be linked even more to general industrial and economic policies than the traditional environmental policy of the 1970s and 1980s. Hitherto, environmental policy was more or less doomed to managing the disposal of prefabricated and ecologically harmful policy products, causing as little damage as possible. This type of environmental policy now should be consigned to the past.

It is important to recall that the Earth Summit in Rio de Janeiro and its Agenda 21 defined sustainable development mainly in terms of using renewable natural resources in a way that future generations still can benefit from them. Traditional environmental policies are not actually natural resource policies (with the exception of nature conservation and landscape protection policies). Instead they are policies aimed at managing one of the many services provided by such resources and consisting in the reception, transformation or dilution of all kinds of emissions or wastes. They do not directly regulate all of the other (more or less sustainable) uses the economy or society makes of one and the same resource. It is high time that traditional environmental policies were transformed into true resource policies governing simultaneously and in a co-ordinated way all of the goods and services produced by natural resources and aimed at the preservation of their reproductive capacity. Such an approach can no longer avoid addressing property rights issues concerning the rights to the use of the related goods and services of resources and the (re-)distribution of these rights among competing user groups. Such a shift in perspective could be productively prepared and accompanied through the joint efforts of political science and institutional economics. This would bring together the necessary analytical tools for both integrated resource-oriented policy designs and appropriately designed property rights arrangements. These interactively determine the institutional regimes of natural resources in a given time and space, which in turn determine the behaviour of a whole range of – competing – resource users and the level of sustainability associated with the way we use our natural resources.

References

Bürgenmeier, Beat, Harayama, Yuko, Wallart, Nicolas 1997. *Théorie et pratique des taxes environnementales.* Paris: Economica

Crenson, M. 1971. The Un-Politics of Air Pollution: A Study of Non-Decisionmaking in the Cities. Baltimore: John Hopkins University Press.

Dales, J.H. 1968. Pollution, Property and Prices. Toronto: University of Toronto Press

Dente, Bruno (ed.), Environmental Policy in Search of New Instruments, European Science Foundation. Dordrecht: Kluwer Academic Publishers

Dierkes, M. Weiler, H.N., Antal, A.A. (eds) 1987. *Comparative Policy Research: Learning from Experience*. Aldershot: Gower.

Eurostat: *Measuring progress towards a more sustainable Europe.* http://www.europa.eu.int/comm/eurostat

Grant, Wyn, Perl Anthony, Knoepfel, Peter (éds) 1999. *The Politics of Improving Urban Air Quality*. Cheltenham UK/Northampton USA: Edward Elgar

Häberli, Rudolf, Gessler, Rahel, Grossenbacher-Mansuy, Walter, Lehmann Pollheimer, Daniel 2002. *Vision Lebensqualität. Nachhaltige Etwicklung: ökologisch notwendig, wirtschaftlich klug, gesellschafttlich möglich.* Ein Synthesebericht des Schwerpunktprogramms Umwelt Schweiz. Zurich: vdf Hochschulverlag ETHZ

Hanf, Kenneth, Jansen, Alf-Inge (eds) 1998. *Governance and Environment in Western Europe. Politics, Policy and Administration*. New York: Addison Wesley Longman

Heidenheimer, A., Heclo, H., Adams, C.T. 1989. *Comparative Public Policy: The Politics of Social Choice in America, Europe and Japan.* 3rd edn, New York: St Martin's Press

Holzinger, Katharina, Knoepfel, Peter (éds). *Environmental Policy in a European Union of Variable Geometry? The Challenge of the Next Enlargement.* Bâle et Francfort: Helbing & Lichtenhahn (série Ecologie & Société, vol. 13)

Horsch, Helga, Ring, Irene, Herzog, Felix (eds) 2001. *Nachhaltige Wasserbewirtschaftung und Landnutzung. Methoden und Instrumente der Entscheidsungsfindung und –nutzung*. Marburg: Metropolis-Verlag

Ingelhart, R. 1997. *The Silent Revolution.* Princetown: Princetown University Press

Jänicke, Martin, Weidner, Martin (eds) 1995. *Successful Environmental Policy. A Critical Evaluation of 24 Cases*. Berlin: sigma

Jänicke, Martin, Weidner, Helmut (eds) 1997. *National Environmental Policies. A Comparative Study of Capacity-Building*. Berlin: Springer

Knoepfel, P., Weidner, H. 1985. *Luftreinhaltepolitik (stationäre Quellen) im internationalen Vergleich*, 6 vol. Berlin: sigma

Knoepfel, P., Weidner, H. 1986. Explaining Differencies in the Performance of Clean Air Policies. An International and Interregional Comparative Study. *Policy and Politics 14*: 71–92

Lafferty, William M., Meadowcroft, James (eds) 1996. *Democracy and the environment. Problems and Prospects.* Cheltenham UK/Lyme USA: Edward Elgar

Larrue, Corinne 2000. *Analyser les politiques publiques d'environnement.* Paris/Montréal: L'Harmattan (Collection Logiques politiques)

Lundqvist, L. 1980: *The Hare and the Tortoise: Clean Air Politics in the U.S. and Sweden.* Ann Arbor: University of Michigan Press

Majone, G.. 1976. Choice Among Policy Instruments for Pollution Control, *Policy Analysis 4*: 589–612

O'Riordan, T., D'Arge, R.C. (eds) 1979. *Progress in Resource Managment and Environmental Planning,* vol. 1. New York: John-Wiley & Sons

Rüdig, Wolfgang (ed.) 1999. *Environmental Policy, volume I, National Environmental Policy.* Cheltenham UK/ Northampton, MA, USA: The International Library of Comparative Public Policy (Elgar Reference Collection)

Schnaiberg, A., Watts, N., Zimmermann, K. (eds) 1986. *Distributional Conflicts in Environmental-Resource Policy.* Aldershot: Gower

SFO/SAEFL/ARE: *Measuring sustainablle development.* Neuchâtel, August 2002. http://www.are.ch

United Nations Commission on sustainable development (UNSCD: *Indicators of sustainyble development.* http://www.un.org/esa/sustdev/isd.htm.

Vogel, D., Kuhn, V. 1987. The Comparative Study of Environmental Policy: A Review of Literature, in: Dierkes, M, Weiler, H.N., Antal, A.A. (eds) 1987. *Comparative Policy Research: Learning from Experience.* Aldershot: Gower

Weidner, H. 1986. *Air Pollution Control Strategies and Policies in the F.R. Germany: Laws, Regulations, Implementation and Principal Shortcomings.* Berlin: sigma

Weidner, Helmut, Jänicke, Martin (eds) 1997 and 2002. *Capacity Building in National Environmental Policy. A. Comparative Study of 17 Countries.* Berlin etc.: Springer

Wildavsky, A., Douglas, M. 1982. *Risk and Culture: An Essay on the Selection of Technical and Environmental Dangers.* Berkeley: University of California Press

Xepapadeas, Anastasios 1997. *Advanced Principles in Environmental Policy.* Cheltenham UK/Northampton USA: Edward Elgar

2
Approaches to an Effective Framework for Environmental Management (1993)

Peter Knoepfel[1]

Forword

The ideas presented in the following pages are based on comparative research – mainly international in scope – which the author has undertaken since the late 'seventies into three areas of environmental policy (maintenance of clean air in proximity to stationary sources of pollution, maintenance of clean air in proximity to mobile sources of pollution, and water protection in agriculture) in countries of Western Europe. Based on the author's postgraduate course on environmental policy and management in the Institut de hautes études en administration publique (IDHEAP) of the University of Lausanne, experience gained from Swiss environmental policy will also be drawn upon. Since the mid-eighties the author has participated in training officials in the Swiss federal, cantonal and communal administrations having responsibilities in relation to the environment. The discussion on observation of the environment is based on the groundwork accomplished since 1988 by the Swiss Commission for Observation of the Environment ("Schweizerische Kommission für Umweltbeobachtung") of which the author is Chairman. Further reading is indicated in the appended References.

An effective environmental policy presupposes the existence, from the outset, of clarity with regard to the precise manner in which participants from social and governmental circles act in furtherance of the policy (Chapter 1). Moreover, such a policy, as compared with those in other areas, must of necessity rest on a relatively broad range of scientific findings – which, indeed, have to be systematically developed. Accordingly, precise

---

1 in : A. Carius, L. Höttler, H. Mercker (Hrsg.), Environmental Management in Kenya, Tanzania, Uganda and Zimbawe, Deutsche Stiftung für internationale Entwicklung, Zentralstelle für öffentliche Verwaltung, Berlin 1995, p. 45–72 = Cahier de l'IDHEAP no 108 (German version) et 108b (English version) (The author wishes to thank Dagmar Kollande for assistance in translating this essay. Copyright: We thank the editor for authorizing the publication of this article in this book.

analyses of the nature and evolution of pollution are essential (Chapter 2). It is only on the basis of these findings that clear operational programmes – within the framework of which the specific administrative measures can be properly coordinated – can be drawn up (Chapter 3). The creation of efficient management structures and of clear divisions of competence between the central government and regional or local organs are of particular importance for the implementation of environmental policy (Chapter 4). It is only when these fundamental conditions have been met that the efficient carrying out of the above-mentioned operational programmes can be put in train. In this connection it is worth noting that some useful experience has already been acquired as to how best to structure such implementation (Chapter 5). Lastly, it must be emphasized that an environmental policy cannot be pursued independently of other policies in the public domain. Some aspects of this desideratum are considered at the end of the present paper (Chapter 6).

2.1 The main participants

2.1.1 The iron triangle

Until well into the late 'seventies, environmental policy was, for all practical purposes, carried out by two parties in a bilateral context. One was constituted by the official administrations concerned with environmental management. From the beginning of the 'seventies onwards, these became integrated – first with the central government apparatus and then progressively with the regional and local authorities, in every country of Western Europe. Facing these administrations was the other party, made up of organized lobbies, which may be divided into industrial and artisanal enterprises having an interest in particular environment-related activities (e.g. protection against pollution of the air and of water supplies, and disposal of waste), private households (sewerage, air pollution, waste), agricultural interests (water and soil pollution, protection of nature) and those concerned with elements of the national infrastructure such as roads, railways and airports (air, noise and soil pollution, protection of nature). A wide variety of relationships came into existence between the State and these various lobbies. Thus, the state authorities regulated pollution levels of concern to the pressure groups by a wide range of procedures involving the fixing of maximum permissible levels of specific types of pollution, as well as outright prohibition: these procedures regularly culminated in the issue of individual and specific orders and prohibitions. On the other hand, the pressure groups participated in drawing up the pertinent regulations and, specifically, in the establishment of the permissible levels of pollution. They were represented in numerous official commissions, where they were in a position not only to contribute their knowledge of the latest technological developments but also to ensure that the standards finally adopted would, to the maximum extent feasible, meet the interest of the lobby in question.

From the mid-'seventies onwards this bilateral cooperative relationship was brought more and more into question. New interests came knocking at the closed doors behind which the discussions between the regulators and the regulated had been taking place on a confidential basis. Since the beginning of the 'eighties these new interests have succeeded in most countries of Western Europe in making a breakthrough: the erstwhile bilateral set-up was expanded into a trilateral arrangement, in which henceforth the organizations dedicated to the protection of the environment were recognized as having an equal right to have their say. This is true not only at the national level in the drawing up of the relevant programmes, but also – and in particular – in carrying them out by means of publicly accessible administrative procedures, in which the environmental organizations came to acquire the status of parties, and today – and rightly – the implementation of every environment-related policy has to be accomplished in harmony with this trilateral reality. This iron triangle, formed by (1) the national administration, (2) the pressure groups and (3) the organizations acting to protect the environment must be accepted as being the basic framework within which the participants in a modern environment policy operate – for the protection of the environment is not a matter for more or less secret negotiations between the State and pressure groups: it concerns everyone whose conditions of work and life are harmed by the pollution caused by the groups' activities.

The following relationships can be perceived within the triangle:

- *Relations between commercial pressure groups and the authorities having responsibility for the environment.* As already mentioned above, the pressure groups bring their technical knowledge, their economic interests and their determination to bear when pollution control programmes are being drawn up, so that as far as possible the official standards which are ultimately adopted do not affect commercial competitiveness. As a general rule, important undertakings and economic sectors, operating on a countrywide basis, tend towards the adoption of detailed standards, while groups representing small-scale and artisanal businesses tend to opt for the adoption of general aims which allow for local variations in application. In such dealings official administrations mostly endeavour to achieve the strictest possible standards, in which context they take account of the practice of the most progressive firms. When doing so, however, they must avoid establishing as an official standard one which is exclusively based on the technology evolved by a single firm which dominates the market: otherwise, governments would, in practice, be furthering the creation of monopolies and cartels in matters of standards, with questionable effects on fair competition in the market place.

For the proper carrying out of an environmental policy, commercial pressure groups must be compelled to furnish the requisite technical and other data. Firms doing so may, in turn, find themselves well placed to influence the administration in formulating the regulations relating to particular standards. Negotiations in this context prove, in practice, to be less concerned with the details of maximum permissible pollution levels than with the fixing of deadlines and with transitional measures. It has, indeed, long been accepted in Western European countries for such bargaining to form part of the process leading to the adoption of environmental standards. The high-handed and unilateral official proclamations of yesteryear are giving way to agreements which in many cases are the outcome of informal negotiations.

- *Relations between the authorities having responsibility for the environment and environmental organizations.* In all countries of Western Europe, environmental organizations (e.g. the World Wildlife Fund, organizations for the protection of nature and of the national heritage, Greenpeace and numerous types of foundations) have become of increasing significance on the national scene and now form an important pressure group in environmental matters. Their political clout stems from their continuously high degree of organization and thus in their capacity to bring public pressure to bear on national governments. It is because of this factor that these organizations are often the most important "clients" of the authorities, whose resolve they strengthen in confrontations with the commercial pressure groups. For their part, the authorities increasingly appreciate the support which they thus receive, and regularly involve these organizations in the preparation of protection programmes and, when deciding on specific aims of environmental policy and on the texts to be adopted for their achievement, draw upon the high-level expertise and the wealth of experience which these organizations have acquired.

Authorities greatly value the contribution which these bodies make in the many instances in which they take the initiative in securing public acceptance of protective measures and in launching publicity campaigns towards that end. What was an initially sceptical, not to say dismissive, attitude on the part of some public authorities towards the environmental organizations has today doubtless become largely a thing of the past.

When it comes to the practical implementation of environmental policy, local groups dedicated to protecting the environment are undoubtedly the most alert watchdogs, in a position to supply the authorities with valuable details of on-the-spot pollution levels. These local bodies also take upon themselves functions – such as alerting public opinion and monitoring the behaviour of major commercial interests – which the official administrations could not carry out on their own; and today, no policy for environmental management, if it is to be really effective, can do without the backing thus afforded by such local groups. More and more, all kinds of groups dedicated to such protection participate in checking on the environment's level of toleration of pollution. In most countries they are allowed to state their case when permissible limits of pollution are being adopted and are accorded right of audience in any subsequent legal proceedings which may be taken to invalidate the limits set. Trouble can, indeed, occur when it comes to practical environmental action, if these bodies are not brought into the procedures at any early stage or if they are not really taken seriously. Today, in negotiations between the authorities, commercial interests and environmental protection bodies, resort is increasingly had to "mediation procedures" with a view to reaching a consensus without following strict and formal legal procedures.

- *Relations between commercial interests and environmental organizations.* Since the end of the 'eighties, direct relations between commercial lobbies and environmental organizations have undergone considerable change in the countries of Western Europe – largely without official intervention. The days of heavy confrontations between, for example, commercial firms, industrial sectors, farmers and the protective bodies may well be past. Firms have come to appreciate the value of a good relationship with the community in which they are located, as well as with these bodies. They also make a point of voluntarily keeping the pollution levels they create below officially-permitted maxima and, within the context of self-regulation, adopt a pro-environmental stance in dealing with any "black sheep" within their ranks. Legal agreements between protective bodies, local authorities and particular firms, providing for the voluntary reduction of noxious emissions are no longer unusual today. The same holds for public monitoring of products, technological processes and plant layout – topics once jealously guarded as commercial secrets. Today, effective management of the environment can scarcely dispense with such cooperation from local communities and protective bodies, even if it to some extent renders governmental actions superfluous. The environmental policies of many large commercial undertakings, as well as some medium-sized firms in Western European countries, now provide as a matter of course for the possibility of the agreed suspension of a process, for moratoria and for agreed compensation, mainly relating to waste disposal, air pollution and to protection of the soil.

2.1.2 Creating a proper framework

The iron triangle already referred to is by no means a pre-existing element of nature. Rather must it be created – and particularly in those countries where national environmental policy is at an early stage – by step-by-step enactment of the requisite measures. The three sides of the triangle can be considered as together constituting the framework for the activities of the three main participants in matters of the environment, and for the evolution of a corresponding culture. The history of environmental management in Western Europe during the 'seventies amply shows that the entry of the protective bodies on the scene of what had until then been the joint preserve of the polluters and the official authorities was not achieved without a fight. There were repeated attempts to exclude these bodies, to blacken them and to deny their expertise. If the lessons thus learned are to be applied today, the authorities must leave nothing undone to provide active support for the creation of such bodies by organizing training programmes to develop the expertise of their personnel, and by fostering the growth of relationships between the bodies and industrial and trade circles as well as with the agricultural sector. And the latter groups must not delay in learning to live with the protective bodies and in meeting the demand of the public for transparent information. They must, moreover, prepare themselves for the new experience of tripartite discussions by organizing training courses for their staff, by building up ecological expertise within the undertaking and by ensuring that the experts concerned are accorded appropriately high status within the ranks of the company hierarchy.

2.1.3 Squares and pentagons

As is well known, government services do not consist exclusively of environment management administrations. In particular, countries with a strong official participation in the national economy are in many instances equipped with separate sectoral bureaucracies for industry, trade and agricultural affairs, each of which naturally seeks to promote the economic sector in question, whether it is wholly or partly state-run, or merely state-regulated. These bureaucracies usually also see fit to ward off what they consider as excessive ecological demands upon "their" sectors. When there are such bureaucracies, the concept of a triangle no longer fits: that of a square or even of a pentagon becomes more appropriate, with the bureaucracies participating with the three parties already mentioned in matters of the environment. As a rule, these bureaucracies tend to be the "natural" allies of the polluter pressure groups, the result being a polarization between two camps. One consists of the bodies dedicated to protecting the environment and of the official administration for environmental management. The other is that of the anti-active pressure groups, who can now mobilize "their" official bureaucracy in support of their stance in the political power game. The stronger such official sectoral bureaucracies become, the more difficult it becomes to implement effective policies for the environment. There is thus in practice a connection between privatization and environmental policy: privatization means the disappearance of the sectoral bureaucracies concerned, which accordingly no longer exist as a protective screen for the polluters. A further outcome of privatization is that all special regulations and exceptions which had been made in favour of a state-run sector of the economy – which, for instance, determined matters in the former socialist republics of Eastern Europe – should lapse.

It must be added, however, that it has not proved possible to prevent the existence of such quadrilateral scenarios, even in the market economies of Western Europe, in a particularly important field – that of the "public works lobby". Entrepreneurs, whether building roads, airports, or waste incineration plants, or in charge of state-run energy production, were in almost all such cases implementing one public policy or another, for which powerful backing was available from within the official administration concerned. The officials in question were – and still are – in a position to shield public works for the national infrastructure from environmental regulations which they deem to be unduly strict. The result has been that in many cases special regulations were made allowing lower standards of environmental protection in such works than those applicable to undertakings in the private sector or in trade. When official administrations for environmental management are being set up, care should consequently be taken that protection vis-à-vis the long-established administrations concerned with public works is effective enough to ensure that the latter will have to meet the same environmental requirements as are applied to the private sector.

2.2 Preparing the scientific basis

2.2.1 Monitoring the environment

Precise knowledge of the nature and levels of pollution is the first prerequisite for an efficient environmental policy: only when it is known which sources give rise to which pollution that any meaningful regulations can be prepared. The mapping of pollution surveys based on verified statements, or if necessary on remote sensing techniques, is essential for the control of air pollution, waste disposal and of water and soil protection in particular. The author considers that the making of such maps is of greater importance than the simple recording of general pollution levels present in an extensive area since the latter merely illustrates the need for remedial action without, however, pinpointing precisely *where* that action is to be taken. In any case, the implementation of established environmental policies will require the creation, in zones of greatest concentration, of networks for monitoring pollution. These serve in particular to locate the priority areas needing remedial action, to identify situations of acute danger (e.g. smog concentrations) with a view to the launching of emergency measures, as well as to provide a check on the effectiveness of the policy followed. There is, indeed, a long-term need to monitor the effects of pollution on flora, fauna, mankind, buildings and the landscape.

Even in Western Europe, such monitoring of the environment is confronted with many problems. Little useful purpose is served if, for fear of a negative reaction in a locality heavily affected by pollution, a figure for the maximum level of pollution to be tolerated in the locality in question, scientifically arrived at, is made to apply to the country as a whole, so that objectively acceptable differences between the levels permissible in different regions simply "disappear". And useful knowledge scarcely results when answers to a specific enquiry relate to different localities, so that their significance as a basis for administrative action becomes virtually nil. Data on the local, regional and national scales must, in the end, be correlated so as to bring out the essential structure of the pollution "scene" and to show just where political and administrative action can best be initiated. Thus, for example, endeavours can be made to link data concerning traffic patterns, urbanization and air pollution so as to develop strategies to maintain good air conditions in the main cities and in peripheral satellite communities in heavily polluted areas. On-the-spot data on the ozone level must be correlated with those for volatile organic compounds and nitrogen oxides so as to locate the areas in which this secondary nuisance originates. Lastly, it must be pointed out that if it is to provide a basis for practical action, environmental monitoring must, rationally, take the whole ecosystem into account. There is no "national quality of the environment": everywhere, different harmful substances affect variously structured ecosystems in quite specific ways.

2.2.2 Monitoring technologies

So that the benefits of the most appropriate pollution prevention technologies may be secured, environmental authorities as well as individual sectors of the economy must be informed of what relevant environmentally sound technologies are on offer. Such "state of the art" knowledge is required not only for the development of environmental programmes but also for their practical implementation, when pollution-creating undertakings have to be made aware of the relevant technologies and techniques available for environmentally sound operations. That such monitoring presupposes research on environmental technologies is just as evident as the fact that the interstices between research and practice must also be covered. The scope of such monitoring embraces not only technology in its narrower meaning, but also observation of behavioural patterns, of the motivations which cause them and of the underlying attitudes of the public. It also covers sociological research on the organizational, sociological and administrative requirements for the meaningful adoption of desirable new techniques and patterns of behaviour. It is, indeed, known that an environmental technology, wrongly applied, often proves more dangerous than no technology at all: the trouble-free application of a specific technology depends on prior awareness of the reactions and behavioural patterns of the social groups affected as well as on appropriate social engineering.

2.2.3 Monitoring of policies

The methodical implementation of an environmental policy presupposes the continuous monitoring of its performance. Programme managers, the governments and the executive authorities politically responsible must have access to the data necessary to enable them to assess just how the policy is working out in practice and what difficulties it encounters. The first questions to be asked relate to precisely where, when and what specific action in furtherance of the policy was taken by the authorities charged with its execution. A compilation of the resulting data enables a map to be drawn for each region of the country, giving an overall view of the impact – or absence of impact – of the policy, and providing a basis for setting new priorities. Such (electronically assembled) databanks, moreover, enable the exact content of current regulations to be reproduced at a later period as an act of political memory. However, even in Western Europe, such policy output mapping systems are still rare.

It is recommended that, at a later stage, the monitoring of environment policy be extended so as to cover the major official policies in other fields (e.g. agriculture, the infrastructure, the national economy, and town-and-country planning), since these policies are often responsible for yet more pressure on the environment. Hence if the environmental policy is to hold its own when conflicts with the other policies arise, its guardians must be aware of the exact implications of the latter for the environment. These implications can be identified by noting the actual steps taken by the authorities charged with carrying out these other policies and by correlating them with the data resulting from monitoring the environment. Such an assessment of the results of the environmental policy itself could, similarly, be undertaken from time to time. In arriving at these assessments, an endeavour should be made to correlate changes in the quality of the environment and in the actions of polluters with the evolution of the results in question. Such a quest often proves, however, to be time-consuming and beset with obstacles: it cannot be carried out on a continuous basis.

2.2.4 Organizing monitoring

Experience shows that monitoring the environment itself and monitoring the policy for environmental management, should not be carried out by the official body entrusted with implementing that policy. If the latter body were to be so entrusted, its independence – and hence its objectivity and credibility – would be placed in doubt. It is preferable that a largely independent agency or authority, not involved in day-to-day administrative concerns, be established to monitor the environment and related matters. Otherwise, the authority responsible for administering the environmental policy would run the risk of not managing to ensure the continuity of monitoring, and might be tempted to prettify results for political reasons. Close collaboration with the administrations directly concerned with the environment must, however, be ensured. These latter administrations should, moreover, be provided with all the data collected, analysed and published by the agency.

2.3 The management programme

2.3.1 Five components of an environmental management programme

Experience has demonstrated the need for the basic legal documents – e.g. laws, decrees and official circulars – relating to environmental management to manifest the greatest clarity when prescribing the solutions to specific problems, in order that uncertainties in execution, and the consequential losses in efficiency may be avoided. These problems arise in the following contexts:

- *Specified quality standards for individual substances (i.e. maximum permissible limits for pollutants):* As a rule these are expressed in terms of the maximum permissible average concentrations of a given harmful substance during a specified interval of time (mostly as μg/m3). They serve as the basis for the qualitative assessment of the environmental component concerned and for prescribing the target levels to be incorporated in the relevant policy. In the countries of Western Europe, such limits have been set for various harmful substances in water, in the air and in the soil, as well as for noise pollution. In general, the limits are arrived at partly by means of empirical eco-toxicological findings as to the levels which can be tolerated by humans, animals, plants and by the ecosystem as a whole, and partly by politically influenced estimates of the risks involved. Due to the incidence of the latter factor, the limits prescribed vary – often considerably – from country to country. A decision as to the limit to be adopted is rendered all the more difficult when subjective considerations of particular groups influence their assessment of the pollution load in question. A typical example of the impact of such considerations is afforded by noise limits. In respect of some fields of application of environmental policy – e.g. natural beauty spots and landscapes – maximum permissible limits are expressed in quite imprecise terms, while in yet other fields specific limits apply to quite small numbers of substances only, limits for the remaining substances being left for fixing in the light of on-the-spot eco-toxicological findings.
- *Measurement techniques and networks; and criteria for evaluation:* The extent to which permissible pollution limits apply in practice depends on how the methods and arrangements for making relevant measurements and on how the criteria for assessing violations of the specified limits have been arrived at. Depending on just how the management programme prescribes these criteria and the requisite meas-

urement procedures, violations which set off remedial action will "happen" with greater or lesser frequency. Thus, the precise manner in which these points are dealt with in the basic texts – and which, despite its technicity, is of major political significance – can tend to strengthen, or alternatively weaken, the practical effect of the limits prescribed. In matters of the environment as in other contexts, the old saying that "he who wants to chance his arm can get away with it" holds true. It follows that, as a matter of political reality, the ostensibly purely technical question of just how pollution limits are to be enunciated has manifestly assumed what might in all innocence be considered an improbable sensitivity.

- *Pollution output limits and product standards:* These stipulations constitute the legal basis for a wide range of official documents and thus establish the legal obligations imposed on the commercial interests concerned. They are accordingly the subject of diverse negotiations between governmental authorities and the pressure groups in question. Standards for pollution output and product standards assume key importance in the setting of maximum limits of pollution input, since they must be expressed in such a way that they cannot be exceeded even if each of a multitude of polluters releases the permitted volume of noxious material. Otherwise, excessive levels of pollution would automatically occur in congested areas – a state of affairs directly due to inherently faulty management programming and not to any shortcoming in the actual execution of the programme. Such a faulty correlation between the standards for pollution output, processes and products on the one hand and, on the other, the core of the programme itself – namely the maximum permissible level of pollution input – must be viewed as a pre-programmed failure, resulting in an inevitable loss of effectiveness in implementing the programme. Each limit prescribed for pollution output must be reviewed to establish to what extent the limits for pollution input can be retained if it is to remain unchanged. It has become the established practice in various countries of Western Europe for national limits for pollution output to be reduced when their application does not, in practice, prevent the limits for pollution input being exceeded.
- *Organization and financing:* Any programme for environmental management, however coherent and comprehensive it may be, with prescribed standards for pollution output, industrial processes and products, will remain a dead letter if it does not specify precisely who is to be responsible for carrying it out. The same holds true if adequate funds for the inevitably costly investigations entailed in its execution are not

provided when the programme is launched. An insufficiently financed executive agency, or one not truly responsive to environmental concerns, will never be able to translate substantial standards into reality. Every legislature concerned with environmental matters must take full account of this interaction between the substantive and organizational aspects of a programme. If a legislature should dodge its responsibilities in this field, failure in implementing the programme is inevitable.

- *Management tools and procedures:* In addition to allocating responsibilities for its implementation, an environmental management programme must, at least, also clearly designate the instruments from which it derives its authority. For a state based on the rule of law presupposes that the carrying out of a programme will be in accordance with procedures which can be ascertained in advance, and that commercial interests as well as environmental organizations will have their say. The legal basis for the management procedures concerned, together with the legal texts to which the parties involved may have recourse, must therefore be codified. This applies to the tools for control of the programme (such as periodic controls by the executing agency itself, controls by independent inspectors, and rights of access), to the procedures leading to the award of licences (such as that for ensuring legal protection against immissions when licences are issued, with or without provision for controlling the environment's capacity to tolerate the proposed pollution, as well as to approvals of plant layout) and to the tools for the initiation of curative measures (such as curative decrees addressed to individual polluters or to entire groups thereof). By the same token, the right of access of the parties to the law courts must also be specified. The clarity with which these various aspects of the procedures to be adopted are set out determines the extent to which the management programme concerned is effectively capable of achieving its goals (constituting the "core" of the programme). Excluding the environmental organizations from the procedure for the award of permits can damage this "core" just as much as it can be strengthened by appropriate intervention on the part of these organizations.

2.3.2 Fixed versus flexible programmes

In the early stages of environmental policies, it is desirable to have a separate policy for the management of pollution in each component affected within the environment, and in particular for air, water, soil, noise and nature. Such programmes may vary from each other when it comes to practical action. The more national regulations are couched in purely general terms, the more it is left to the executing authorities to decide just how to interpret them. Undertaking such a task presupposes the existence of the necessary expertise among the personnel of the regional and local authorities concerned. It also calls for political ability on the part of those authorities to resist pressure from powerful individual polluters who endeavour to influence the process of implementation in their favour. It must also be said, however, that vaguely worded official texts have the obvious advantage of facilitating the carrying out of the particular environmental policy in the context of other policies at the local level. The executive authorities are thus presented not only with scope for decision-making in the political field, but also with political responsibility for their actions – which in many ways often proves to be a healthy state of affairs.

It will be seen, therefore, that there is no general rule as to how far central environmental authorities should go in the management of the individual components of the environment. Several countries have, in this connection, opted for the solution adopted by the European Community – namely, the central state authority concerned establishes a series of norms for immissions, and a (generally longer) series for emissions, and the regional authorities are left free to set additional norms for any component not already covered by a national standard. In some countries as, for example, in Switzerland, the regional executive bodies may set norms higher than those adopted at the national level.

2.4 Management structures

The experience gained in building up environmental management bodies in countries of Western Europe may be summarized in the terms of the following nine principles, which have been more fully elaborated by the author elsewhere.

2.4.1 First principle: keep monitoring separate from administration

As has already been pointed out under 2.4. above, monitoring the performance of a particular agency should be entrusted to another body having a relatively independent legal status. Such an arrangement ensures that the monitoring cannot be suspected of a self-indulgent bias. It also guarantees the continuity, calm and professionalism necessary for collecting and analysing data – activities which should not be disrupted by immediate political considerations. Monitoring the environment is by definition a long-term exercise, requiring continuity and scrupulous adherence to procedure. The trend to separate monitoring agencies from the general administrations responsible for environmental management has been evidenced both at the level of the European Communities (where an environment agency has been set up) and at the national level (as in France, Germany and Switzerland).

2.4.2 Second principle: structure the environmental authorities on a component-by-component basis

Since the early 'seventies the concept of structuring environmental management authorities on an environmental component-by-component basis has become generally accepted in most Western European countries. This fosters the progressive accumulation within the authority of the requisite expertise with regard to such different components as the air, noise and water. In most cases this development took place in the wake of the progressive enactment of environmental legislation which was itself component-specific, generally starting with water-protection and, as from the mid 'seventies, successively dealing with the other components – air, soil, the biota (including fauna and flora) – as well as with the landscape. Administrative departments were next set up to deal with pollution by noise, radiation and chemicals. The validity of this basic tenet for the structuring of environmental management has proved itself in practice. It has been the traditional, component-by-component approach to environmental regulation in the majority of industrialized countries in Western Europe. This approach led to the adoption of different environmental policies in response to the impact of emission-creating activities on the different environmental components. As these activities, when they impact on the components of vital importance to human beings, animals and plants, become immediately evident, governments enjoy ever-greater political support for protecting the components concerned.

2.4.3 Third principle: a responsive administrative structure focused on emissions

This component-by-component approach first becomes evident on the immission side, and is not primarily concerned with the causes of the pollution in question. There exist totally immission-oriented strategies, in the pursuance of which component-specific remedial activities are undertaken, without going back to the emissions themselves or to those responsible for them. Several concrete examples may be cited: a phosphate-polluted lake is revitalized by an influx of oxygen; the quality of nitrate-polluted drinking water is improved by mixing with water low in nitrate content; dying trees in a forest are replaced by more acid-resisting species; the concentration of harmful substances in the main sewer of a local community is reduced by the mixing-in of rainwater. Such immission-oriented strategies were central to the growth of component-specific environmental management administrations during the 'seventies.

Today, however, the validity of these strategies is being queried: after all, they do nothing to reduce the pollution load; they merely distribute it differently. When these strategies gave way to emission-oriented environmental policies, corresponding changes in management structures became necessary. Within component-specific administrative divisions, emission-oriented subdivisions were set up. Thus, for example, in a water protection division, subdivisions for "community drainage (sewage treatment)", "industrial effluents" and "agriculture" are now to be found. Similarly, in a clean air division, subdivisions for "industry and trade", "households" and "traffic" have come into being. Such a responsive structure of administrative divisions fosters increased knowledge of the groups responsible for the emissions in question, and of the technologies available for avoiding the emissions. It also facilitates close and on-going communication between the members of those groups and the specialized administrative unit concerned; and in general it provides those responsible for protecting the particular environmental component with greater insight into the causes of the pollution affecting it. Depending on the component, separate subdivisions should be envisaged to deal with industry and trade, households, agriculture, traffic and public works.

2.4.4 Fourth principle: link activities relating to different components of the environment (intra-policy cooperation)

Once a separate management structure, focused on the groups responsible for pollution, does come into being for each component of the environment, the linking of the different components through the groups is only a matter of time. Such integration is, indeed, essential for the efficient control of the different groups responsible for emissions, and facilitates cooperation with them. Instead of each polluter having to deal with a different official in connection with controls or discussions relating to each specific component of the environment, it makes more sense to group the regulations applicable to each category of polluter, so that air, water and the landscape are dealt with together. The jump from tackling matters for each component separately to dealing with them collectively on an integrated basis, which took place in environmental administrations in most countries of Western Europe towards the end of the 'eighties was, indeed, a requirement of rational, ecologically responsive environmental policy. Here, priority is given to controlling and managing the pollution loads on entire ecosystems, rather than to controlling and managing individual components of the environment. Ecosystems, after all, interact with each of the permanent components of the environment, and their management accordingly calls for an integrated approach. Ecological cycles cannot be split on a component-by-component basis.

The networking of component-related environmental management here recommended presupposes the incorporation, within the administrations, of "horizontal" structures for cooperation between the different groups of polluters or corresponding to the track of particular noxious substances through successive stages. This can be achieved by the creation of a number of "horizontal" divisions as part of the administrative matrix. Thus, component-specific divisions would be supplemented by the setting up of "horizontal" subdivisions corresponding to the importance of the main polluter groups. The author recommended such a structure, for example, for the progressive establishment of the Swiss Federal Environment Authority – which is mainly concerned with programme management, and plays only a limited role at the level of programme execution. Every practician in this field knows that at that level, forms of intra-policy cooperation almost inevitably appear. This is because the entire range of environmental components to be affected by a proposed industrial plant must necessarily be taken into account in each case, when the tolerance of the environment is being tested and the issue of a permit is under examination. It is in the light of this fact that considerations of intra-policy cooperation have long had a bearing on the internal routing of files when applications for licences are being processed within authorities charged with implementing environmental policy.

2.4.5 Fifth principle: networking environmental authorities with those dealing with other policies relevant to the environment (inter-policy cooperation)

As has been indicated at 1.3. above, government policies in many cases share responsibility for pollution creation. Thus, an agricultural policy aimed at maximizing production can result in soil and surface pollution; the all-out implementation of an energy policy may be accompanied by significant air pollution; road construction policies fostering the development of one branch of transport regularly cause air and noise pollution; while sector-specific industrial policies can lead to heavy water and soil pollution. Since the beginning of the 'nineties, those concerned with environment protection in Western Europe have become increasingly aware of the extent to which practical policies for such protection are thus subject to other policies which are not primarily concerned with the environment. Rather than attempt the interminable task of coping with the side-effects of these other policies, the environmental authorities are becoming more and more determined on open confrontation with those policies at the national programming level, and likewise at the implementation stage. Towards this end, within official administrations bridgeheads were – and rightly continue to be – established so as to promote such inter-policy cooperation. Among the possibilities being considered in this connection, mention may be made of interdepartmental working groups, stationing "satellite" officials of the environmental authority in other ministries, and of incorporating elements of non-environmental policies in those for the environment.

In each design for a new environmental management authority, in-depth consideration should be given to the incorporation of structures which will facilitate inter-policy cooperation: an environmental policy cannot be ecologically successful unless these other policies are "ecologized" and are made subject to the overriding exigencies of ecology. An ecological policy is therefore not to be considered as merely one governmental policy among many: as was noted in the Single European Act of Luxembourg, 1987, the requirements of the protection of our ecological balances must form an integral part of all other policies, such as those of the European Communities.

2.4.6 Sixth principle: networking of structures by means of established procedures

The management structures described above lend themselves to the detailed specification of each task which is to be undertaken by the different administrative subdivisions, component-specific and "horizontal" divisions. The required networking of a component-oriented with a polluter-oriented stance, or with an overall approach having special regard to the ecosystem can, by contrast, hardly be accomplished by specification of the administrative structures to be adopted for discharging the tasks in question. Networking must permeate procedures: it cannot be ordained merely by what a text may prescribe. There is, accordingly, a need for this networking concept to be put into practice through down-to-earth measures to ensure that appropriate administrative procedures are always followed. The famed "beaten paths", daily trod by all concerned, have demonstrated their indispensability. Such mechanisms for the routine correlation of policies on a day-to-day basis can be arrived at in a variety of ways – by built-in procedures for the circulation of files relating to important matters through the different divisions and other appropriate administrative units, by periodic meetings of those involved in a given project, by setting up joint working parties to deal with particular topics, or by prescribing procedural rules covering matters such as obligatory consultation, exchange of information and discussion. Experience has also shown the value of other informal channels for cross-communication in strengthening such in-house networking of different policies. Lastly, mention must be made of how very important it is for the authority as a whole that it should acquire and maintain its own corporate identity, that all its services should give good example, and that its basic strategies should be periodically reviewed.

2.4.7 Seventh principle: separation of programme planning from execution

In the course of the sorting-out of environment-related functions, an administrative separation took place during the 'seventies in most[2] countries of Western Europe between the authorities responsible for programming and those charged with carrying out the programmes adopted. This even holds true when a single authority is responsible for the two functions. Despite some shortcomings (such as insufficient feedback between the two functions, or unrealistic and impracticable task-programming), such separation has widely justified itself and the concept should not lightly be abandoned. It has the advantage of relieving the front-line executive authorities, in a politically objective manner, of the manifold political pressures to which they would otherwise be exposed, and of providing similar political relief to the programming authorities which, when they are engaged in preparing programmes, are no longer harassed on each particular point. They are thus rightly freed to take a broader view and to take objective account of scientifically valid criteria. A further advantage lies in the fact that they can work on the basis of really important parameters relevant to the country as a whole.

It must be added, however, that – as discussed in connection with management programmes at 3.2. above – such separation cannot in all cases be

[2] Notable exceptions: the United Kingdom and, to some extent, France. These are characteristically unitary States, marked by a high degree of centralization.

achieved in an orderly fashion. In spite of the increasing volume of regulations governing the relevant management programmes, there will in practice always be some cases which are not covered. In such cases, which were frequent during the early years of these programmes, the executive administration has to take an ad hoc decision on just what has to be done – in other words, to do the requisite programming itself. This means that, for such individual cases, the executive authority must be in a position to acquire the knowledge necessary for reaching an ecologically meaningful solution – knowledge which may well prove useful in dealing with future cases of a similar order. This, indeed, often happens when new harmful substances have to be dealt with. While, theoretically, it should be possible to submit such cases to the appropriate programming body, in reality there is often pressure for them to be dealt with quickly, so that in most instances an ad hoc solution is indicated.

2.4.8 Eighth principle: implementation on a regional basis

Programme formation is manifestly the responsibility of the central environment authority. In almost all the countries of Western Europe,[3] however, the principle of regional implementation of programmes has become accepted. Implementation by the central authority is now mainly confined to chemicals (approval of substances), transport (vehicle equipment and fuels), air pollution (the importation and production of propellants), radiation protection, and control of the internal and cross-border transport of dangerous substances and wastes.

The advantages of implementation at the regional level, in line with the principle of "subsidiarity", are manifest: it facilitates fitting in national aims with the requirements of the regional or local environment, and taking account of the economic and social realities of the regions concerned. To these advantages must be added: close contact with the population, specialized knowledge of the local scene; and the possibility of associating the networks existing for the implementation of other, non-environmental, state policies ("political carpets") with the carrying out of the environmental programme. But the assignment to the regional level of responsibility for the implementation of these programmes is generally said to entail the risk of greater vulnerability to local or regional economic and political pressure. As has been pointed out under 4.7. above, such pressure can be

[3] Sole exception: the United Kingdom. Today, even in Italy and France, the national environment policies envisage substantial regional participation in their implementation.

countered by the manifest separation of programming and execution, by the clearest possible wording of management programmes when dealing with contentious matters (see 3.1. above), and above all by mobilizing pro-active support from the ranks of the environment protection movement (see 1.1. above).

It should be noted that regional implementation is not communal implementation. The high degree of technicity involved scarcely warrants local authorities (other than those of major cities) being saddled with this responsibility. A communal authority set up to cope with it would be inefficient and all too costly. This should not be taken to mean, however, that there is no role for communal authorities in the carrying out of environmental policy. On the contrary: without the active collaboration of these authorities, no regional body could cope. In those cases in which, in the interests of local industry, a communal authority actively obstructs a regional body, there is little that can be done. A pro-active communal authority can, on the other hand, alert the regional body concerning any nuisances within its territory rather sooner than a passive one may do. The role of a communal authority lies above all in the assiduous monitoring of the environment, without which the regional body cannot implement the policy. Communal authorities can – and should – contribute their detailed knowledge as well as their political convictions to the practical implementation of environmental policy. Their cooperation is therefore of vital importance if such a policy is to be effective. In many cases their role is similar to that of the environmental organizations: not infrequently they act as the catalyst for what subsequently emerge as popular local movements which show their mettle in on-the-spot protective action.

2.4.9 Ninth principle: staff training

Environmental policy is manifestly technical in nature. Neither lawyers nor administrative functionaries can conceive or implement it. Thus it has come about that in Western European countries since the 'seventies such policies have led to a significant trend towards the professionalization of those bearing responsibility for conceptualizing and carrying them out. This in turn led to the creation of new academic courses and to options for further education at the university and non-university levels. New profiles came into being for new professions – those of the environmental chemist, the environmental biologist, the environmental engineer and the environmental economist. Today, without such staff, no environmental policy could be carried out. It is therefore not only a scientific imperative but also – and in particular – a managerial necessity that such courses be set up in the various university and non-university institutions. This provides yet another example of how public policies have contributed to the creation of new professions and new educational opportunities.

This professionalism must not, however, be confined to the ranks of environmental management: it must also permeate the other two sides of the "iron triangle" mentioned in section 1.1. above. It is, indeed, in the interest of protecting the national environment that such ecology-oriented professionalism should be made welcome in industry, in trade, in agriculture and in the bureaucracies in charge of public works. Because of their key importance (also referred to at 1.1. above), the same holds for the environmental organizations.

2.5 Implementation

The experience gained by the countries of Western Europe in implementing environmental management programmes may be summarized under the following four headings.

2.5.1 Implementation means politics

The implementation of environmental policy is a technical, carefully focused operation *and* an act of political will. Both these elements are present – and not infrequently combine to create an explosive mixture. Both are, indeed, equally necessary: if sheer political pigheadedness is allowed to carry the day against technical good sense, the consequences can be just as bad as when implementation is based on exclusively technical desiderata but fails to obtain general acceptance by the authorities and the parties concerned within the community. Effective implementation is essentially a matter of repeatedly reconciling these two elements with one another. Both as an act of will and as a technical, carefully focused operation, it depends above all on the creation of robust regional executive authorities in conformity with the nine above-mentioned principles, on adequate financing of these authorities and on the decisions, indicated in the following sections, which have to be taken in regard to the plan of action and to the individual decrees (both formal and informal), as well as in connection with evaluation.

While, thanks to the desirable homogeneity of the staff entrusted with implementation, the technical, carefully focused operations themselves vary little from one part of the country to another, there are marked differences from region to region in the quality and emotional implications of the acts of political will. It may therefore be assumed that, even in centralized states, there will similarly be differences in actual implementation as regards the assiduity with which emissions are monitored, and in particular in connection with the progress of remedial action both in time and in different parts of the country. This is confirmed by experience in Western European countries as well as by considerations of plausibility, as presented hereunder. For the act of political will is the net outcome of a variety of estimates, within local and regional *milieux*, of the risks run, as well as of differences in environmental awareness which demonstrably exist within a given country, as also conceivably between North and South or across linguistic frontiers. Whether one likes it or not, the fact remains that these differences can never be entirely eliminated, even when the most firm and explicit injunctions for programme execution have been laid down by the central state authority. They are, moreover, exacerbated by the fact that, at the regional level, environmental politics are of necessity always acted out within the context of – and must be interwoven with – existing national, regional or local politics, which also differ considerably from place to place. This interweaving is, furthermore, subject to the conditions resulting from the existence of a particular amalgam of the technical, carefully focused operations and the acts of political will. In the light of these realities, it would be naïve to expect "uniformity" in the implementation of environmental programming.

2.5.2 The regional plan of action

Before the regional authorities tackle the drafting of individual decrees, they must first set their priorities: no authority can be continuously omnipresent when it has to discharge its task in a scientifically sound manner and if it is not to engender political antipathy on all fronts at one and the same time. In the countries of Western Europe it has in most cases proved feasible to establish territorial priorities by identifying areas of heavy pollution, as revealed by immission measurements. At all events, it is a fact that when excessive immissions create acute danger for human beings, animals and plants in a given area, the authority is well advised to concentrate its (generally modest) analytical, financial and staff resources on the area in question. Such territorial concentration, as compared with attempts at simultaneous country-wide implementation, has the advantage that it lends itself to the creation among interested parties of local networks which can enable informal understandings (in the sense of 1.1 above) to be reached between the parties, thus leading to the speeding-up of implementation.

As a rule, territorial priorities are linked to priorities in time. The concentration of administrative attention on a particular polluted area implies that measures to remedy emissions outside the area are initially deferred. Such chronological staggering can, however, also take place independently of the setting of any territorial priorities – when, for example, a decision is taken to deal in sequence with remedial action for different sectors of the economy. Such a course is, indeed, to be recommended, particularly because investment in any given sector tends to follow a cyclical pattern: remedial action can conveniently be timed to coincide with the writing-off of plant or with the launching of new technologies on the market.

The last case to be discussed in the present context relates to the setting of priorities for remedial action on a function-by-function basis, as is also feasible in connection with other sources of emissions – e.g. agriculture, households, public works. As in the case of chronological priorities, functional priorities may be disrupted due to the political niceties of what might be termed "parallelism in sacrifice". For in many instances, action affecting a given sector of the economy can, in practice, be taken only if at the same time corresponding action is taken with regard to other sectors. Well-known examples of this phenomenon are those relating to water protection and to the maintenance of clean air in the context of transport: farmers are reluctant to consider reducing their use of fertilizer as a contribution to water-protection, until their communal authorities stop discharging their sewage untreated into the waters in question. Automobile associations – which enjoy much political clout in the countries of Western Europe – insist on action being taken against excessive Nox emissions by local industry in return for acquiescence in traffic restrictions aimed at improving the quality of air.

Action plans entailing such "parallelism in sacrifice" can result in earlier willingness to change and thus in greater efficiency: they stem from an act of political will, which in certain cases may not mesh with purely scientific considerations. Such a case shows that the "monitoring of technology", referred to at 2.2. above, must always accompany the preparation of action plans and the setting of priorities. It also poses the question of how ready the interests concerned are to change their attitudes and, consequently, to agree to the curative measures proposed. The preparation of action plans will accordingly be seen to require not only technical expertise but also the ability to achieve a public consensus. Thus, for the implementation of environmental policy, and in addition to such technical expertise, there is a growing need for social engineering.

If the construction of an action plan is not to meet the same end as the building of a sand castle, the environmental authority must endeavour to bring into a consensus all the other authorities which will be affected by the set priorities. The wide-ranging and early participation of these other authorities in the planning process will make it easier to obviate any resistance on their part when it comes to drafting individual decrees. Thus, for example, when priorities are being set with a view to curbing air pollution – largely caused by traffic – in the worst affected zones of Western European cities, it is essential to involve the authorities in charge of road construction, traffic control and town planning in the process. The same holds for corresponding curative measures in heavily polluted areas in Eastern Europe, for which close cooperation with the regional authorities responsible for industry, energy and town and country planning is required. In order to deepen the involvement of the "other" participating authorities in the plan of action, it is advisable for a binding document to be prepared, setting forth explicitly how individual measures will in due course be implemented. The structure of the body which prepares and takes these implementation decisions should, in so far as practicable, be the same as that having elaborated the plan of action. By this means, any subsequent opting out on the part of a particular "other" authority is at least made more difficult.

2.5.3 Executing individual decrees

When the plan of action has been drawn up, the authority responsible for its execution proceeds to implement the various measures set forth therein. Here the authority will respect the agreed procedures for implementation, as well as the entitlement of the commercial interests and of the environmental protection bodies (see 1.1. above) to their legal rights. Official instruments for implementation may take the form of the classical act of management – i.e. an individual, concrete official decision, including injunctions and prohibitions applicable to every potential polluter and set forth in permissory terms and/or in decrees requiring the taking of specific remedial measures. When several polluters are concerned by a single decree, or by similar ones, a "general-concrete" plan may constitute an appropriate instrument. Such a plan enables the labour-intensive preparation of several separate decrees to be replaced by a collectively applicable decree conforming to the general requirements of legal protection, thus allowing implementation to get ahead faster. Plans and collective injunctions of this kind are particularly apposite in dealing with agriculture, for example, when several farms within the catchment area of a ground-water course are required to reduce their use of fertilizers. Similar collective decrees are in common use with regard to town and country planning and (in the form of generally applicable detailed requirements concerning traffic signs) in the context of an environment-oriented transport policy.

It must be said that no authority will survive a campaign of curative measures which is exclusively confined to the promulgation of formal decrees: for one thing, the sheer volume of work would be overwhelming. In most countries of Western Europe, therefore, practical implementation is increasingly effected by informal means. These include, for example: verbal agreements between the authority, the environmental protection bodies and the commercial interests concerned (both collectively and at the level of an individual firm); offers of advice (particularly in agriculture and in the waste-disposal sector); the organization by the authority, in collaboration with industry-specific associations, of events to impart information and training; and the dissemination, by means of informal procedures, of texts of regulatory agreements likely to be accepted by a consensus – e.g. within the context of events held with a view to mediation. It is clear, however, that such low-key methods are applicable only when there is a practical prospect of securing the voluntary collaboration of the commercial interests and of the environmental organizations, and thus the uncontested implementation of the legal requirements. If informal procedures should, however, fail, the authority can of course always fall back on the promulgation of a formal decree. Depending on the legal system in a given country, such a formal decree may be subject to judicial review in an administrative court.

2.5.4 Evaluation

The final stage of any procedure for implementing an environmental policy should consist of an assessment of the real value of the various measures taken. Towards this end the regional authority responsible for implementation examines, on the basis of monitoring data furnished by an independent body (environmental- and policy-monitoring as discussed at 2.1. and 2.3. above), to what extent the measures prescribed in the plan of action and put into effect by means of specific decrees have actually led to changed behaviour by the industrial interests in question (e.g. as regards emissions) or even to an actual improvement in the quality of the environment (e.g. as regards immissions) or in the effects of immissions on the recipients of the policy in question (human health, animals and plants). As has already been pointed out at 2.4. above, such evaluations are in practice hampered by numerous obstacles. Experience reveals that environmental authorities tend to ascribe to their own doings any improvement in the environment which has happened as a by-product of another, non-environmental, policy, or which has come about quite independently – e.g. as the result of a recession. Such evaluations should therefore be carefully scrutinized. It is, indeed, in the manifest interest of a regional authority to avoid any such exaggeration of the impact of environmental policy. Outside experts could therefore usefully be associated with the making of such assessments – or could in fact be put in charge of the exercise.

The findings of such assessments must be made accessible to the public. This is a matter of major interest to the non-environmental authorities which have participated in the programme, as well as to the commercial interests and the environmental protection bodies. Moreover, evaluation reports present useful environmentally relevant data, such as the results of monitoring. These are, indeed, supposed to be published within the European Communities, pursuant to the guideline[4] of 7 June 1990 on free access to environmentally relevant information. In the light of the public debate thus set in motion, the regional authority will be able either to rectify its plan of action or to set other priorities in the context of its on-going preparation of a new plan. If an evaluation should reveal any shortcomings in the central government's environmental programme, the programme management must be brought into the picture.

2.6 Conclusion

The recommendations set forth in the present paper have all been evolved in the light of the most recent experience of the environmental policies of the countries of Western Europe. The author has presented them as a whole, in response to numerous questions from Eastern European colleagues about how to organize an effective environmental policy in their countries. In all the discussions which he has had, however, he has been at pains to stress that his limited knowledge of the countries in question does not enable him to assess the extent to which Western European experience can be applied in the context of Eastern European realities. He wishes to reiterate this reservation at the conclusion of this paper. It may well be that, in view of the dramatic levels of pollution in these countries, resort may, as a matter of principle, have to be had to environmental policies other than those thus far applied in Western Europe. The author does not consider himself at the present juncture to be in a position to put forward any concrete proposals for such policies which – in the light of the gigantic social

[4] E.C. Guideline 90/313/EEC dated 7 June 1990 on free access to information concerning the environment, AB1 1990 L 158/56.

upheavals accompanying the transition from a "socialist" to a market economy – will presumably have to bite deeper before accomplishing a transformation (thus far by no means achieved in the capitalist countries of Western Europe) into a truly ecological market economy. Fundamental rethinking along such lines would imply getting away from the emission-focused programming which has hitherto held sway, and embracing instead the vision of an entirely new regime, based on the concept of dedicated resource-management. A discussion of this topic would, however, exceed the scope of the present paper.

References

Dente, Bruno (ed.) 1993. Environmental Policy in Search of New Instruments. London: Gower

Knoepfel, Peter, Weidner, Helmut 1986. Luftreinhaltung (stationäre Quellen) im internationalen Vergleich. Berlin: sigma (6 vol.)

Knoepfel, Peter, with the collaboration of Imhof, Rita 1991. Les cycles écologiques et le principe de légalité – De la nécessité d'assouplir les liens de causalité, Cahier de l'IDHEAP no. 72. Chavannes-près-Renens: IDHEAP

Knoepfel, Peter 1991. La Protection de l'environnement en proie aux problèmes d'acceptation et aux déficits de mise en oeuvre – Mise en oeuvre de la législation environnementale: situation actuelle à la lumière des secteurs industrie et artisanat, Agriculture et Politique infrastructurelle de l'Etat, Cahier de l'IDHEAP no. 70a., Chavannes-près-Renens: IDHEAP

Knoepfel, Peter, Imhof, Rita, Zimmermann, Willi 1993. Luftreinhaltung im Verkehr: Analyse von Vollzugsprozessen In der städtische Umweltpolitik mittels Massnahmenplänen gemäss Luftreinhalte-Verordnung Art. 31 ff., Final Report of Project, Nr. 40228112, of the (Swiss) National Research Programme on Town and Traffic, (Nr. 25)

Knoepfel, Peter, Zimmermann, Willi 1993. Gewässerschutz in der Landwirtschaft. Evaluation und Analyse des föderalen Vollzugs. Basel: Helbing & Lichtenhahn (Reihe Oekologie & Gesellschaft, Bd. 7)

OECD 1989. Instruments économiques pour la protection de l'environnement. Paris

OECD 1991. Politiques de l'environnement. Comment appliquer les instruments économiques. Paris

Schweizerische Kommission für Umweltbeobachtung (Swiss Commission for Environment Monitoring) 1993. Konzept für die Einführung eines

Beobachtungssystems, Final Report. Berne: Schweizerische Akademie der Naturwissenschaften

Tsuru, S., Weidner, H. (eds.) 1989. Environmental policy in Japan, Berlin: Wissenschaftszentrum, Research Area Technology, Work, Environment, Berlin: sigma

Voig, N.J., M. Kraft, E. 1990. Environmental Policy in the 1990s. Towards a New Agenda. Washington: a division of Congressional Quarterly Inc.

Weidner, Helmut, Zieschank, Roland, Knoepfel, Peter (eds) 1992. Umweltinformation: Berichterstattung und Informationssysteme in zwölf Ländern. Berlin: sigma

3
Formulation and Implementation of Air Quality Control Programmes: Patterns of Interest Consideration (1982)[1]

Peter Knoepfel, Helmut Weidner[2]

3.1 Introduction

Recent studies from the field of implementation research (e.g. Downing/ Brady, 1980; Majone/Wildavsky, 1979; Barrett/Fudge 1981) have indicated clearly that a strict division according to content cannot be made between processes of programme formulation and those of implementation from a theoretical-analytical standpoint, and that such a division further leads to unrealistic presentations of empirical policy processes. "Implementation does not assume a fully articulated policy decision: it creates and recreates it". Furthermore, "implementation ... must reformulate as well as carry out policy" (Majone/Wildavsky, 1979). This view could hardly be debated on the abstract level. However, it neither aids a practice-oriented formulation of theory concerning the specific methods of function of administrative policy enforcement, nor does it serve the practical "enforcement policy" built upon this.

---

[1] The paper is based on experiences gained by the authors in the preparatory phase of two research projects – funded by the Volkswagen Foundation and the German Research Foundation – in the area of "international comparative analysis of standard setting and implementation processes in SO2 air pollution control policy of the EC countries and Switzerland" and which is referred to in Knoepfel/Weidner, 1980. The available literature and the pertinent standards and documents were analyzed, and numerous interviews with representatives of central and enforcement agencies were evaluated. Further bases include the results of pilot studies in Bristol, Emilia Romagna and Zurich, which were carried out by "national research teams". The phase of broader empirical study started in the summer of 1980. The observations presented, therefore, have a temporary and in some places heuristic-hypothetical character.

[2] in : Policy & Politics (1982), vol. 10, no 1: 85–109. Copyright: We thank the editor for authorizing the publication of this article with some minor linguistic adaptations proposed by Susan Cox.

This unsatisfactory temporary result is probably due in no small measure to the fact that research up to now has operated with a concept of programme which is too narrow and not sufficiently thought through. Despite the recurring indications of their relevance to implementation (cf. Poland, 1974; McGill/Wooten, 1975; Williams, 1975), only a few studies are concerned with programme elements and structures. Only when it has been established that varying programme-structures lead to clearly divergent results under similar policy goals can the thesis of the relevance of programme structure be confirmed. The reason that the treatment of the programme has been neglected in comparison to that of other influence variables[3] probably lies in the fact that American studies, which are limited to the local scene, can usually include no such variants because these do not occur within one and the same country to any significant degree. As soon as studies aim toward an international comparison, it becomes necessary for them to concern themselves with programme structures. Thus, in the first part of this study, we will present several considerations in the area of sulphur dioxide (SO_2) pollution control policy.

The consideration of programme structure brings with it the recognition, particularly when this is done with attention to the interest-specific perspective, that it is very difficult to correctly formulate the question of differences in content between programme formulation and implementation processes. Whoever operates with a realistic concept of programme, which along with internal goals also includes process, organization and resource decisions, will understand the programme less in the light of its contents than in that of its procedural linkage of performance standards, prescriptions for action and corresponding level of control. Programming is "fixing the game" (Bardach, 1977: 268), and therefore consists of the more or less rigid, explicit or implicit establishment of patterns for the consideration of interests for later policy processes under conditions of already established infrastructures. This will be demonstrated on the basis of three different programme structure types in the second part of this study. Given this background, the inclusion of essential programme elements in implementation studies in the sense of integral policy process analyses will be urged.

---

[3] E.g. implementation field structures or implementation structures. Cf. the survey by Mayntz, 1977: 5lff.

3.2 Problem structure and programme needs in SO_2 pollution control plicies

3.2.1 Global and regional of SO_2 pollution control policy strategies

SO_2 pollution control policy has as its object the control of pollutant concentrations (ambient air or "immission" conditions) to which certain selected objects (human beings, animals, plants, buildings or ecosystems) are exposed. The points of departure for administrative action are the combination of raw materials which go into manufacturing and combustion processes *(input control),* the technology of production or combustion facilities *(process control)* and the distribution of emissions occurring in a specific area *(transmission control)* by regulating the distribution of local sources or control over chimney heights.

The SO_2 emissions which are more or less administratively controlled result in ambient air conditions which are technically measurable, and in the case of the pollutant SO_2 these have been measured in European agglomerated regions for some time. With the extension and the continual refining of monitoring networks, a control of ground level pollution is possible. A comparison of the data thus obtained with the goals of clean air policy indicates whether reasons exist for further measures. On the basis of trend analyses, conclusions can be made concerning the effectiveness of the strategies applied (evaluation). In comparison to other policy and pollutant areas, this one presents probably the most favourable preconditions for evaluation.

In clean air policy, we differentiate between *global and regional strategies* (Stamer, 1976): the former are directed at the control of the total amount of pollutants emitted in the entire country. Their local distribution thus does not come into consideration. The principal means of such strategies are standards for products (the sulphur content of fuels), technologies (e.g. prescriptions for the effectiveness of furnaces) and activity-specific emission norms (limitation of amounts Of SO_2 emission). By contrast, regional strategies concentrate on the *pollutant concentration in the surrounding air,* and thus on the control of emission activities of individual emitters under the specific local distribution conditions. Besides the control of transmission, the same instruments are used as those employed in global control. Here, however, they are differentiated according to the varying regional ground level concentrations. Thus, in many European countries, stricter sulphur content standards for fuel, sharper technology prescriptions or more extensive limitations on emissions apply to highly polluted areas than to other regions of the country. It is also usual to characterize global strategies as emission-oriented, and regional strategies as ambient air quality-oriented (air quality management approach or, in German terms: "immission-oriented approach"). We find the differentiation of global versus regional to be more appropriate, since it more precisely clarifies the importance of the regional concept in air pollution control policy for the structure of programmes.

Certainly all global control measures alter the ground level concentration most particularly in regions with a high number of sources. Nevertheless, on the basis of structural and/or regional economic policy as well as varying additional pollution tolerances of the regions, they are not oriented to the highest level of pollution. For this reason, most countries have additional regional strategies.

3.2.2 The trend toward environmental quality standards

Global and regional strategies are both aimed at the regulation of SO_2 concentrations. Nevertheless, they have differing programme structures. Global structures can do without ambient air oriented quality criteria and corresponding air quality standards both from the standpoint of decision-theory and that of administrative practice, since they are not selectively oriented toward defined heavily polluted areas. Regional strategies, on the other hand, are only fully capable of functioning in the realm of enforcement practice if they include a programme of established quality criteria and of air quality standards which are not to be exceeded. Air quality standards prescribe the highest allowable concentration of a particular pollutant in yearly, daily, hourly or half-hourly mean averages. Often, short and long term standards are established[4]. The practical demands on administrative programmes in regional strategies have extensive consequences: since there are heavily polluted areas in every highly industrialized country, regional strategies are always necessary as a supplement to global strategies. Thus additional orientation standards for air quality exist everywhere. Under pressure from measurement data collected and publicized, as well as from the recognition of the effects of pollution, global strategies are also legitimated by, or forced to take into account, ambient air quality standards. This is important, for example, to the extent that different administrative officials are responsible for regional and global strategies: the Ministry of the Interior[5], which is in many European states responsible for the regulation of sulphur content of fuel, often finds itself in a situation, in

[4] Cf. the survey by Weidner/Knoepfel, 1979: 161.

[5] As in Belgium (Labour Ministry), Italy and in Switzerland.

which it must base a regulation of sulphur content on air quality standards which were established by the Environmental Ministry. Furthermore, the necessity of air quality standards for the implementation of regional strategies results in the fact that the air quality question becomes a political theme from the standpoint of social, medical and risk policy considerations. Global strategies are generally conditioned by the technical status (more seldom: the latest advances) of abatement technologies (in Britain: best practicable means). They also take economic justifiability into account. The costs of prevention can already be calculated at the point that standards are established. The focal point of the policy process lies in the setting of standards, which usually appears as a relatively closed "bureaucratic-industrial co-operation and negotiation process"[6]. This bipolar interaction system is interrupted by the appearance of air quality standards: a new factor, the costs of which are difficult to calculate, comes into play. Medical experts and biologists are not interested in the particular production conditions of air pollutants, but rather in their effects[7]. When levels are regarded as unsatisfactorily high, emission reduction is often urged without regard to the costs of prevention. The entry of such "pollution-effects-researchers" (especially epidemiologists) into the system of standard setting which is occupied by bureaucrats, technologists and firms, signals the inclusion of the affected parties into the standard setting process which then becomes the subject of increased pressure toward politicization (Knoepfel/Weidner, 1979: 167).

Harder bargaining processes must also reckon with air pollution control policies operating on the basis of air quality standards after the phase of standard setting. Wherever the establishment of air quality standards occurs as a quasi contractual agreement between administration and influential emitter groups, difficulties in programme implementation will "only" occur when smaller and medium-sized emitters cannot keep up with the standards agreed by the "big" emitters. "Implementation deficiencies" can reflect in such cases intramural and intergroup disagreements between parties. Where the standpoint of affected parties is taken into account aside from production considerations, and therefore room is provided for explicit political decisions, resistance to enforcement reflects conflicts between individual and collective goals which can hardly be anticipated in the phase of standard setting. The results of such conflicts are considerably

---

[6] A good example for this is Belgium, which is the only European country operating with emission norms without providing for corresponding ambient air quality standards. Cf. the chapter on Belgium in Knoepfel/Weidner, 1980.

[7] Cf. Antweiler, 1978.

influenced by constellations of interests and enforcement opportunities within the various implementation fields.

The perception of the political importance of processes which have environmental quality standards as their object rises correspondingly among emitters and affected parties. The demand for their further political legitimation, for legal developments[8], and the corresponding attempt by organizations of affected parties to expand their activities beyond the implementation process back to the stage of standard setting shows this clearly. This increasing political importance is also expressed in the growing significance of enforcement practice: environmental quality standards provide the central orientation standards for the formulation of licensing conditions, abatement ordinances and for - sometimes quite radical - temporary interventions in the production process or in personal freedom in alarm situations. Furthermore, they are a basis for the identification of highly polluted regions, and are finally used for the evaluation of the clean air activities of officials. They can therefore considerably strengthen the bargaining position of organizations of affected parties and officials in their debates with emitters[9].

Against this background, it is no wonder that only a few countries operate with explicit environmental quality standards[10]. Particularly France and Great Britain rigorously oppose the inclusion of such standards in air pollution control programmes. In official government statements, despite assurances to the contrary, there is a certain fear of showing one's colours in the realm of environmental and health policy. As will be shown below, France at least operates with environmental quality criteria.

3.3 Structures of actors and interests

Standard setting, understood as the totality of decisions which lead to environmental quality standards, criteria or action-controlling equivalents, can occur according to the type of programme in the actual policy formulation, as well as in implementation processes. It can proceed either as a relatively isolated process with its own communication network on one of the two

[8] Particularly in the Federal Republic of Germany; cf. Fall STEAG/Voerde in DVBl, 1978, p. 594ff, and Volume 1 of the Zeitschrift für Parlamentsfragen, 1979.

[9] Arguments on this point in Mayntz et al., 1978, and for the USA in Shiffman, 1973.

[10] E.g. the Federal Republic of Germany, Italy, the Netherlands and Switzerland. Cf. the corresponding national sections in Knoepfel/Weidner, 1980.

levels, or as a more or less specialized exchange process between programme formulation and implementation. As a rule, in policy formulation and implementation processes there are to some extent disparate actor groups and action calculations to be found. The basis of this may be, for example, purely objective (the need for other information), political (higher chances for influence in one of the two areas), bias in the giving of concessions or material (lack of resources). Mere ignorance can also play a role (a lack of consciousness of the importance of various process stages). The factual interdependence of both processes makes it necessary for empirical studies to form a complete picture of actors and interests for the total policy area. It is therefore necessary that the behaviour of "related" actors be analysed throughout the entire policy process (Borghorst, 1979). The bias in an analysis limited to the implementation process, for example, consists of the danger of mistaking petty infighting (i.e. disputes on a lower order of importance or "implementation wrangling") for the processes whereby standards are established. Whoever wishes to have more than peripheral influence upon the implementation process can seldom avoid extending his activities to the phase of programme formulation.

In air pollution control policy, standard setting in the sense of institutional establishment of environmental quality standards is a central element of programming, and thus of the policy formulation process. If one wishes to estimate the difficulties involved with implementation effects by means of alterations of such programme elements, then it is not sufficient simply to know the pattern according to which interests are taken into account with regard to effects of such standards, which is central to this study. The pattern of accounting for interests in the process of its formulation, which does not always amount to the same thing, must also be analysed.

Potential actors can be the beneficiaries or the victims of a situation which was previously unregulated or "otherwise" regulated, previous or new political-administrative actors, or other participants by virtue of their own or their appointed "right". A differentiation between interest and actor groups is possible by means of an analysis model which divides a societal system into three areas with regard to the principally differing functions within the total system (Habermas, 1973: 15; Offe, 1973: 213):

- In the *socio-cultural (legitimatory) system*, we find the parties affected by air pollution, whose interests at least consist of living under environmental conditions which cause no injury to health. Measures aimed toward this goal should have a high degree of generalizable power of legitimation.

- In the *economic system,* it is the emitters who are interested on the basis of profit calculations at least in a certain degree of allowable air pollution. However, it is difficult to legitimize measures which publicly embody this goal; strategies will be adopted which reduce transparency and which work against evaluation.
- In the *politico-administrative system* it is the clean air authorities who are at least interested in the "management" of the two above named interest spheres as well as other interest conflicts which can be traced back to the same basic pattern, in order to make them "capable of coexistence" (Grottian, 1973: 20), so that no serious disturbances occur in the total system or within the area for which it is responsible. The legitimation effect is precarious because of the conflicts of interest, so that here the utilization of strategies which hinder transparency must be anticipated to some extent.

This method of analysis which attempts to depict the influences and calculations of the actors against the background of structurally fixed social-power relationships, is oriented to a very abstract concept of interest, which is as yet too amorphous in its depiction of interest groups for the purpose of an empirical investigation. Even within the individual systems, it is well known that there are conflicts of interest which have an exceedingly high political effect. In order to cover the actual variety of these interests within the three sub-systems in the framework of an empirical study, the analysis grid must be fine enough (cf. in detail Knoepfel/Weidner, 1978: 32ff). This is possible if the procedure is oriented towards pollutants: which groups would have an over-riding interest, brought about by their being specifically affected by sulphur dioxide, in additional emissions or in their reduction? This is possible only through an empirical analysis. Prior theoretical considerations have the purpose of assuring that no relevant interest group is ignored, and that a realistic interest preference can be discovered. Such an extensive analysis of interrelationships can be carried out in the case of the pollutant SO_2 with the fine detail that we have postulated here, since SO_2 air pollution primarily occurs through combustion processes in the course of energy production. The emitter groups in question, the energy policy and economic implications of various possible clean air measures, and the effects of the pollutant are sufficiently known and observable that the theoretical claims would not be undermined by the over complex demands of an empirical study[11].

[11] Cf. in detail: Downing, 1979.

3.4 programme structure

The need for programmes presented here is regarded as a necessary control dimension for an effective clean air policy. Programmes of regulative policy usually have as their core concrete performance standards, which translate the general goals contained in laws or in governmental programmes into guidelines for action. This programme core - in the case of clean air policy, the "naked" ambient air quality standard - requires an additional dimension which is evaluative (measuring systems, basis for the evaluation of measurements) and emission controlling (emission product and process standards), in order to be at all administratively feasible. As in other areas of policy, the actual standard at the core of the programme functions only as privilege/discrimination, the conversion of which into process oriented administrative guidelines for action demonstrates a variety of political contingencies. For studies oriented to interests, it is appropriate and realistic to analytically separate this additional dimension from the programme core, and to conceive it as the first group of "programme shells" surrounding the core *(inner programme shells)*. In this way, tensions between performance standards based on results and action standards based on processes can be made evident. As in many areas of regulative policy, the performance standards formulated in the programme core require an *organizational and instrumental foundation* which must be similarly formulated in the programme and will be designated *"outer programme shells"* in the following outline.

The principal difference between the individual countries, when one compares problem formulations and strategies utilized, lies not in the control dimension itself but rather in the *control level* which is explicitly or implicitly laid down in detail according to the control dimension in the respective national programmes. It will be the task of the explanations of the three programme structural types presented above to demonstrate that in air pollution control policy as well[12], formal and actual control levels quite often differ from one another. Such factual distortions occur particularly when co-ordination efforts (because of central "under-control") or practitioners (because of central "over-control") manage to bring about an extension of local contingencies. Wherever explicit rules are lacking, modes of administrative behaviour arise, preferably on a low control level, to "cover up" the corresponding (intentional or not intentional) loophole in the programme.

[12] Cf. also Bohnert/Klitzsch and Garlichs.

Practically every one of the nations surveyed demonstrates a division of the individual control dimension which is already institutionally differentiated according to national, regional and local *control levels*. This variety is also retained when the typification proposed below is utilized. This permits a choice of programme structures of air pollution control policy suitable to a comparative sketch for the Federal Republic of Germany, Italy, France and Great Britain, as well as the pattern of interests contained therein. These structures and interest patterns vary from one extreme (national control level for practically all dimensions) over a medium solution (national control level for inner and outer programme shells) to the other extreme (national control level only for the outer programme shells).

The following diagram presents this schematically.

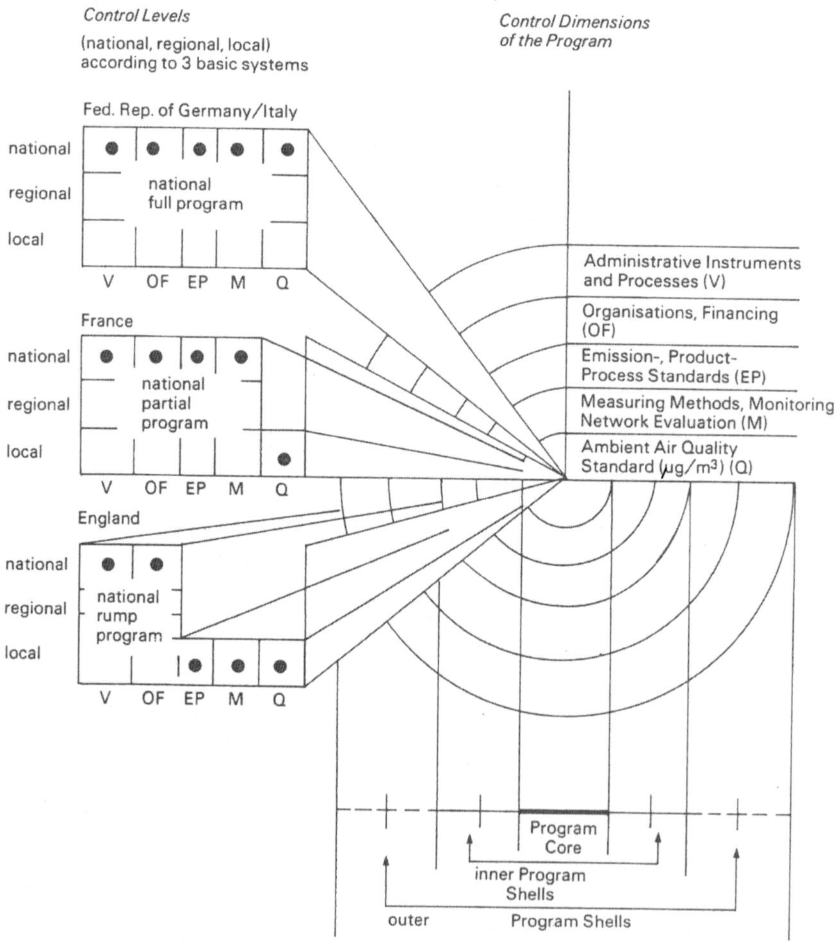

3.5 Full national programme (type 1): handling of interests through the formulation of inner and outer programme shells[13]

3.5.1 Programme shells as relevant to interests

Centralized standard setting in the programme types found in the Federal Republic of Germany and in Italy are characterized by the fact that they often take place on the presumed sidelines. Representatives of economic interests cannot afford in the present situation to publicly attack the health demands which are based on medical-scientific considerations. Thus, the interest is concentrated less on the ambient air quality standard as such (programme core) than on its programmatic surroundings: attempts are made, by pulling the teeth out of the enforcement section of the programme, to partially checkmate the programme core itself. Often support of environmental interests is demonstrated by a neutral behaviour toward ambient air quality standards (programme core), which are standing in the spotlight, while at the same time it is tried to weaken their impact by manipulating essential elements of the peripheral programme shells safe from public control. An analysis of interests must proceed with this in mind. It is evident that emitter organizations often use such "shell strategies". Since they often are the only actors who are systematically active *both* on the norm formation and the implementation fronts, they have usually built up an effective system of information transfer by means of which they are to gain immediately applicable knowledge concerning the implementation relevance of central control dimensions.

With regard to the *inner programme shells,* the following strategies can be established:

- *Percentile Reduction Strategy:* An ambient air quality standard can be defused through the establishment of a low percentile[14]. By this means, the number of those peak values which do not have to be included in calculations of the mean average can be increased. If, for example, the percentile of the mean average value can be reduced from 95% to 92%, then instead of 5% as previously, 8% of the peak values can be struck

[13] Cf. on the clean air policies of the Federal Republic of Germany and Italy the corresponding national sections in Knoepfel/Weidner, 1980. Also for Italy: Dente/Lewanski, 1979.

[14] A good example is provided by the Belgian discussion concerning the establishment of the percentile. Cf. Knoepfel/Weidner, 1980. Italian monitoring data do not give the percentile.

from the calculation of the average. Such reductions are particularly effective in areas in which, because of unfavourable topographic and climatological conditions, a small number of high peak values occur.

- *Measurement System Control:* The value of an ambient air quality standard for enforcement practice can be considerably varied by means of manipulation of measuring systems, monitoring areas, evaluation methods and the degree of publicity of monitoring results. Thus, an extension of the area to be monitored allows the "measurement" of a more favourable result than in systems with relatively small measurement areas[15]. Of further importance are the organization, type, and particularly the periodicity of data publication. Monitoring data which - as in Italy[16] - are made public only after a lengthy processing and/or authorization process in a manner which can hardly be deciphered, have a lower mobilizing effect upon the activities of parties affected, which otherwise appear spontaneously as a reaction to acute peak loads.

In the full programme type, the importance of the outer *programme shells* - in contrast to partial and rump programmes - often lies in a weakening of the programme core. This purpose is served by the following strategies:

- *Implementation Structure Control:* At the centre point of debates which can become quite intensive is, first of all, the formal distribution of planning, approval, supervision and enforcement instruments among local, regional or national bodies, which is explicitly defined in the full programme. Such options are often affected by calculations of interest strategies. One strategy which should be mentioned is the "worry" which is often dragged into the field of air pollution control policy by industrial groups regarding federalist or decentralized concepts of democracy (Borghorst, 1979; Knoepfel, 1977: 194ff). In fact, they hope to gain far stronger bargaining positions in such decentralized regulations - above all in small and medium-sized industry - than is possible with regional or even national authorities[17]. This striving for localized protectionism meets with resistance within the industrial organizations. Large industries in the Federal Republic of Germany and Italy see their interest in the harmonization and unification of markets endangered[18]. In view of the high degree of technologization of air pollution control,

[15] Cf. on the general problem area: Krämer, 1978: 93ff.
[16] The newest data available are from 1975.
[17] Cf. in general also Downing, 1979.
[18] These are expressed e.g. in the demand for legal establishment of ambient air quality standards by the BDI (Federal Association of Engineers).

the personnel and material equipment of the responsible administrations and technical service agencies becomes a political theme. Its great importance for the chances for implementation of ambient air quality standards is included in the calculations of the participating actors. As the example of Italy demonstrates, economic interest groups can maintain a pro-environment facade by accepting relatively stringent standards, whereas in the budget debate, which is undertaken elsewhere, they seek to prevent the voting of the necessary administrative funds[19].

- *Control of Administrative Instruments and Processes:* In air pollution control policy, the basis for approval, the approval process, and the administrative supervision and abatement instruments stand in the foreground. The structure of the formal administrative procedural law, which still regulates behaviour, is particularly important as an instrument for the distribution of bargaining power (Knoepfel, 1979: 72ff). It corresponds to the professional qualifications of a still large portion of the enforcement officials and makes possible a high degree of routinization of implementation processes. Such formalization, combined with the almost loophole-free legal protection system in the Federal Republic of Germany, brings about a relatively centralized control of the relevant dimensions of the process. This is particularly true with regard to the duty of secrecy and the exclusion of third parties, which exists to a great extent in Italy[20]. In contrast, the inclusion of third parties[21] postulated on a programme level in the Federal Republic of Germany may have essentially weaker chances for implementation. As has been shown in corresponding studies (Mayntz *et al.,* 1979), the outcome of the approval or abatement decision is often established by the end of the preliminary hearings which are legally regulated without participation of parties affected. Citizens who participate in the main process face in such cases an already hard and fixed, quasi contractual decision; their subordinate function is in fact institutionally assured.

[19] A good example is provided by Italy, the provinces of which are denied the means for the establishment of monitoring systems by the national government. Cf. Dente/ Lewanski, 1979: 153ff.

[20] Third party participation in approval processes is not provided; the law indicates for third parties the possibility of legal recourse following the granting of the approval decision, which occurs under exclusive co-operation of the emitters (Legge 13, luglio 1966, n. 615: Provvedimenti contro l'inquinamento atmosferico).

[21] Federal Air Pollution Control Law of March 15th, 1974 (BGBl.I. p.721), §§ 10ff.

3.5.2 Pattern for the handling of interests

The mere presence of national full programmes, particularly the existence of environmental quality standards and decisions on measuring methods as well as emission and product standards, by no means allows the conclusion that, despite the sharp differences in content and results between the West German and Italian air pollution control policies, there is in the case of the former, any concern for those affected. Full programmes certainly force a thematization of the total programme dimensions and put them at the disposition of the national authorities. However, by this very process they are subject to legitimation through organizations of affected parties. This pressure toward thematization which is assured by the programme structure translates into actual favouring of the interests of affected persons in the content of the programmes when this pressure is applied systematically by powerful organizations of affected parties along the entire line, particularly in the area of the programme shells. The example of Italy makes this dilemma clear; such organizations, in comparison to the Federal Republic of Germany, are almost non-existent up to now. Dominant industrial organizations succeeded in protecting their interests even in national parliamentary decisions in spite of a conclusion of arrangements with agricultural and tourism organizations of the highly polluted northern regions. This was particularly the case in the areas of measurements, product standards and approval processes (Mannozzi et al., 1979). The latest debate on the revision of the "Technical Instruction on Air Pollution" in the Federal Republic of Germany has met with comparatively high resistance to change the programme which was developed in the period of high economic trends. What part of this is due to the insistence on the chosen programme structure and its particularly high pressures toward justification before a clearly more sensitive public (Fietkau/Kessel, 1979) can only be estimated on the basis of the planned revision of the Federal Air Pollution Control Law and the discussion which will probably arise therewith concerning the concrete programme structure. Here the starting point is a pattern of interest consideration in the framework of full programmes which is precarious both for emitters and for affected parties, because the pressure for legitimation is so high.

3.6 Partial programme (type 2): policy of flexible control relationships[22]

3.6.1 Programme shells as the central national control dimension

In France, regional ground level concentration controls in highly polluted regions are expected to be achieved not only through the "side effects" of global strategies, but also by means of stricter approval, clean-up and supervisory practice in the enforcement of clean air and industrial construction laws. The national administrative programme explicitly requires the enforcement practice to take into account the local air quality conditions. As the basis of the French system is the classic liberal concept of case to case interventionism in policing of business, concrete standard setting thus develops according to the views of the inspector, to some extent as a reflection of his implementation behaviour. Standard formation is a type of case-wise actualization of common sense, i.e. of the notions of the official concerning air quality regulations. The goal of intervention can only be described negatively; its purpose is the absence of disturbances of all kinds. Therefore, the programme core of central control is generally included. This discretionary power of the regional inspector is surrounded by an increasingly tight web of pragmatically and additively developed central programme amendments, without itself being affected *per se*. The central administration attributes to the discretionary power a great deal of importance for the efficiency of environmental strategies just of the flexibility involved. Bargaining processes concerning quality criteria in the area of enforcement thus belong to the programme. Their acknowledgement as part of the programme promotes them out of the darkness of "machinations" into the "bright light" of a pluralistically legitimized social contract. They acquire the ethos of democratic rationalistic micro policy – a specifically Giscardian mixture of rationalizing attitudes and functionalistic-technocratic administrative management.

In the varied arsenal of these programme amendments to national air pollution control policy following control dimensions are found:

- *Control of Technological Regulations Specific to Facilities:* In the framework of global strategies, France is increasingly transferring to a sector-absed or activity-based unification of the technological aspects of national standard prescriptions. These simultaneously effect a partial standardization of leeways, which exist for overlapping regional strate-

[22] Cf. the chapter on France in Knoepfel/Weidner 1980.

gies in agglomeration areas for the sake of a stricter treatment of increased prior loads.

- *Creation of Special Protected Zones:* A particular variant of central controls of inner programme shells by means of special product and emission standards is presented in the setting up of special protected zones by order of the national Minister of the Environment. No quality standards are set for these zones – in contrast to Belgium. Their creation permits a stricter regulation of fuel, more intensive supervisory activities and more comprehensive approval and abatement practice than is customary elsewhere in the country. In the five protected zone[23], which have been thus set up, unified product and supervisory standards apply; comparable conditions are thus provided for an administrative practice which is usually not more strictly defined.
- *Central Organization Control:* In recent years, a clear upward evaluation of the interdepartmental offices for industry and mining so important for approval and supervision in the regions has occurred through increased grants of money, personnel and material by the central administration which are oriented toward increased performance standards. At the same time, the authority for applications was extended to the department prefect, who still had formal authority over decisions in air pollution control policy, but also in economic promotion policies. Through this increasing concentration of professional expertise at the regional level, which includes a number of *départements,* the decentralized foundation of prefectoral business policing continues to lose importance. The same is true for the development and extension of regional monitoring networks, which at present is aided by the central administration with considerable financial and personnel resources; the establishment and operation of these monitoring networks is generally State-supported but privately organized by the associations. The establishment of regional evaluation areas effects an actual regionalization of the control level, even without formalized coupling mechanisms between monitoring and implementation administrations.

The harmonization of enforcement was also supposedly assured by the establishment of a national air agency *(Agence de lair)* in 1978, which was to have provided technical counselling for the regional offices in problems of enforcement. Particularly high hopes were placed on the planned central training courses for factory inspectors. It is certainly conceivable that in the framework of the instruction material, equivalents to environmental quality standards are developed, which could ultimately develop into more

[23] Presently: Marseille, Paris, Rouen/Le Havre, Rhône (Dijon).

rigid and practice-relevant standards rather than compromise formulas arrived at politically and administratively. In comparison to national standards such "technical standards" are often more applicable and are conceived more practically. The political costs of such "norm formation via teaching and course materials" can be estimated as extremely low. The formula for the estimation of chimney heights, which was formally adopted in 1970 and 1974 as a ministerial regulation, as well as the air quality guideline included therein, has more the character of a technical aid. Detailed foreign regulations are thus brought into application throughout the country. In Rouen, for example, you can find a quite well-thumbed copy of the Federal German Technical Instruction on Air Pollution.

- *Central Control of the Implementation Process:* Energetic attempts at harmonization can be seen in the increase of central control of the implementation process. In 1974, the approval process was subjected to extensive regulation, in 1976, the associations' standing to sue was extended, and in the same year the "environmental impact assessment" *(Etude d'impact)* for all facilities subject to approval was introduced[24]. The innovations permit a total increase of the importance of information concerning environmental risks and expected alterations in local air quality conditions among approval and supervisory authorities in the entire country. Simultaneously, an increased participation by citizens could be thus mobilized without sacrificing the legally better position of the industrialist. If a case of conflict is not politicized so that it becomes a national issue - which would constitute an exception - the new participatory regulations will probably lead generally to a consolidation of loyalties within an industry-favouring practice.

3.6.2 Pattern of interest consideration

With the increasing importance of air pollution control policy and the corresponding increase in the costs of prevention, the local protective action by the highly decentralized business regulatory agencies lost its attractiveness. Instead, the advantages of a harmonizing practice for the industrial associations which had in the meantime become strongly integrated, and for the large concerns, took the foreground. This development from the previous dominance of technologically outdated small and medium-sized firms into modern large concerns, supported by an effective investment policy on the part of national and regional sectors, which was consciously

[24] Through the nature protection law (Loi du 10 juillet 1976 relative à la protection de la nature, no. 76/629, art. 2).

aimed at regional structural policy, made a partial unification necessary. In the natural course of events, this first affected global and then also regional strategies for industrial agglomeration areas.

With the increasing integration of the industrial sector and its incorporation into industrial associations with cross-local functions a loss of power and influence occurs among medium-sized and small industrial firms within the associations. The increasing development of national programme elements constitutes this transfer of power which takes place within the private and the national economic systems.

The "coreless" national programme type allows the retention of a low politicization level of standard setting in micro-political implementation processes through the suppression of themes in the area of technical-legal matters which could cause conflict. An invasion of conflict-laden processes of politicization from the implementation front to the centre can be prevented by the immediate decoupling of the - unique - individual front from the total system. This system of flexible control conditions between the centre and the region allows both authorities a largely independent management of legitimacy. While legitimacy under more rigid control systems tends to be indivisible, in this system it can presumably be produced practically autonomously in parallel both by the central administration and the prefectures. This constellation enables a control of the degree of integration of the entire system by the central administration in close co-operation with the industrial system and the regional technical services, which tends to be independent and removed from politics. The central administration makes continuous announcements to the nation of new pragmatically additive programme elements, and *"Monsieur le Préfet"* can fall back on his formal right of decision in cases of local protests or industrial intervention and drop the demands of the regional inspector in individual cases like a hot potato, for political reasons. This system is very attractive for the total economy: on the one hand, a certain amount of external control of implementation agencies is possible on the part of those to whom the standards are addressed, but such control is also available to affected parties. At the same time, environmental quality standards are prevented from entering the political arena and thus the area of bargaining on the national level, thereby avoiding the risk of infiltration into the central control level by environmental organizations, the results of which would be difficult to calculate.

3.7 Rump programme (type 3): central minimal programme - policy of quasi autonomous programme formation and harmonization at the base[25]

3.7.1 Basic situation: lack of a national programme core and inner programme shells

Great Britain has always been different, and this also applies to the area of clean air policy. Here we find the paradox that an explicitly postulated locally oriented air quality control is supposed to be established without operationalized legally binding target limits for ground level concentrations. National air quality standards or even guidelines do not exist. The common approach for large emitters operates in all areas on the basis of a case-wise quality standard for each site resulting from an emission control oriented to the individual case according to the principle of the *best practicable means* (bpm). Under the Health and Safety at Work etc. Act, registered works have to use the bpm for preventing emissions of noxious or offensive substances into the atmosphere and "for rendering harmless and inoffensive such substances as may be so emitted". This vague and therefore flexible goal is made even more imprecise by the requirement that the financial implications of abatement requirements and local conditions are to be taken into account in the concrete formulation of the status of technology. The required inclusion of environmental conditions, however, makes it practically unavoidable for implementation to take account of ambient air quality criteria.

In its capacity as implementation agency of the central administration "HM Alkali and Clean Air Inspectorate" (here referred to as Alkali Inspectorate) has set so-called "presumptive (emission) standards", which serve as the official concretization of the best practicable means principle and as a national minimal standard. The standards can be made more strict by the respective district inspectors according to local conditions. Whether and how they use this discretion is hardly estimable "from outside", since no binding rules or duties of information exist. The only central control dimension is the nationally unified formula for chimney heights. This technical guideline, however, does not include - as, for example, in France - a "hidden" ambient air quality standard. The local ambient air situation is taken into account in the calculation of the height of chimneys only indirectly and approximately, in that different domestic and industrial struc-

[25] Cf. the chapter on England in Knoepfel/Weidner, 1980 and Hill, 1980.

tures may be taken into account. The concrete definition of "low" or "serious" air pollution is left largely to the implementation officials. Nevertheless, this chimney height regulation has an important part in the generally relatively favourable SO_2 situation in Great Britain since in particular the parameter "maximal capacity of the combustion facility" which is applied leads to relatively high chimneys. A considerable "purification effect" of the pollutants thus emitted results further from the favourable prevailing winds, which carry a good part of the pollutants to Scandinavia. The "policy of high chimneys" is the core of British SO_2 policy.

3.7.2 The central political position of outer programme layers

The focal point of central control lies on the organizational level: the implementation structure shows a dual order of organization and responsibility. The central Alkali Inspectorate is responsible for industrial facilities, and the local authorities are primarily responsible for domestic heating and small businesses.

- *Industrial Emissions:* For all problematic firms and processes, the Alkali Inspectorate is the responsible authority. 80% of the industrial SO_2 emissions come from these sources. The Alkali Inspectorate belongs to the Health and Safety Executive, but can be regarded practically as a quasi autonomous institution: the law allows it a high degree of discretion, and the Department of Environment generally refrains from actively using the means of intervention remaining to it. Thus, the Royal Commission on Environmental Pollution stated in its 5th Report (1976) that bpm is generally interpreted as "best practicable means to the satisfaction of the Alkali Inspectorate." The following factors are probably responsible for this uniquely strong position from an international perspective: the long tradition, the high degree of specialization (exclusively air pollution), the (internationally recognized) high reputation of the experts; and the fact that the inspectorate has a long history as the main authority for systematic air pollution control in the country. A further factor is the successful clean air policy, particularly in the early years, which through technological innovation sometimes led to positive economic effects in regulated firms. The close, mutually trusting co-operation with industry, as well as the legal assurance of an extremely restrictive information and co-operation policy regarding affected third parties of the public gives further support. Public conflicts between regulatees and the inspectorate are as good as non-existent. The inspectors regard themselves as partners of the emitters who are specialized in technical environmental prevention; one points specifically to the necessity of good relationships.

The organizational isolation of the inspectorate from politics at the district level and from third parties is the programme shell relevant to control. The clientele relationship, which is a partnership and is as free from disturbance as possible, replaces a legal standardization over wide areas. A "surprise-free" implementation practice which is oriented toward co-operation and equal treatment is thus assured nation-wide. The implementation process can then insure the high flexibility required and at the same time satisfy the (cost saving) need for standardization of branches of the same industry. Technical standards defining bpm generally proceed from a norm formation process, in which the affected firms, their trade association, and the Alkali Inspectorate are the authoritative participants. Endogenous conflicts are only to be expected where disagreements break out within the industrial lines. However, in view of the very high homogeneity of the regulatee group, this will presumably occur very seldom. Furthermore, the regionally responsible inspector retains a certain flexibility for individual variations.

Affected third parties as well as local authorities are to a great extent excluded from the local decision-making process not only by law, but also as a result of the close relationship between emitters and "their" authority. Exogenous conflicts are thus made more difficult by national programme decisions, and in practice they are sometimes blocked out in a manner which almost smacks of conspiracy. This is favoured by the fact that the inspectorate is at the same time the supervising and enforcement agency. Under these conditions, implementation difficulties almost never occur.

- *Small and Medium-Sized Firms, Domestic Heating:* Almost as high a degree of discretion with regard to standard setting, organization of enforcement and supervisory measures is given to the local authorities, who are responsible for the remaining, not registered emitters. The central programme also restricts itself in this area to the establishment of general maxims (best practicable means, prevention of damage to health) and general competencies. It leaves the "nasty field work" largely to the local authorities. The implementation system is strongly decentralized: there are no regional control authorities, and there are no institutional control and supervision lines from the central authority to the local authorities. The harshest measure which the central administration can use consists of "calling in" an approval process to itself. The regulation of the emitters falling within the local authorities" area of competence is accomplished by decentralized and splintered processes. Standard setting in the sense of taking into consideration ground level concentrations occurs to a great extent as immediate reaction to local air quality and power conditions. The emitters have a stronger position as the affected parties in that they are familiar with the relatively high bargaining potential of the implementation authorities in the concretization of national programme elements, and are in a position to better utilize these. Furthermore, they are procedurally better armed for defending themselves when claims are raised against them.

3.7.3 The need and the efforts toward harmonization at the basis of the system

In this conception, two relatively autonomously acting implementation authorities clash with one another in the same administrative area (Alkali Inspectorate and local authorities) without the availability of sufficient co-ordination mechanisms in the national programme. Thus, conflicts sometimes arise between local implementation officials and the responsible District Alkali Inspectorate, because the politically acceptable room for manoeuvre in questions of air pollution has already been used up by the clientele of one of the two implementation authorities. Such "transfer problems" also occur between neighbouring local authority areas. Particularly on the part of the emitters, there is a need for harmonization concerning common standards. This may be an essential reason for the fact that in regions which form an economic and/or climatological unit, similar or equal notions of environmental policy are found among the various officials. In Bristol and its environs, there also exists a *trans-local administrative co-operation network,* which undertakes some co-ordination functions. In other cases, such networks are also found between local authorities and the responsible District Alkali Inspectorate. Such forms of co-operation can be understood as *informal regional implementation systems,* which could constitute a *"quasi autonomous spread of standards"* among the participating actors. Intermediate studies indicate that such standards usually originate from local administrative units which are particularly well equipped in personnel and material.

3.7.4 The pattern of consideration of interests

Decentralized orientation toward individual cases among the institutional conditions described results in an overall isolation of the implementation process from interventions oriented toward environmental protection by affected parties. Correspondingly, the position of the affected parties as far as the process is concerned is quite weak, and their possibilities for intervention are essentially limited to the pointing out of direct pollution effects. They are dependent to a high degree on the "goodwill" of the Alkali Inspectorate as well as the advocacy roles of their local political institutions. Strict limitations are set on legal recourse, since certification of the compliance with regulations for emitters is considered sufficient redress. Mobilization of local political institutions is largely dependent on the respective political culture of the locality. Such mobilization can be favoured whenever an open conflict relationship exists between the inspectorate and the local authorities. Such situations do appear now and then in "stronger" local authorities whose political and economic intentions are thwarted by the Alkali Inspectorate. Isolated examples demonstrate that in such cases the central administration (Department of the Environment) actively supports the Alkali Inspectorate.

In the framework of such confrontation strategies, the local authorities set up their own ambient air quality standards, which are also proposed to the nation. The Greater London Council agreed on (not legally binding) SO_2 ambient air quality guidelines, which lie at the level of the "strict" WHO proposed standards. The main purpose in this is political mobilization in favour of a movement toward standards which should thematize the great gap between implementation performance and the demands of public health. Parallel to this, an "issue-supporting" emission inventory is being built at the same time. Such political measures could open up the access to the hard core of the "implementation cartel" by means of legitimation pressures, and could then bring about stronger environmental effects than those possible through additional legally binding programme elements.

The stubborn resistance of the central administration to official ambient air quality standards, which is very difficult to understand because of the relatively favourable national ambient air situation, is probably explainable for two primary reasons: first, there are still "high load islands" with partially increasing pollution trends. Secondly, a "domino effect" is to be feared; i.e. the introduction of environmental quality standard strategies for *one* pollutant would mean that such standards could hardly be prevented for pollutants other than SO_2, which are usually more problematic. Policy which operates with explicit ambient air quality standards must be supported above all in "high load islands" with new patterns of co-operation, which could easy destroy the system structure as a whole; as this has existed up to now. Traditional and accustomed relationships with specific patterns for consideration of interests on the whole must be placed in question. Less dangerous by contrast are the informal *ad hoc* co-operation networks between the implementation authorities which are based on autonomous bargaining processes. The local policy aspect then does not systematically affect the national strategic concept; the option for co-operation primarily remains with the Alkali Inspectorate, for which reason this body steers a pronounced course against national ambient air quality standards.

This exclusion of health oriented quality standards, combined with the above described exclusion of the influences of affected third parties or the ordinary public from the standard formation and implementation process and the close co-operation of the Alkali Inspectorate with its industrial clientele, allows conflicts with important firms which could be called up by environmental regulations to be prevented to a great extent. This strategy may be ideal from the standpoint of national economic policy. The interests of large and medium-sized firms are to a great extent taken into account in the formation of standards and in implementation practice.

In local districts with unfavourable air quality conditions, the situation of the small emitters, which are under local authorities' regulation, is more dependent upon implementation field structures: the interests of affected parties can be brought to bear more strongly by means of the local political mechanisms. Particularly unfavourable is their situation where they must compete for scarce resources in the same area with the large emitters regulated by the Alkali Inspectorate. Their supervision is generally separate from the local power structures to the extent that the local authorities can be forced to fall back upon the nearest group that they can influence whenever citizen protests arise against excessive air pollution. The "screws of the standards" can then be tightened by means of approvals for new or expanded facilities. Through the dual implementation and standard setting system, it is the affected parties which are disadvantaged above all. By their activities at the local level, at least for the mid-term only a limited sector of emitters can be regulated (those under control of the local authorities). The Alkali Inspectorate is driven to a stricter practice only under external pressure, which is not easy to bring about, because there exist sufficient "safety valves". Especially in the area of SO_2 clean air policy, the position of those affected parties who because of their participation in domestic heating belong at the same time to the emitters to whom the implementation measures are addressed, is relatively more favourable. This lies in the special strategy of the "smoke control areas", within which the addressees are allowed extensive possibilities for participation. Implementation problems are alleviated further by the principle of coupling measures with subsidies. Conflicts therefore are almost restricted to mining areas where larger parts of the inhabitants heat with so-called concessionary coal.

3.8 Conclusion

This outline of the three varying programme types on the national control level (national total programme, national partial programme and national rump programme), made on the basis of a typology of clean air policy programme structures, and taking into account the standpoint of interests, demonstrated that by means of a practicable, extensive concept of programme, an analysis of programme structures brings to light varying patterns for the handling of interests which occur as particular hindrances or aids to implementation in the enforcement system, the enforcement process or the implementation field, and which enter the "implementation game" more or less from outside. Corresponding privilege and discrimination patterns are already laid down institutionally in the programme itself, the use of the concept of implementation deficit is relativized. Such presumed deficits often represent nothing more than the influence, in accordance with the programme, of programme shells that intrude consciously or unconsciously on the programme core, which is often falsely referred to as the only central standard for evaluation. Implementation research should thus yield hypotheses by means of programme structural analysis against the background of these considerations as to what should actually be implemented, particularly in regulatory policy areas.

This is often possible only when, following an analysis of the presumed patterns of handling of interests in the programme, an investigation of the formation of the programme is undertaken under specific consideration of implementation relevant aspects. Implementation research will thus have to become an integral policy process analysis, integrating implementation and program formulation processes.

References

Antweiler, H. 1978. Voraussetzungen zur Aufstellung von Maximalen Immissionsschutz-Konzentrationen (MIK) Air Schwefeldioxid zum Schutze des Menschen, *VDI-Berichte 314: Sauerstoffhaltige Schwefelverbindungen*. Düsseldorf

Bardach, E. 1977. *The Implementation Game: What Happens After a Bill Becomes a Law?* Cambridge, Mass., London: MIT Press

Barrett, S., Fudge, C. (eds.) 1981. *Policy and Action.* London: Methuen

Bohnert, W., Klitsch, W. 1980. Gesellschaftliche Selbstregulierung und staatliche Steuerung. Steuerungstheoretische Anmerkungen zur Implementation politischer Programme, in: Mayntz, R. (ed.), *Implementation politischer Programme. Empirische Forschungsberichte.* Königstein/Ts: 200-215

Borghorst, H. 1978. *Die wechselseitige Anhängigkeit von Bund und Kommunen in der Stadtsanierungspolitik der Vereinigten Staaten von Amerika.* Berlin

Downing, P., Brady, G. 1978. *The Clean Air Act is Dead.* Typescript. Virginia, Miami, Florida

Downing, P. 1979. Implementing Pollution Laws: Observations from the American Experience, *Zeitschrift für Umweltpolitik*

Dente, B., Lewanski, R. 1979. *Implementation of SO_2 Air Quality Control Policy in an Italian Region, the Case of Emilia Romagna.* Paper presented at the workshop on Standard Setting and Implementation of SO_2 Air Quality Control Policy. A Comparative Analysis of EC Countries and Switzerland. Berlin: International Institute for Environment and Society, unpublished manuscript

Garlichs, Dieter 1980. Politikformulierung und Implementierung im föderativen Staat, in: Mayntz, R. (ed.), *Implementation politischer Programme. Empirische Forschungsberichte.* Königstein/Ts: 20-35

Grottian, P. 1974. *Strukturprobleme staatlicher Planung.* Hamburg: Hoffmann & Campe

Habermas, J. 1973. *Legitimationsprobleme im Spätkapitalismus.* Frankfurt: Suhrkamp

Hill, M. 1980. *The Role of Brush Local Government in the Control of Air Pollution – A Growing Gap Between Policy and Implementation?* Paper presented at the ECPR Workshop on Environmental Politics and Policies, Florence, March 1980

Knoepfel, P. 1977. *Demokratisierung der Raumplanaung.* Berlin: Duncker & Humblot

Knoepfel, P., Weidner, H. 1978. *International vergleichende Analyse von Normbildungsprozessen im Bereich von Umweltqualitätsnormen.* Eine konzeptiononelle Vorstudie. Berlin: International Institute for Environment and Society, unpublished manuscript

Knoepfel, P. 1979. *Öffentliches Recht und Vollzugsforschung.* Bern: Paul Haupt Verlag

Knoepfel, P., Weidner, H. 1980. *Handbuch der SO_2-Luftreinhaltepolitik,* vol. I and 2. Berlin: Erich Schmidt Verlag

Krämer, H. J. 1978. Probleme im Zusammenhang mit der Messung und Beurteilung von SO_2-Emissionen. *VDI-Berichte 314, Sauerstoffhaltige Schwefelverbindungen.* Düsseldorf

Majone, G., Wildavsky, A. 1978. Implementation as Evolution: Exorcising the Ghosts in the Implementation Machine, *American Journal for Political Science*

Mannozzi, S., Melloni, F., Tozzi, S., in co-operation with Cassese, S. 1979. *The Law 319/76 for Water Protection Against Pollution*. Paper presented at the workshop Standard Setting and Implementation of SO_2 of EC Countries and Switzerland. Berlin: International Institute for Environment and Society, manuscript

Mayntz, R. 1983 Die Implementation politischer Programme - Theoretische Überlegungen zu einem neuen Forschungsgebiet, *Die Verwaltung*

Mayntz, R., Bohne, E., Hesse, B., Hucke, J., Muller, A. 1978. V*ollzugsprobleme der Umweltpolitik. Empirische Untersuchung der Implementation von Gesetzen im Bereich der Luftreinhaltung und des Gewässerschutzes*, Materialien zur Umweltforschung, herausgegeben vom Rat von Sachverständigen für Umweltfragen. Köln: Kohlhammer

McGill, E. M., Wooten, L. 1975. Management in the Third Sector, *Public Administration Review*, Vol. 35

Offe, C. 1973. *Strukturprobleme des modernen Staates*. Frankfurt: Suhrkamp

Poland, O. F. 1974. Program Evaluation and Administrative Theory, *Public Administration Review*, Vol. 34

Shiffmann, M. A. 1973. The Use of Standards in the Administration of Environmental Pollution Control Programs, *Air Pollution and the Politics of Control*. New York: MSS Information Corporation

Stamer, P. 1976. Niveau- und strukturorientierte Umweltpolitik, *Schriftenreihe der Kommission für wirtschaftlichen and sozialen Wandel,* Nr. 79. Göttingen

Weidner, H., Knoepfel, P. 1979. Politisierung technischer Werte - Schwierigkeiten des Normbildungsprozesses an einem Beispiel (Luftreinhaltung) der Umweltpolitik, *Zeitschrift fur Parlamentsfragen*

Williams, B. 1975. Implementation Analysis and Assessment, *Policy Analysis*. Berkeley

4
Explaining Differences in the Performance of Clean Air Policies: An International and Interregional Comparative Study (1986)

Peter Knoepfel, Helmut Weidner[1]

4.1 The concept

4.1.1 Basic questions

On the basis of a large international comparative study of implementation policies from 14 different regional implementation systems (RIS) from the Federal Republic of Germany[2], United Kingdom[3], France[4], Italy[5] and the Netherlands[6], we first tried to identify the impact of regional policies in terms of their capacity to influence the behaviour of individual emitters

---

[1] in : Policy and Politics, (Bristol), Vol. 14/N.1 1986, p. 71–91. Copyright: We thank the editor for authorizing the publication of this article in this book.

[2] See Knoepfel and Weidner, "Die Durchsetzbarkeit planerischer Ziele auf dem Gebiet der Luftreinhaltung aus der Sicht der Politikwissenschaft. Ergebnisse aus einer internationalen Vergleichsuntersuchung", in Zeitschrift für Umweltpolitik, (Frankfurt), Nr. 2, 1981, p. 87–115; as well as the national report mentioned in note 1.

[3] See M. Hill "The Role of the British Alkali and Clean Air Inspectorate in Air Pollution Control", in: P. Downing and K. Hanf (eds.) op. cit., 1983, p. 87–106 and the British report mentioned in note 1.

[4] See R. Darbera, C. Larrue and H. Magalhaes, Politique nationale et mise en oeuvre du contrôle de la pollution atmosphérique par le SO2 Ìe S02 en France, Berlin (IIUG rep 84–9), 1984; P. Knoepfel and C. Larrue "Distribution spatiale et mise en oeuvre d'une politique publique: le cas de la pollution atmosphérique", in: Politiques et Management Public, No. 2/1985, p. 43–69 and the French report mentioned in: note 1.

[5] See B. Dente and R. Lewanski "Implementing Air Pollution Control in Italy: The Importance of the Political and Administrative Structure", in: P. Downing and K. Hanf (eds.) op. cit., 1983, p. 107–128; B. Dente, P. Knoepfel, R. Lewanski, S. Tozzi, S. Mannozzi, Il controllo dell'inquinamente atmosferico in Italia: analisi di una politica regolativa, Roma (officina), 1984, and the Italian report mentioned in: note 1.

[6] See the Dutch report mentioned in: note 1.

within selected local implementation areas (LIA). For each of the 14 regions we selected two to three LIAs with different structures of industrial and domestic SO_2 emitters. We aimed at having one of the three LIAs as a metropolitan area, the second one as a heavily industrialized area and the third one with a somehow mixed emitters structure[7]. Within each of these local areas we tried to compare changes over time in the local ambient air quality ("immission" data in the field of sulphur dioxide – SO_2), relating these to the total amount of emission produced by local emitters. In a second step we tried to find out, by means of interviews conducted with the main emitters, the different motives behind observed changes in their behaviour.

One of the most striking results was that, as a whole and independently of the countries' national programmes, emitter behaviour was more decisively influenced by such factors as the production situations of companies, relative price differences on the fuel market, and energy saving measures undertaken by firms on their own initiative, than by specific public control measures. In the field of SO_2 pollution there is an inseparable relationship between production and emission volumes, and this is only marginally affected by environmental policy activities. Thus impacts of environmental quality policies in the field of air pollution control are much more limited than initially assumed.

As we will show in this contribution, there are, however, significant differences amongst our examined regions which nevertheless can be attributed to policy activities. Within this paper we would like to stress these differences in the performance of the regional policies, although keeping in mind that even the most effective policy in our terms cannot completely explain the whole of changes in emitter behaviours. Politicians and agencies tend to considerably over-estimate the impact of their policies.

The common feature of the selected RISs relies on the overall responsibility of regional or local agencies for the control of the main emitters within their jurisdiction. There is only one exception to this: the case of the

[7] The study included the following Local Implementation Areas (LIAs); Italy: Bologna, Casalecchio di Reno, Piacenza (Emilia Romagna); Sesto San Giovanni, Villasanta, Cassano d'Addo (Lombardia); Turin (district Nr. 16), Chivasso, Moncalieri (Piemonte); France: Fos-Etang de Berre-Martigues, Marseille (Provence-Côte d'Azur); Paris, Vitry, Creteil (Ile de France); Lille, Dunkerque (Nord-Pas de Calais); Federal Republic of Germany: Berlin; Nürnberg, München (Bayern); Köln, Duisburg (Nordrhein-Westfalen); England: Greenwich, Westminster, Brent (Greater London). Sheffield, Barnsley (South Yorkshire); The Netherlands: Ijnmond, Amsterdam (North-Holland); Gelderland and Arnhem.

two regions selected in Great Britain. Here the formal programme formation and implementation responsibility lies with the national Industrial Air Pollution Inspectorate (known at the time of the study as the Alkali Inspectorate). Nevertheless also in the case of this country we found significantly different interaction networks within the two selected metropolitan areas (South Yorkshire and Greater London). These differences justified the identification of at least informal RISs also in the case of Great Britain.

The most fascinating aspect of the comparison of the impacts of the selected regional policies was to identify different patterns of the distribution of regulatory activities amongst the examined LIAs. So we could find some regions within which regulatory activities were concentrated upon one single area whereas in areas equally or even more polluted regulatory activities were much more limited or even absent. Given these varying distributional patterns of scarce administrative resources we decided to concentrate our comparative impact evaluation also on this distributional aspect. Thus our dependent variable in the comparison of the regions became both a combination of the efforts and the relative intensity of regional regulatory activities as a whole as well as the different patterns of the distribution of regulatory activities amongst the different areas compared to what we called "the problem pressure" (the extent of the exposed population together with the absolute immission levels). Our comparative question was therefore the following: which factors coming from inside or from outside of the RISs affect, with what weight, the activity levels, the intensity and the distribution of regional regulatory activities? On the basis of our empirical data we could distinguish the following three complexes of independent variables: the national programme structure the political choices of regional implementation policies and the "problem structure" (the actual region-wide distribution of the problem pressure). Table 1 shows the concept of the inter-RIS comparison.

Table 1

Independent variables		*Dependent variables*
national programme		overall activity level
political choices of regional implementation policies	regional regulatory activity	overall impacts impacts in highly polluted area
problem structure		

4.1.2 Possible characteristics of the dependent variables (performance level)

As in the case of most international comparisons it does not make much sense to compare impacts in terms of absolute figures[8]. It is evident that those figures might be the result of various country-specific factors which are of no interest for an international comparison. Therefore we produced, on the basis of all relevant data, statements concerning the overall regional policy performances. These statements take into account the following three dimensions:

- *The overall activity level:* this level has been defined by comparing the number of all emitters of the region with the number of individual policy out-puts and/or the annual budget (personal, financial means etc.) spent by the agency for the purposes of SO_2 control. Given the fact that, with the exception of Germany, our data do not indicate significant differences amongst the related averages of the involved countries, we could base our statements upon a national frame of reference. This country-specific frame of reference has been finally selected also for the case of the German RISs because, on the one hand, we could find a significant difference between the three regions (Nordrhein-Westfalen vs. Bayern and Berlin) and because, on the other hand the highly formalized German administrative procedures need more personnel than the more informal ones of the other countries involved[9]. Furthermore we took into consideration the equity of the distribution of the outputs with regard to the distribution of the problem pressure within the region. A high overall activity level was attributed to those regions where the number of outputs and/or the agencies" budgets were relatively high (mainly compared to the other RISs of the same country) and where most or all of the problematic areas were affected by regulatory activities.
- *Overall impacts:* this dimension was determined on the basis of our interviews with the main emitters in the local implementation areas. Again we took into consideration the spatial distribution of the impact: regions were only characterized as having high overall impacts where we could observe a relatively high impact of regulatory activities in all

[8] See L. Lundqvist, The Hare and the Tortoise: Clean Air Policies in the United States and Sweden, University of Michigan Press, Ann Arbor, Michigan, 1980.

[9] Cf. H. Weidner and P. Knoepfel, "Innovation durch international vergleichende Politikanalyse. Dargestellt am Beispiel der Luftreinhaltepolitik", in: R. Mayntz (ed.), Implementation politischer Programme II. Ansätze zur Theoriebildung (Opladen: Westdeutscher Verlag, 1983), pp. 221–255.

three selected LIAs. Again the main frame of reference consisted of the national context; this turned out to be more reliable than the largely uncontrollable cross-national comparisons of simple facts stemming from very different regulatory traditions of the countries under comparison.

- *Impacts in highly polluted and densely populated areas:* as already mentioned, we could find regions where in the most polluted metropolitan areas the impacts of regulatory activities were high, whereas for instance in industrial pools we could only barely observe visible effects. Using this criterion we introduced a particularly immission-oriented evaluation dimension. This choice turned out to be necessary because of the predominantly immission-orientation of the national programmes of all of the included countries. According to this orientation clean air regulation is considered as a means to control local ambient air quality trends mainly in the surroundings of air pollution sources. The objective of such policies is in fact achieved whenever local concentrations of SO_2 (and not of H_2SO_4 e.g.) can be kept below a certain immission level. This policy concept only calls for a limitation of emissions when local air quality is affected. High chimney policies are therefore one of the most recognized and most visible consequences of these concepts. In the seventies no national programme of the included countries provided a systematic control of the total amount of the emissions of the country. We all know that from the point of view of the eighties (dying forests, trans-boundary air pollution, acid rain) an emission-orientation seems to be the only reasonable concept for an actually preventive air quality control policy. We are convinced, however, that any serious implementation evaluation of the policies of the seventies must take as its frame of reference the explicit policy intentions and not the point of view of the researcher, it is, nevertheless, also true that no national programme explicitly limited air quality regulations to highly polluted areas. Furthermore any immission-orientation by definition implies the mandate to observe and to control air quality trends in less urbanized rural areas ("non deterioration clauses"). Thus, even a successful concentration of regulatory activities on metropolitan areas, leaving apart the other areas, must be considered as a half way victory even under an immission-oriented policy concept. Therefore we had to take care of the already mentioned overall impact within the whole region.

By combining these three dimensions (1: overall activity level; 2: overall impact; 3: impact in highly polluted areas) we developed the following eight types of performance levels:

1: 1-, 2+, 3+

2: 1+, 2+, 3+

3: 1-, 2-, 3+

4: 1+ 2-, 3+

5: 1-, 2+, 3-

6: 1+, 2+, 3-

7: 1-, 2-, 3-

8: 1+, 2-, 3-

The proposed scale ranging from 1 to 8 corresponds with a decreasing performance level. This range takes account of the relation between costs and benefits, because the patterns with the even numbers reflect high activity and the odd numbers low activity, and administrative costs are a function of activity level.

1. *High effects with visible improvements in all areas of the region (including highly polluted areas)*

   low administrative costs: type 1

   high administrative costs: type 2

2. *Visible improvements in the highly polluted and densely populated areas within the region*

   low administrative costs: type 3

   high administrative costs: type 4

3. *Diffuse effects in all the region which, however, do not lead to significant improvement in the polluted areas*

   with low administrative costs: type 5

   with high administrative costs: type 6

4. *Regulation without any significant impacts*

   with high administrative costs: type 7

   with low administrative costs: type 8

4.1.3 The independent variables

Our results permitted us to reduce the complexity in the field of our three selected independent variables (programme structure, political choices of regional implementation policies, and problem structure) in the sense that we could in fact reduce them to two for each case: the mix of variables which turned out to mainly explain the differences between the selected

RISs either consist of a combination between the variables "political choices of regional implementation policies" and the specific "programme structure" or between "political choices of implementation policies" and the "problem structure". This reduction in the number of variables highlights the most significant elements of policy even if it may partly underestimate the weight of the eliminated variable. In this sense it must be remembered that even in the case of those regions where the national programme did not significantly shape the implementation policies, the fact of its existence will undoubtedly have had a certain influence on the regional agencies.

The different variables can be characterized as follows:

- *The national programme structure:* its explanatory capacity is high where most of the substantive, procedural and organizational elements of these programmes as well as their specific linkages are reflected within the regional policy. By definition this is only possible in the cases where national programmes contain relatively concrete elements (as in Federal Republic of Germany, Italy and the Netherlands)[10].
- *Problem structure:* the explanatory capacity of this variable is high where the arrangements of outputs and their impacts can be interpreted as an immediate reflection of the problem pressure within the region. Both the spatial distribution and the intensity of regulatory activities in such cases can be mainly explained by the problem structure itself; the programme structure or specific political policy choices are of limited importance for the explanation of such policy effects. A good example is the case of the Paris region (Ile de France): here our data show that policy implementation activities are distributed almost in complete compliance with the trends shown by data as produced by the SO_2 measurement network. It must be stressed that this considerable correspondence is achieved due to the use of collective rather than individual outputs, committing a whole range of emitters within a certain area to emission reductions (special protection areas coupled with "alarm networks" which automatically "order" changes in fuels in critical periods).
- *Political choices of regional implementation agencies:* if the implementation results neither reflect the observed problem-structure nor can be reasonably explained by specific priorities already set within the national programme, this third variable complex becomes important. We were in fact able to find regional policies where the main explanation

[10] See the more detailed analysis in: P. Knoepfel and H. Weidner: op. cit. 1983, p. 91ff.

was specific (mainly political) policy decisions, setting particular priorities for the region. These priorities not infrequently turned out to be even in opposition with the national programme and moreover inconsistent with the given problem pressure distribution. The main representative of this type could be found in those situations where the regional implementation policy seemed to be an immediate reflection of political pressure calling for or effectively preventing regional regulatory activities within the concerned localities.

Table 2 shows the different possibilities of combining characteristics of the selected independent variables:

Table 2

A.	National programme	high	middle	low
B.	Political choices of regional implementation policies	high	middle	low
C.	Political choices of regional regulatory activities	high	middle	low
D.	Problem structure	high	middle	low

Before presenting the results of our study we should stress the fact that, against our initial hypothesis, one of the most frequently discussed[11] variables turned out to be of very limited importance: the degree of formalized access of the public to the policy implementation processes. We could not find significant differences in the contents of outputs between those countries which, by means of their procedural laws, guarantee a formalized participation of the public (the Federal Republic of Germany, France and the Netherlands) and those countries where the procedural arrangements must be characterized as relatively closed (Italy and Great Britain). The possibilities of influencing choices of regional implementation policies (against or in correspondence with the problem pressure) seem to depend much more on possibilities and capabilities of local governments to mobilize regional resources, or alternatively, to block the regional attempts to effectively control local emitters. This observation at least partly has to do with the fact that such policy decisions consist of general priority setting and planning decisions rather than individual clean up orders or permits. For-

[11] See Kunreuther, H. and Linneroth, J., Risk Analysis and Decision Processes: The Siting of LEG Facilities in Four Countries, IIASA, Laxenburg, Austria, March 1982

malized public participation is normally guaranteed within individual procedures.

Again, contrary to our initial assumption, the administrative and technical capacity of the regional agencies as such turned out to be of a very limited importance. Varying administrative capacities within one and the same country affected the activity and, more especially the impact of the agency policies much less than we assumed. Such variations only lead to significant differences in the impacts when the corresponding interaction networks with local governments were equally developed. It seems that regional administrative and technical capacities can only be fruitfully used if interactions with local government are intensive and more or less in harmony.

4.2 The different performance levels of the 14 regional implementation systems

Table 3 shows the different performance levels (ranging from 1: high performance, to 8: low performance, according to the typology presented above) of the 14 regional implementation systems for the five countries included in our research. The ranking position is calculated by means of a simple addition of the positions of the individual regional systems indicated by the type-number we have attributed to these systems. The lower the total for a country, the better the country ranks in the comparison.

Table 3

FRG	*England*	*France*	*Italy*	*Netherlands*
Berlin 7	London 2	Ile de France1	Lombardie 2	Nord-Holl. 3
Bayern 7	South Yorkshire 7	Nord-Pas de Calais 4	Emilia Romagna 6	Sud-Holland4
NRW 8		Provence-Côte d'Azur 6	Piemont 7	Gelderland 7
Total 22	**9**	**11**	**15**	**14**
Average 7.3	4.5	3.7	5	4.7

Table 4

Group	*Type*		*Regional Implementation Systems: (:types)*
1 high performance overall	1	2	London (2) Lombardie (2) Ile de France (1)
2 high performance in polluted areas within the region	3	4	Nord-Holland (3) Sud-Holland (4) Nord-Pas de Calais (4)
3 low performance	5	6	Emilia Romagna (6) Provence-Côte d'Azur (6)
4 without effects	7	8	Berlin (7) Bayern (7) Nordrhein-Westfalen (8) South Yorkshire (7) Gelderland (7) Piemont (7)

These results suggest the following comments:

With the exception of the Federal Republic of Germany (FRG) we can observe significant differences between the performance levels within one and the same country. On the other hand we can find similar performance levels in a cross-national inter-regional comparison.

For several observers at least the two extreme positions will not be surprising: the first place of the French implementation systems will not surprise those scholars familiar with comparative policy analysis. Studies of other policy areas (industrial development[12], telecommunications[13], public

[12] See W. Neumann, H. Uterwedde "Industriepolitik in Frankreich und der Bundesrepublik Deutschland", working paper, presented at the German-French Institute of Ludwigsburg (Germany), Dec. 1984; F. Jenny "L'évaluation de politiques industrielles", in: Nioche, J.P. et Poinsard (ed.) L'évaluation des politiques publiques, Paris (economica), 1984, p. 187–204; C. Durand "Les politiques industrielles de l'Etat", papier présenté dans le cadre de la direction de la recherche du M.E.N. du C.N.R.S. dans l'action spécifique sur "politiques gouvernementales et entreprises publiques", le 9/10 mai 1985 à Lille.

[13] See B. Jobert et E. Brenac: "La contribution des entreprises publiques et l'élaboration des politiques gouvernementales: La D.G.T. et le plan cable." papier pré-

transport[14]) have indicated evidence of effective implementation, which is attributed to the high professionalization of French policies ("les corps"), strong vertical interaction-networks amongst the individual policies, and the high "technicity" of French policies with their concomitant "campaign character". On the other hand there are several reports which show the relatively low performance of German public policies (in the field of labour market policies[15] or environmental policies[16] as a whole). As in our case, this low performance has to do with the structure of national programmes[17]. It is surprising that Italy is not ranked last; furthermore one of its three regions even ranks amongst the best performers. As we will see, this exceptional position of the region of Lombardy is due to the low degree of politicization of its regional policy as well as to the relatively developed technical and managerial capacity of the regional agency.

Last, but not least, we must stress the fact that both our observations within the investigated regions and our comparative evaluative statements

---

senté à L'Atelier Séminaire à Lille (see note 12): J. Autin: 7 défis audiovisuels, Paris (economica) 1984; Guillou B. "Les stratégies multimédias des groupes de communications", La Documentation Francaise, Paris 1984.

[14] See J. C. Thoenig, L'Ere des technocrates. Le cas des Ponts et Chaussées, Paris (Editions d'organisation) 1973; C. Montrade "Financement et politique de développement de la RATP: analyse historique et prospective de la dialectique internalisation d'objectifs de la politique économique – adaptation à une demande évolutive", papier présenté lors du Premier Congrès du journal Politiques et Management Public (Séance 3–1) du 27 au 28 Sept. 1984 à Paris.

[15] See F.W. Scharpf, "Strategy Choice, Economic Feasibility and Institutional Constraints as Determinants of Full-Employment Policy During the Recession", in: K. Gerlach, W. Peters, W. Sengenberger (eds.): Public Policies to Combat Unemployment in a Period of Economic Stagnation, Frankfurt, New York (Campus), 1984, p. 67–114; G. Schmid: "Equal Opportunity Policy: A Comparative Perspective", in International Journal Manpower, vol. 5, No. 3, 1984, p. 15–25; F.W. Scharpf "Economic and Institutional Constraints of Full-Employment Strategies: Sweden, Austria and Western Germany, 1973–1982", in: J.H. Goldthorpe (ed.), Order and Conflict in Contemporary Capitalism, Oxford (Clarendon Press), 1984.

[16] See J. Hucke, State and Local Relations on Environmental Regulations in the Federal Republic of Germany, in: P. Downing, K. Hanf, op. cit., 1981, p. 129–142.

[17] See: J. Feick, R. Mayntz et al., Regulative Politik und politisch-administrative Kultur. Ein Vergleich von fünf Ländern und vier Interventionsprogrammen, mimeo, Köln 1982, reported in: J. Feick "Internationale Vergleichbarkeit staatlicher Interventionsprogramme – Konzeptionelle und methodische Probleme", in: R. Mayntz (Hrsg.), Implementation politischer Programme II, Ansatz zur Theoriebildung, Opladen (Westdeutscher Verlag) 1983, p. 197–220.

might be incomplete. Therefore the established ranking order for the included countries might be problematic and even dangerous. As we will show in the following, our research is actually much more interested in the cross-national regional comparison and especially in the attempt to compare the patterns of explanatory variables for different performance degrees than in the comparison of the countries as such[18].

Table 4 ranges the 14 regional implementation systems according to the four main performance levels from 1 (= high performance) to 4 (= without significant effects).

It is interesting to notice that the three high performing regions come from three countries which are extremely different with regard to their general constitutional, political and economic conditions. The three regions, however, each represent one of the most important economic and political metropolitan areas of their respective country. The fact that two of the three Dutch regions ranked in the second class which, according to our typology, is characterized by a concentration on highly polluted areas, shows the still predominant immission-orientation of the Dutch clean air policy, in contrast to the official national government's declarations. According to the official statements by the Dutch Government and to central policy goals established in governmental air pollution abatement programmes, the main objectives are to prevent a deterioration of ambient air quality in general (non-deterioration clause) and to achieve a total national SO_2 emission load of no more than 500 kt/a[19]. The same tendency also seems to be true for the Nord-Pas de Calais of France (the Ile de France is also a region where there is this strong immission-orientation; the reason it is in the first local group is that more or less the whole policy-relevant territory can be considered as a highly polluted and densely populated area). Four of the five countries are represented by at least one of their regions within the group of the mostly ineffective implementation policies. This last group contains six of the 14 regions. Furthermore it is interesting to notice that all three German regions are in this last group.

---

[18] See P. Knoepfel, C. Larrue, “Les politiques publiques comparées: tourisme intelligent ou vrai progrès? Le cas des politiques comparées de l'environnement”, in: Politiques et Management Public, Paris, no 2, 1984, p. 561s.

[19] Cf. S02-Beleidskaderplan, Tweede Kamer, zitting 1979–1980, no 1–2.

4.3 Cross-national comparison of the independent variables

4.3.1 In high performing implementation systems

Table 5 shows the explanatory capacity of the three independent variables for the three highest performing regional implementation systems:

Table 5

Variables	*London*	*Lombardie*	*Ile de France*
National programme	--	--	--
Political choices	middle	low	low
Problem structure	middle	high	high

This table tells its own tale: what is striking one is the almost complete absence of any significant role of the national programme for all three implementation policies in spite of their different structure. Both the British and the French programme, which are characterized by a very vague, incomplete (absence of concrete ambient air quality standards and general abatement equipment standards) and mainly organizational set of rules (e.g. the British Clean Air Acts) and the extremely detailed Italian programme (immission and emission standards etc.) did not actually guide regional implementation activities. On the other hand, we observed in all three regions the explanatory importance of the dimension "problem structure". Regional implementation priorities as well as the content of individual outputs were significantly shaped by the actual pollution at the beginning of regulatory activities, by the observed pollution situation at the beginning of regulatory activities by the observed pollution trends within most of the local areas of the region and by the technical regulation requirements of the different groups of sources to be controlled. Correspondingly we found little evidence of explicit political policy choices concerning regulatory activities. In all three cases well staffed agencies seem to have succeeded in conceiving regulatory activities in compliance with the varying problem pressure within the controlled areas. Notice that all three agencies had to deal with political pressure coming from one or more local areas but were strong enough to use this pressure in order to identify problematic activities without becoming dependent on it. The agencies could maintain their independence by partly managing and dominating the network within which they had to act. One of the most important conditions for this success was the strong position of the agencies within this network but also their capability to use the pressure for their own purposes: namely the setting of priorities completely according to technically-perceived problem structures. One of our best examples is the Lombardian air pollution control agency (CRIAL), which explicitly set problem oriented priorities (extension of the monitoring network within the problematic areas of the province of Milan and concentration of regulatory activities upon emitters out of these areas following a schedule "branch per branch"). These often had to be implemented by postponing individual municipalities' requests for further regulations. By this means the agency could achieve a relatively equal distribution of its regulatory resources which also took care of the pollution risk of those parts of the population living in municipalities, the political weight of which was not sufficient to mobilize regional interventions[20].

4.3.2 In regions where high performance was present in high polluted areas

Table 6 shows the explanatory capacity of the independent variables for those regional implementation policies which rank in the second group:

Table 6

Variables	*North-Holland*	*South-Holland*	*Nord-Pas de Calais*
National programme	middle	middle	--
Political choices	middle	high	high
Problem structure	--	--	low

The concentration on the most critical areas to the disadvantage of smaller industrial zones or rural areas seems to reflect, in all of the three regions, a corresponding political pressure articulated by the local governments of the major urban areas. It becomes evident that the explanatory capacity of the problem structure is much more limited than for group 1. This becomes evident if one considers the increasing pollution trends outside the metropolitan areas which, according to the immission-oriented air pollution control concept laid down in the legislation of both France and

[20] See B. Dente, R. Lewanski, op. cit., 1983 (note 5), p. 123ff.

the Netherlands, would have called for further regional implementation activities. We can observe therefore a partial discrepancy between the distribution of the problem pressure and the distribution of the regulatory activities within the region. This group is particularly interesting for those scholars interested in distributive effects of regulatory policies.

4.3.3 In low performing regions

Table 7 shows the explanatory capacity of the two relatively low performing regional implementation systems, Emilia-Romagna and Provence-Côte d'Azur:

Table 7

Variables	*Emilia Romagna*	*Provence-Côte d'Azur*
National Programme	--	--
Political choices	middle	middle
Problem structure	middle	middle

It is difficult to interpret this table. However, we can stress the fact that these two policies within the period of investigation have to be characterized as typical transitional policies. Within their countries they underwent a remarkable change from an immission-oriented towards an emission-oriented clean air policy. This change involved a concentration on a selected set of industrial activities within specific plants which were cleaned up independently of their location within or outside highly polluted metropolitan areas. The selection of these plants (Emilia Romagna: iron and steel works; Fos Etang de Berre Martigues within Provence-Côte d'Azur: new large-scale industrial plants) was influenced by political pressure working in both a positive and negative sense. This group is therefore interesting for those scholars mainly studying conditions for such change in clean air policies. The increasing concern with the issue of acid rain, with its impact upon the forests of central and northern Europe, indicated a need for a change in policy emphasis of this kind, focusing attention upon emission rather than entirely upon local ambient air quality (immission).

4.3.4 In regions where implementation had no significant impact

Table 8 shows the explanatory capacity of the independent variables of the six regional implementation systems which belong to the last group comprising implementation policies with almost no actual impacts.

Table 8

Variables	*Berlin*	*Bayern*	*NRW*	*South Yorkshire*	*Gelderland*	*Piemont*
National programme	high	middle	high	--	middle	low
Political choices	low	middle	middle	middle	low	high
Problem structure	--	--	--	high	--	--

This table shows in general the inverse picture of Table 5 above concerning the most performing systems. This is especially true as far as the low explanatory capacity of the variable "problem structure" is concerned. The only exception is South Yorkshire. Here, compared to the RIS London, the influence of the problem structure must be rated high because regulatory activities have been decisively influenced by the specific problem situation (strong position of the coal mining industry, unfavourable economic situation). Thus, with a certain exception of the LIA Sheffield due to its rather effective Smoke Control Programme (domestic heating), industrial emission sources (and in the case of the LIA Barnsley also domestic sources) have been treated in a way which clearly reflects the prevailing problem structure (minor enforcement activities, especially concerning old industrial emission sources)[21].

Less significant but still evident is the high explanatory capacity of the programme structures. In the reports on regulatory activities in these regions we found many observations pointing out the constraining role of the national programme (for example legal clauses concerning the "economic arid technical feasibility" used by emitters as a successful weapon against

[21] See the British report mentioned in: note 1.

regional agencies; or the legal provision that every emitter should be regulated (FRG); or the formal clean up procedure in the Italian legislation). Not infrequently we also found programme related regulatory priorities which turned out to be far from the actual regulatory needs. Such priorities often led to inefficient outputs. In the case of the region of Piedmont the application of the national programme by the legalistic regional agency actually hampered the development of adequate technical control rationality. This also held true for the three German regions investigated. The German air pollution control programme as established by laws and regulations of the national and state governments is a highly complex one because almost every minor aspect of enforcement activities is covered by formal, very detailed regulations and directives to be observed by the responsible (implementation) authorities. Yet, on the other hand, the legal provisions have not led to the establishment of obligatory control requirements: there is a lack of clear and binding objectives as well as effective administrative instruments directed to the control of major emitting sources. Thus, to give just one example, the formal requirements have brought about rather sophisticated regional Clean Air Plans which were highly cost-intensive and required large numbers of staff. However, they soon turned out to be quite ineffective because the law did not provide for specific means to enforce these plans. Instead of eliminating this obvious shortcoming with the help of new, enforcement-oriented legal provisions, another development occurred: the tremendous increase in regulations with more detailed requirements concerning specific elements of the Clean Air Plans (e.g. emission inventory, monitoring requirements).

Table 9

	High performance	*High in polluted areas*	*Low*	*Without effects*
National Programme:				
- high				Berlin (7) NRW (8)
- middle		Nord-Holland (3) Sud-Holland (4)		Bayern (7) Gelderland (7)
- low				Piemont (7)
Political choices:				
- high		Nord-Pas de Calais (4) South-Holland (4)		Piemont (7)
- middle	London (2)	Nord-Holland (3)	Emilia Romagna (6) Provence-Côte d'Azur (6)	Gelderland (7) NRW (7) South Yorkshire (7)
- low	Lombardie (2) Ile de France (1)			Berlin (7)
Problem structure				
- high	Lombardie (2) Ile de France (1)			South Yorkshire (7)
- middle	London (2)		Emilia Romagna (6) Provence-Côte d'Azur (6)	
- low		Nord-Pas de Calais (4)		

In general, the highly formalized German control system with its emphasis on secondary aspects and neglect of the core issues not only reduces the flexibility and latitude of the responsible authorities but also increases administrative costs and – owing to the limited budget – also swallows up resources badly needed for effective measures (e.g. supervision). All in all, the predominance of very detailed, formal regulations fits the political-administrative culture of this country and supports the attitude of public administration which is altogether in favour of legalistic approaches[22].

4.4 Conclusions: comparison of the explanatory patterns of the 14 regional implementation systems

Table 9 shows the influence of the three independent variables on the performance level of the regional implementation systems studied.

A highly programme-controlled clean air policy seems to have negatively influenced the performance level of regional implementation activities in the case of the Federal Republic of Germany but also, to a lesser extent, of Italy. In the latter case the relative success of the region of Lombardy has actually to be attributed to the fact that the regional implementation policy has been developed partly against the national programme. Highly programme-controlled clean air policies seem to be less problematic in the case of the two Dutch regions North-Holland and South-Holland (as well as in the case of two Swiss regional implementation[23] systems studied but not included in this report). On the contrary all three regions of the apparently "optimal" case of France turned out to be much less programme-controlled than others.

One could argue that the extent to which public policies arc programme-controlled is rather more dependent on the overall national political-administrative culture than on policy-specific features[24]. So we can find multiple public policies in the FRG as well as in Italy sharing the characteristic of being over-controlled by highly detailed programmes. This situation often hampers the development of adequate implementation policies on the regional and local level. Also, French public policies often

---

[22] Cf. H. Weidner, "Schwachstellen in der Luftreinhaltung. Die Bundesrepublik im internationalen Vergleich", Umweltmagazin No. 9, September 1983, pp. 22–24.

[23] See M. Peteis, "Standard setting and implementation in S02-Air quality control policies", Fallstudien Schweiz, Zürich. (MS) 1982.

[24] See on this "classical" question D. Ashford, "The structural analysis of policy or institutions really do matter", in: D. Ashford (ed.), Comparing Public Policies, New Concepts and Methods, Beverley Hills, London (Sage), 1978, p. 81 97. An interesting contrast: J. Feick, op. cit., 1983 (note 16), 202ff.

share common features: national programmes are often formulated in vague terms leaving a lot of discretion to regional or departmental implementation agencies which, in turn, are often both managerially and technically well staffed. Swiss and Dutch policies again seem to share relatively detailed programmes which anticipate a decision on those conflicts which in other countries normally arise only in the implementation phase. Therefore we can find more conflictual programme formation processes and less controversies in related implementation activities.

In spite of such cultural-specific explanations we can learn from our study that, independently of these political-administrative environments, clean air policies tend to be more successful if they leave to the regional implementation agencies a considerable leeway of political and administrative or technical discretion. This necessity stems from the nature of clean air policies themselves which have to control extremely different problem constellations. Furthermore such policies need a large consensus amongst local target groups which cannot be anticipated by detailed but still general national programmes.

Politically controlled regional implementation activities should not be automatically assumed to be low performers. Much depends on the position of the regional implementation agency within the regional policy network, and whether it is strong enough to countervail local political pressure likely to privilege selected areas to the disadvantage of others. However, we also found politically controlled misallocation of implementation resources to the advantage of less polluted areas in the two regional implementation systems of Emilia Romagna and Provence-Côte d'Azur[25].

If combined with the characteristics of programme-controlled policies, explicit political regional implementation policies, however, seem to be condemned to ineffectiveness: in the case of the region of Piemont the political control over the regional implementation policy was able to mobilize all elements of the national programme in a way which could check or even block regulatory activities, to the advantage of the most important emitters of the city of Turin[26]. Those agencies however, which conceived their implementation policies mainly according to problem pressure data either by following a more technocratic concept (Ile de France and Provence-Côte d'Azur) or a more participatory approach (intensive cooperation with local governments as in the case of the region of Lom-

[25] See the more detailed presentation in: P. Knoepfel, "Distributional Issues in Regulatory Policy Implementation: The Case of Air Quality Control Policies", in: A. Schnaiberg, N. Watts, K. Zimmermann (eds.), Distributional Conflicts in Environmental-Resource-Policy, 1986, pp. 363–379.

[26] See B. Dente, R. Lewanski, op. cit., 1981, p. 115f.

bardy) turned out to be most successful. In both cases implementation could profit from the absence of a significant deviating political control articulated through local governments' pressure as well as from the strong technical staff of the regional agency and its predominant role in the regional network. One should stress, however, that political pressure in the case of the region of Lombardy was not absent as such, but that the agency was successful in canalizing this pressure in a positive way. The same held true for the more informal regional implementation system of London, where both the Greater London Council and some local governments were able to mobilize political pressure in order to support locally (domestic heating control) and nationally (industrial air pollution control) determined policies.

4.5 Recommendations

On the basis of the results we have reported here in a very concentrated way, we have tried to formulate some recommendations for future modifications of clean air policies. These recommendations are presented in separate publications which did not only take care of comparative aspects but made an attempt to contribute further to different national policy debates according to the individual conditions of these countries[27]. Given the problem of long range air pollution, leading to the well known phenomenon of acid rains and dying forests, we recommended first of all a radical reorientation of the objectives of all policies. The recommended changes from immission towards emission oriented clean air policies require drastic measures in the field of technological innovation concerning all combustion activities, independently of their location in more or less polluted areas. This reorientation concomitantly should lead to a certain shift of the burden of implementation within the regional agencies in the form of standardized national requirements for smaller plants as well as for large scale combustion facilities (such as power plants and different types of refineries). According to all our results it will still be important to limit these standards to a set of minimal requirements; regional agencies should have the possibility of fixing further technical conditions as well as branch-wide and/or locally different timetables for compliance. Given such guidelines for these to specific emitter groups, regional agencies will have resources enough to start clean-up activities in the large field of middle sized plants where at least in the immediate future they need considerable leeway of

[27] See the reports mentioned in: notes 2 to 5.

discretion. In order to exclude time consuming court suits about the extent of this discretionary power, parliamentary acts should explicitly clarify the legal leeway of regional implementation policies.

Given the fact that the French regional implementation systems turned out to be the most successful ones our recommendations are mainly based upon the following characteristics of these systems:

- a relatively subtle structure of the national programme, leaving important discretion to regional agencies;
- an adequate technical and managerial staffing of regional agencies;
- the capability of these agencies to defend their independent positions within the region-wide interaction network, as well as managerial capacity to use local political pressure for the purpose of a problem-oriented setting of priorities;
- the existence of collective outputs oriented towards a whole group of emitters ("alarm networks" and "special protection zones") the application of which is based upon contractual procedures including the main interest groups (local governments, industrial associations, regional agencies and environmental protection groups).

Nevertheless, we share the common scepticism with regard to any simple policy transfer proposition. As pointed out above, important elements of the French approach are rooted in the overall French political administrative culture and their transfer to another country risks failure because these conditions do not exist in the new country. Two illustrations of this are: the strongly organised "corps" mainly responsible for French clean air policies, and the relatively low priority given to equity within the French political culture allowing, amongst others, a relatively high degree of flexibility in decisions on individual outputs. It would be outside the scope of this presentation to discuss each of these recommendations. Nevertheless, we would like to emphasise that for the development of an internationally co-ordinated environmental policy national and regional features and preconditions of air pollution control policies must be given due consideration. As yet international policies have neglected the importance of regional aspects to an even greater extent than national policies. Especially when it comes to "programming the programme", regional features, as we tried to demonstrate in this article, may have a great influence on the actual achievements of general pollution control programmes[28].

[28] A striking example of the shortcomings of an international air pollution control policy neglecting the basic national and regional features and their influence on implementational achievements is the EEC directive on ambient air quality

References

Ashford, D. 1978. The structural analysis of policy or institutions really do matter, in: Ashford, D. (ed.), *Comparing Public Policies, New Concepts and Methods*. Beverley Hills, London: Sage: 81–97

Autin, J. 1984. *7 défis audiovisuels*. Paris: Economica

Darbera, R., Larrue, C., Magalhaes, H. 1984. *Politique nationale et mise en oeuvre du contrôle de la pollution atmosphérique par le SO_2 en France*. Berlin: IIUG rep: 84–9

Dente, B., Lewanski, R. 1983. Implementing Air Pollution Control in Italy: The Importance of the Political and Administrative Structure, in: Downing, P., Hanf, K. (eds.), *Internatianal Comparison in Implementing Pollution Laws*. Boston, The Hague, Dordrecht, Lancaster: Kluwer-Nijhoff: 107–128

Dente, B., Knoepfel, P., Lewanski, R., Tozzi, S., Mannozzi, S. 1984. *Il controllo dell'inquinamente atmosferico in Italia: analisi di una politica regolativa*. Roma: Officina

Dente, B., Lewanski, R., Prittwitz, V., Schreiber, H., Brammer, R., Hill, M., Garrard, P., Larrue, C., Darbera, R., Magalhaes, H., Hanf, K., van der Tak, T., Toonen, T. 1982. Formulation and Implementation of Air Quality Control Programs: Patterns of Interest Consideration, *Policy and Politics*, vol. 10 No. l: 85–109; and: Implementing Air Quality Control Progmms in Europe: Some Results of a Comparative Study, in: Downing, P., Hanf, K. (eds) 1983. *Internatianal Comparison in Implementing Pollution Laws*. Boston, The Hague, Dordrecht, Lancaster: Kluwer-Nijhoff: 191–211

Durand, C. 1985. *Les politiques industrielles de l'Etat*, papier présenté dans le cadre de la direction de la recherche du M.E.N. du C.N.R.S. dans l'action spécifique sur "politiques gouvernementales et entreprises publiques", le 9/10 mai 1985 à Lille

Feick, J. 1983. Internationale Vergleichbarkeit staatlicher Interventionsprogramme – Konzeptionelle und methodische Probleme, in: Mayntz, R. (ed.), *Implementation politischer Programme II, Ansatz zur Theoriebildung*. Opladen: Westdeutscher Verlag: 197–220

standards for S02 and particulates of 1980. We have tried to draw attention to these shortcomings in an ex ante assessment shortly after the directive was enacted: H. Weidner and P. Knoepfel, "Implementationschancen der EG-Richtlinie zur S02-Luftreinhaltepolitik. Ein kritischer Beitrag zur Internationalisierung von Umweltpolitik", Zeitschrift für Umweltpolitik, vol. 4 No. 1, March 1981 pp. 27–67.

Feick, J., R. Mayntz et al. 1982. *Regulative Politik und politisch-administrative Kultur. Ein Vergleich von fünf Ländern und vier Interventionsprogrammen*. Köln, mimeo

Guillou, B. 1985. Les stratégies multimédias des groupes de communications, *La Documentation Francaise*, Paris

Hill, M. 1983: The Role of the British Alkali and Clean Air Inspectorate in Air Pollution Control, in: Downing, P., Hanf, K. (eds.), *Internatianal Comparison in Implementing Pollution Laws*. Boston, The Hague, Dordrecht, Lancaster: Kluwer-Nijhoff: 87–106

Hucke, J. 1981. State and Local Relations on Environmental Regulations in the Federal Republic of Germany, in: Downing, P., Hanf, K. (eds.), *Internatianal Comparison in Implementing Pollution Laws*. Boston, The Hague, Dordrecht, Lancaster: Kluwer-Nijhoff: 129–142

Jenny, F. 1984. L'évaluation de politiques industrielles, in: Nioche, J.P., Poinsard (eds), *L'évaluation des politiques publiques*. Paris: Economica: 187–204

Jobert, B.. Brenac, E. 1984. *La contribution des entreprises publiques et l'élaboration des politiques gouvernementales: La D.G.T. et le plan cable*, papier présenté à L'Atelier Séminaire à Lille

Knoepfel, P., Weidner, H.. Hanf, K. 1980. *International Comparative Analysis of Program Formulation and Implementation in SO_2 Air Pollution Control Policies in the EC Countries and Switzerland. Analytical Framework and Research Guidelines for ihe natianal research teams*. Berlin: IIUG, Sciencecenter

Knoepfel, P., Weidner, H. 1981. Die Durchsetzbarkeit planerischer Ziele auf dem Gebiet der Luftreinhaltung aus der Sicht der Politikwissenschaft. Ergebnisse aus einer internationalen Vergleichsuntersuchung, *Zeitschrift für Umweltpolitik*. Frankfurt, Nr. 2: 87–115

Knoepfel, P. 1982. Conceiving Comparative Policy Analysis. Implementation of Air Quality Improvement Programs, in: *Architecture and Bevaviour 1 (1980/1981)*. St. Saphorin: Georgi: 287–305

Knoepfel, P.. Larrue, C. 1984. Les politiques publiques comparées: tourisme intelligent ou vrai progrès? Le cas des politiques comparées de l'environnement, *Politiques et Management Public*. Paris, no 2/1985: 561s

Knoepfel, P., Larrue, C. 1985. Distribution spatiale et mise en oeuvre d'une politique publique: le cas de la pollution atmosphérique, *Politiques et Management Public*, no 2/1985: 43–69

Knoepfel, P.. Weidner, H. 1985. *Luftreinhaltepolitik bei stationären Quellen*, 6 vol. Berlin: sigma

Knoepfel, P. 1986. Distributional Issues in Regulatory Policy Implementation: The Case of Air Quality Control Policies, in: Schnaiberg, A.,

Watts, N., Zimmermann, K.(eds.), *Distributional Conflicts in Environ-mental-Resource-Policy*. Hants: Gower: 363–379

Kunreuther, H.,Linneroth, J. 1982. *Risk Analysis and Decision Processes: The Siting of LEG Facilities in Four Countries*. Laxenburg: IIASA

Lundqvist, L. 1980. *The Hare and the Tortoise: Clean Air Policies in the United States and Sweden*. Ann Arbor: University of Michigan Press

Montrade, C. 1984. *Financement et politique de développement de la RATP: analyse historique et prospective de la dialectique internalisation d'objectifs de la politique économique – adaptation à une demande évolutive*, papier présenté lors du Premier Congrès du journal Politiques et Management Public (Séance 3–1) du 27 au 28 Sept. 1984 à Paris

Neumann, W., Uterwedde, H. 1984. *Industriepolitik in Frankreich und der Bundesrepublik Deutschland*, working paper, presented at the German-French Institute of Ludwigsburg (Germany)

Peteis, M. 1982. Standard setting and implementation in $S0_2$-air quality control policies, *Fallstudien Schweiz*. Zürich (MS)

Scharpf, F.W. 1984. Economic and Institutional Constraints of Full-Employment Strategies: Sweden, Austria and Western Germany, 1973–1982, in: Goldthorpe, J.H. (ed.), *Order and Conflict in Contemporary Capitalism*. Oxford: Clarendon Press

Scharpf, F.W. 1984. Strategy Choice, Economic Feasibility and Institutional Constraints as Determinants of Full-Employment Policy During the Recession, in: Gerlach, K., Peters, W., Sengenberger, W. (eds.), *Public Policies to Combat Unemployment in a Period of Economic Stagnation*. Frankfurt, New York: Campus: 67–114

Schmid, G. 1984. Equal Opportunity Policy: A Comparative Perspective, *International Journal Manpower*, vol. 5, No. 3: 15–25

Thoenig, J. C. 1973. *L'Ere des technocrates. Le cas des Ponts et Chaussées*. Paris: Editions d'Organisation

Weidner, H., Knoepfel, P. 1981. Implementationschancen der EG-Richtlinie zur $S0_2$-Luftreinhaltepolitik. Ein kritischer Beitrag zur Internationalisierung von Umweltpolitik, *Zeitschrift für Umweltpolitik*, vol.. 4, No. 1: 27–67

Weidner, H. 1983. Schwachstellen in der Luftreinhaltung. Die Bundesrepublik im internationalen Vergleich, *Umweltmagazin*, No. 9: 22–24

Weidner, H., Knoepfel, P. 1983. Innovation durch international vergleichende Politikanalyse. Dargestellt am Beispiel der Luftreinhaltepolitik, in: Mayntz, R. (ed.), *Implementation potitischer Programme II. Ansätze zur Theoriebildung*. Opladen: Westdeutscher Verlag: 221–255

Section II: Institutional Change

In the European Union and in most other European countries, the most important bulk of substantive environmental regulations came into force during the 1980s. This was also the period of the creation of new environmental agencies, especially at the regional level. At the same time, sectoral environmental agencies of centralised states reorganized to integrate regulatory activities into new aspects of environmental protection, such as toxic substances, reinforced nature and landscape protection, energy saving, noise abatement, and soil protection. Hence, in Switzerland and in many other European countries towards the end of this decade and at the beginning of the 1990s, policy analysts were confronted with the question of adequate administrative structures for managing an increasing number of sectoral environmental policies. Thanks to important progress in ecosystemic research, it became rapidly clear that administrative structures had to take into account more intensively horizontal linkages between specific environmental policies (air, noise, etc. – intrapolicy co-ordination) and between environmental and non environmental policies (interpolicy co-ordination). Having been charged by the Swiss government to systematically conduct case studies on the regulation and implementation of environmental policies in important fields of mostly conflicting interpolicy relationships (such as between environmental policy and agricultural, energy, military, tourism and economic development policies) we had the opportunity to analyze the institutional weaknesses of environmental policies with regards to other well established strong policy communities, which were eager to protect themselves against intrusions from environmental protection actors.

During the 1990s, we never made the distinction between policy implementation and policy evaluation research in our practical work. Implementation analyses always were considered research that should include policy implementation and their effects in the real world. According to that concept, studying administrative implementation structures only made sense when looking closely at the various policy effects the different implementation arrangements produced. Hence, efficacy was the predominant criterion for selecting suitable administrative structures. In my opinion, this research concept is still valid. It is demonstrated in *Evaluation of the Federal Office of Environmental Protection: Across Two Levels of Government (1996)* by Willi Zimmermann and myself, where we present theoretical insights into our practical work which sought to restructure the Swiss Federal Office of the Environment in 1992 and 1993 on behalf of the federal parliament. Indeed, restructuring this large federal environmental policy agency with the objective of reinforcing environmental policies inevitably required knowledge on the decisive dimensions of this structure for cantonal implementation activities and their effects that, at the time, varied extensively according to the existence or absence of adequate mechanisms of horizontal interaction within the federal agency.

New Institutional Arrangements for a new Generation of Environmental Policy Instruments: Intra- and Interpolicy Co-operation (1995) is an article written within a collective book of the above mentioned international study group financed by the European Science Foundation. The two basic concepts of intra- and interpolicy co-ordination have been applied frequently since to theoretical studies and practical administrative reforms in Switzerland and abroad. They are still valid for shaping conflicts amongst environmental policies as well as between environmental and non environmental policies. The presented conceptualization becomes even more important when considering environmental policy as being part of comprehensive resource management policies, as we tend to do nowadays.

Different international comparative studies realized within a stable and friendly community of policy analysts during the 1990s[1] gave us the opportunity to work on such institutional aspects of intergovernmental relationships and administrative reforms in a comparative manner. These beneficial working conditions allowed us in turn to actively participate in many processes of environmental administrative reforms in Switzerland on the cantonal and federal level as well as abroad. Some of the crucial principles having shaped what we today can call the second generation of political institutional arrangements of environmental policies are presented in the three contributions of this section. In *Regulatory Change and Institutional Rearrangements: Building of New Policy Arenas for Ecologization of Agriculture. A Comparative Analysis of Programmes in France and Switzerland (1987)*, we present the results of in-depth comparative research on the environmental regulations regarding water pollution caused by agriculture in selected French and Swiss regions. It can be taken for granted that policy studies, conducted at that time by an increasing number of scholars and practitioners, contributed to a radical change in agricultural policies towards new ecologically-sounder subsidy systems in many countries of the European Union, although it is also true that major driving forces for such changes came from international trade regulations (World Trade Organization). Such implementation studies demonstrated in a very concrete way that environmental policy would never have had the chance to be implemented against a very strong incentive mechanism of heavily state subsidised agricultural policies, which encourage increasing agriculture production by using water-polluting fertilizers.

---

1 See forward note 1

5
Evaluation of the Federal Office of Environmental Protection: Across Two Levels of Government (1996)

Willi Zimmermann, Peter Knoepfel[1]

In the beginning was a conflict between the Swiss parliament and a federal ministry. In the end, an institution, a federal agency was evaluated. At the very beginning, the question was whether the merging of two federal agencies would mean the creation of a higher organisational unit, for which parliamentary approval would have been necessary, or whether it is simply a question of merging the two agencies to form the Federal Office for Environmental Protection (FOEP). At the end, this new federal agency was the subject of an evaluative study. The present study is about this evaluation.

5.1 Point of departure: policy orientation replaces the business management approach

The authors of this essay were appointed by the Federal Parliament auditing commission to analyse the implementation of environmental policy in Switzerland and clarify the following specific issues:

- the ability of the structure of the FOEP to guarantee homogeneity and effectiveness of environmental policy;
- the execution of environmental protection measures from the point of view of the organisation of the federal agency. FOEP structures were to be analysed in the light of this execution;

---

[1] in : O. Rieper, J. Toulemonde (éds.): Politics and Practices of Intergovernmental Evaluation, New Brunswieck, New Jersey, (Transaction Publishers), 1996, p. 133–150. Copyright: We thank the editor for authorizing the publication of this article in this book.

- the difficulties facing cantonal authorities in the execution of environmental policy measures.

The strict business management approach (i.e. cost-benefit analysis etc.) was abandoned in favour of policy-oriented administrative analysis, on the basis of which external and internal organisational structures are judged in the light of the quality of the administrative output. The parliamentary auditing commission decided to abandon the business management approach because one of the two agencies and their merger had previously been evaluated using such an approach. It opted instead for our policy-oriented approach because this approach involved the evaluation of internal and external organisational structures in the light of their administrative output, and the evaluation of the relations between the two agencies. This interaction takes place at both federal and cantonal level; our IGE is of the managerial type (as defined in the introduction). The evaluators, therefore, worked on the basis of the assumption that there is an inherent link between an administrative product and the administrative structures and processes involved in its production. In the case of the FOEP, a hypothesis was proposed to the effect that the production of a consistent and coherent administrative product (...) necessitates organisational structures and decision-making process with a specific capacity for coordination. It is important to note here that "policy-oriented" administrative analyses are based on a policy concept which is not restricted to parliamentary or government decisions. The concept of "public policy" as used in the Anglo-Saxon context (French: politique publique; Italian: politica pubblica) is actually more accurate here as this concept covers the sum of public decisions and measures used in the solution of collective problems, which are legislative, administrative, legal and sometimes even social or associative in nature. These decision-making processes involve political, administrative and even legal instances at federal, cantonal and even local authority level. The entire contribution of a particular authority or administrative body to such a public policy is termed as its "policy" in the area in question (according to the traditional designation: "practical" or "management"). Thus, for example, in the area of clean-air policy, it is possible to distinguish between a "FOEP policy", a "policy of canton XY" or a policy of the federal government or parliament. This policy concept is purely analytical in nature; it has nothing to say about the contradictory relationship for example between a federal agency policy and that of the relevant federal minister or the federal government. This does not broach the (always controversial) issue of the extent to which a federal agency has the right to make "political" decisions, or the extent to which it actually makes such decisions of political value independent of legal competence.

In this study, "FOEP policy" is understood as the sum of measures and decisions implemented and taken by this agency in the solution of specific environmental problems on the basis of existing legal provisions.

In accordance with the task set by the Federal Parliament auditing commission, the aim of this study was to evaluate the quality of the FOEP's entire administrative output in different environmental areas between 1985 and 1991 from the point of view of the experience of the cantons, and the reactions of other important federal agencies. A secondary aim involved the identification of the "organisational" or procedural causes of possible "errors" within the FOEP.

The main focus in the evaluation of the FOEP's activities was on clean air policy. In addition to this "policy field", which is linked in many ways to other areas of environmental policy, activities in connection with environmental impact assessment were also studied, and thirdly, measures in agricultural water protection were also examined. It should be noted that other important FOEP activities such as soil protection and waste management etc. were not included in the evaluation.

The evaluation was, therefore, undertaken from the perspective of the effective implementation of the legal aims of environmental policy. Other policies, which on the basis of their specific tasks or specific interpretations by their ministries and federal agencies involve aims which tend to contradict those of environmental policy, are seen as potentially restrictive factors. The analysis of effectiveness focuses on services provided by the various offices of the federal agencies which are directly involved in environmental policy, in particular, however, the FOEP. Together these offices constitute an open socio-political system. Such systems are characterised firstly by the way in which they must accept and adequately deal with impulses from their environment (other agencies, cantons, international contexts) and by the way in which they must be capable of indicating their aims to their environment in a way which enables the observation of these aims.

Few economic studies have hitherto been undertaken on administrative bodies in Switzerland. Those that are available do not provide any clear guidelines on the methods which should be used. For this reason we have based our approach in part on our experience in research and in part on new methodological elements.

5.2 Execution of the evaluation

To ensure optimum accuracy and comprehensiveness in the execution of this evaluation we availed of a specific combination of quantitative surveying (questionnaires) and qualitative information (analysis of documentation, interviews). Moreover, we took a "bottom-up" approach in the interviews which involved first interviewing those staff members in "subordinate" positions within the hierarchy of the agencies in question. The interviews provided an opportunity to go into the answers provided by these persons in their questionnaires in more detail and to discuss and their experiences relating to environment policy using examples from actual cases. The most important results of these intensive interviews were presented for verification/opinion to the employees at the next level in the hierarchy etc. (so-called cross interviews). The latter were also questioned on their own questionnaires and on subjects of relevance to their work and experiences.

The first step in the evaluation involved the implementation of a quantitative survey in the 26 cantons with a closed-question questionnaire. The relevant cantonal ministers were requested to fill out a questionnaire themselves and distribute questionnaires among colleagues responsible for environmental protection (usually the environmental protection offices). Of the 130 questionnaires dispatched, 89 were returned to us. At 68%, this response is deemed satisfactory.

A total of sixteen cantonal ministers and all twenty-six directors of the cantonal environmental protection offices participated in the quantitative survey. Longer discussions about the results of this survey were held with officials from four cantons. One of the additional aims of the evaluation was to discuss the nature of concrete co-operation between the FOEP and the specialised cantonal bodies. Prior to the discussions, we reconstructed this co-operation with the help of documents. Three relevant cases were prepared for each canton. These cases involved: i) the redevelopment of a waste water unit and water protection in agriculture, ii) environmental impact assessment, iii) the production of the cantonal plan for clean air policy. The FOEP was and is strongly involved in all of these cases. We evaluated the nature of the co-operation between the cantonal and federal agencies on the basis of this dossier and presented our theories to the cantonal officials for their views. The lack of clarity in the administrative output of the FOEP was also touched on. The interviews lasted approximately 90 minutes.

The methodological approach used for the cantonal organisations was also used for FOEP itself. A comprehensive questionnaire containing closed questions on the main subject was distributed to all persons (318) working in the agency. The questions dealt with the organisational and management structure of the agency, the capacity for innovation in the agency, communication inside and outside the agency, specialised knowledge and perception of problems. The three cases listed above were also referred to and those employees involved in clean air policy, waste water disposal, water protection in agriculture and environmental impact assessment were asked special questions. 248 of the questionnaires (78%) were suitable for statistical assessment. The qualitative survey was then implemented on the basis of these results. This involved one and a half-hour conversations with a total of 44 employees of the FOEP.

The auditing commission also requested that important federal agencies be included in the evaluation and questioned about their experience in co-operation with the FOEP. We selected those agencies defined as "difficult" partners in the written questionnaire completed by the FOEP employees. The agencies in question were those whose policy areas can be described, among other things, as extremely "harmful to the environment" and include: the police, public transport, Federal Agriculture Office, highway construction and external commercial relations. These interviews lasted 60 to 150 minutes and were normally conducted with at least two persons. At the very end of this phase in the evaluation, discussions were also held with the minister and general secretary of the Ministry of the Interior responsible for environmental protection. The following points were discussed in this order during all of these talks: the results of both questionnaires, the vertical and horizontal co-operation within and outside the FOEP, the quality of standard material, models and visions, informal structures, the FOEP corporate identity and proposals for reorganisation.

5.3 Difficulties encountered in the implementation of scientific evaluation in federal contexts

Before going into some of the difficulties encountered in the execution of this evaluation, it is necessary to give a short description of the Swiss political system.

Switzerland is a country with many special features. Its political system does not neatly fit any of the current theories of the modern state. Some of the most important elements of the political system include: i) New problems are, if possible, solved by private means, the state takes second place. ii) If political action is required, the closest possible co-operation is sought in many political fields. Social partnership is well established. iii) The system includes institutions such as referendums and popular initiatives.[2]

Another distinctive feature of Swiss political organisation is the federal system. This system consists of three political levels: i) The Confederation with the Swiss government, the Federal Administration which has its own budget and the Federal Assembly or Parliament with its two chambers, the National Council which represents the Swiss people and the Council of States which represents the cantons[3]. ii) The 26 cantons, each with its own cantonal government, parliament, administration and budget. iii) The 3029 local (political) communes with their own governments, modes of representation, administration and budget[4], which, however, always act within the scope of their respective cantonal context.

Federalism is achieved in a number of ways. According to the Swiss Constitution, the 20 cantons and 6 half-cantons are members of the federal state. They are not sovereign states as they were in the past; however they are distinct political entities with their own traditions. Each canton has its own constitution and a set of laws to regulate its own internal affairs. The cantons have considerable scope for participation in the formulation of legislation and decision-making at federal level through their participation in the Council of States, the federal referenda and the pre-parliamentary consultation procedure.[5] The cantons also play a decisive role in many political domains when it comes to the execution of federal legislation. For this reason, the system is often referred to as one of *Vollzugsföderalismus* (executive federalism). As a general rule, it is possible to say that the Confederation defines the aims of a public policy and the cantons choose the means to be used in its implementation and execute it in the form of individual projects. This structure should be kept in mind when reading the following paragraphs.

We would now like to discuss the difficulties encountered in the execution of the evaluation. As a first step, a quantitative survey with closed questions was undertaken in the cantons. The questionnaires were sent to the responsible cantonal ministers with the request that they be passed on

---

[2] These paragraphs are based on Kloeti Ulrich, *Swiss Democracy – Exception or Model?*, Forschungsstelle für Politische Wissenschaft, University of Zurich, Zurich 1987.

[3] The members are not instructed by the cantons on how to vote.

[4] While the size of the communes can vary considerably, each commune has just one administration, one parliament etc.

[5] A short overview of the history of Swiss federalism is provided by Knapp, Blaise, *Etapes du fédéralisme suisse*, in: *Manuel Système politique de la Suisse*, Vol. 3, Berne, 1986, pp. 31.

to those responsible for environmental issues within the canton administration. For political reasons, it would have been inconceivable for experts appointed by the parliamentary auditing commission to approach the cantonal offices directly without first approaching the cantonal governments. This means that, inevitably, the experts did not have complete control over the entire survey. All of the cantons responded, however, one canton only returned one questionnaire. Despite this however, the willingness of the cantons to participate in the evaluation can generally be described as positive.

Another difficulty arose from the fact that it was almost impossible to discuss the organisational problems in isolation from the general structural, management and personnel problems within the entire FOEP and its political environment. The administrative-scientific analysis of the FOEP output and its identification with key individual departments within the FOEP was not very productive. This was due to several reasons. Firstly, the legal analysis revealed few contradictions between the different environmental decrees. Thus, this analysis of origins proved to be not absolutely necessary. Also, the output was shown to have a predominantly natural science and engineering content. The number of non-technical, unsubstantiated reports and decisions was much lower than we expected. As political scientists we are unable to make judgements on the necessity and adequacy of technical standards, calculations and formulas; the commissioners, once informed of this "shortcoming", accepted it. It was not, therefore, possible to evaluate the extensive technical output. A third problem lay in the fact that the responsibility for the processing of a document can usually be traced to a specific department, however within the internal development of objectives within the agency, it is difficult to reconstruct the number of actors involved in retrospect. Other departments, the administration, the general secretariat and the department (Ministry of the Interior) were involved in different capacities in the subsequent consultation. The concrete contributions and influences of all of these actors, thus, largely elude empirical documentation.

The method described here was implemented using the rules of accuracy we have applied in other projects. It is a combination of a conventional quantitative approach, which provides a broad statistical overview, and a qualitative approach, which gives more in-depth insight into the problems and objects analysed. In our interviews with the civil servants we discussed the results of the survey, the questionnaire they themselves completed and possible significant discrepancies. Questions arising from the analysis of documents were also dealt with. The experience we had already gained from several previous evaluations of this nature had taught us that we had to avoid confusion and ambiguity (in answers, documents etc.) at all costs, and that there was always a risk of obtaining subjective accounts which do not tally with the expected responses, or what we already knew etc. In such cases, we requested that the civil servants provide written proof in support of their opinions in the form of either empirical data and/or a written statement on official paper. This helped us to obtain factual statements on the practice of administrative units. In some instances, however, we were deliberately denied the relevant information; it is, however, also possible that we simply did not receive the requested documents, due to a lapse of memory. There were also cases where it was made clear to us that such information was under "cantonal lock".

There is no "objective" measure for the evaluation of statements concerning specific policy and executive services of the FOEP. Our own experience and our (certainly limited) knowledge of the history and development of the relevant policy area were used as a basis for the evaluation. This was combined with the insight gradually gained from the repeated questioning in the discussions with our interviewees from the cantons, FOEP and the representatives from other federal agencies in the course of the survey. However, despite the utmost vigilance, there is always a degree of conjecture on the part of the evaluators which after lengthy discussion they claim to managed and controlled to the best of their knowledge.

Another limitation is ultimately that the social scientist working on an empirical study must also document things that are "objectively untrue". From the perspective of the specialist or politician, some of the answers we have recorded may be untrue: they do, however, represent "objective" content for the social scientist because they can be assumed as "true" for groups which must be taken seriously which is most likely why they influence their behaviour. This must be made clear to several politicians from the outset.

5.4 Some results

The questions set by the auditing commission for the experts and the answers obtained shall not be provided in detail here. We will concentrate only on aspects which are of interest to the Inter-Governmental evaluation. All that shall be said about the organisational structure of the FOEP is that it is not suited to guaranteeing the homogeneity and effectiveness of environmental policy. The agency should be restructured and the new structure should be created on a step by step basis. The question concerning the difficulties which the cantonal authorities have, for example, in the implementation of clean air measures covers the following points:

- resistance to traffic-related clean air policy by relevant interest groups (particularly: local sections of the Swiss Touring Club or the Automobile Party). It is interesting to note that trade and industry registered significantly less resistance to the air-purity clean-up measures;
- resistance by the cantonal infrastructure agencies (particularly the highway authorities and cantonal police); resistance from communal civil engineering, planning and police offices;
- the nature and intensity of the resistance varies from canton to canton; it is more intensive and widespread in French Switzerland than in the German Swiss cantons.

We would now like to return to the questions of interest to the Inter-Governmental evaluation. We have seen that the federal system includes several politically autonomous units. These include the federation and the cantons. In general, the cantonal actors gave the FOEP quite good grades. For example, the results of the survey of the cantonal officials reveal that they consider that the FOEP clean air department deals successfully with the variety and complexity of the problem. The cantonal officials also report that they make considerable use of the recommendations, statements and reports provided by the FOEP. It is interesting to note that only a small minority (4% or 6% of cantonal answers) are of the opinion that the FOEP speaks a contradictory language in relation to technical or political principles. The quality of the – extremely frequent – contact with representatives of the FOEP clean air department was described as good to very good. A third of those questioned had absolutely no problems with respect to co-operation with the FOEP, 64% admit to having "some" problems. The problems mentioned were based on "conflicts of interest between the federation and the canton", "internal problems of co-ordination within the FOEP" and "lack of availability of FOEP officials".

The cantonal actors, thus, gave the FOEP a rather positive evaluation. They were less positive, however, in their evaluation of the "Confederation". The latter includes the federal government and parliament. The questions we asked about the activities of these bodies related to their definitions of legal concepts and norms, their fixing of deadlines, their statements in relation to specific decisions (words versus deeds) and, last but not least, the way in which they made allowance for too many or too few exceptions. From the point of view of cantonal environmental officials, the Confederate bodies behave in a far more contradictory fashion than the FOEP. This discrepancy can mainly be explained by the functional specialisation and professionalization which pervades all three levels in the federal system (federation, canton, bigger towns).This "cartelisation" has the advantage that it facilitates the relatively simple and rapid establishment of consensus. This process is simple and rapid because the actors involved usually share the same interests and perception of problems, and share the same definition of "reality".[6] With few exceptions, the cantonal actors attested the good will and excellent specialist knowledge of their FOEP colleagues.

However, certain disadvantages within this professionalised sector should not be overlooked. Because they are specialists and are obliged to sell their "specialised product" to other sectors, the capacity of these actors for action is basically limited. Moreover, increasing specialisation heightens the potential for conflict and the frustration of the officials at the "front". This is due on the one hand to the very technical emphasis in the environmental protection policy of the FOEP, which takes an end-of-the-pipe direction. It is also due to the increasing difficulties faced in the execution or policies – we assume that this mainly results from the individualisation of society and is, thus, is not directly connected with the activities of FOEP. The more detailed discussions on the results of the survey (questionnaire) and the individual cases with the officials from four cantons revealed that from their point of view, the technical approach to environmental protection, and particularly its control, have reached its limits. For this reason it is imperative that it now be complemented with economic measures.

By drawing on the information we obtained from the civil servants, it is possible for us to observe several social developments which indicate a trend towards individualisation in highly developed West European states: higher levels of income and increased leisure are giving rise to greater individual liberty; the growth of the third sector is leading to large-scale mobility (and upward mobility) and to the diversification of lifestyles; improved educational opportunities mean greater freedom for the individual to pursue his/her chosen career and greater scope for general "personal development". Mass consumption of television etc. help to promote highly individualised lifestyles whereby the individual considers himself/herself at the centre, free to exert complete control over the planning of his/her life. Under such circumstances, public policies lose their "steering capacities", it becomes more difficult to implement public laws and protect the environment. The technical end-of-the-pipe solution no longer suffices, for, to put it bluntly, it allows pollution to occur and expects the public authorities to clean up the mess. Pollution must, therefore, be prevented in

[6] Cf. Scharpf F. W. et al., *Politikverflechtung*, Kronberg, 1976

such a way that "individualists" are also included", that is using means which prevent pollution and involve all members of society. These ways and means are mainly economic (internalise the cost of pollution). This, at least, was the opinion of most of the civil servants we interviewed. Thus, a clear need was identified for a change in strategy in the area of economic steering and also in the case of clean-air policy.

A positive observation was that the clean air concept – an target-oriented strategic paper – also provided federal clean air policy with a razor-sharp gauge which was very useful to the cantons in the production of their cantonal plans for clean air policy and remains so. In addition, the corresponding decree with its unambiguous limit values and the mechanisms for achieving them provides a totally suitable executive instrument. In contrast to the noise pollution decree, the clean air decree only contains a few exceptions which are very clearly formulated. The most significant point about which lack of clarity prevails is the, politically speaking, highly sensitive question of what should be done with new installations in critical areas (polluted areas).

The potential for conflict can, therefore, only be reduced by a change in strategy. The cantonal peers of the FOEP are not the only source of potential conflict; the office is also involved in relatively "substantial" conflict with numerous federal actors! As an example here we have chosen traffic-related environmental policy. This area is a particularly clear example of the difficult situations which the agency (FOEP) must master. In principle, much depends on the decisions taken by other agencies and ministries and the conditions surrounding other policy areas. A typical example here being the Federal Police Office which is responsible for the provision of road signalisation and the type testing of vehicles. The work of the FOEP, which must guarantee the safety and continuity of execution, was not made easier when, for example, a high official or representative of several cantons declared in 1991 that in view of the difference in the strict limit values stipulated in the clean air ordinance as compared with those in other countries, it would be advisable for the federal government to re-examine these limit values...

Similar inconsistencies also exist between the FOEP clean air policy and public statements made by the Federal Office for National Highways. In a lecture in June 1991, a high official from this office made the following mysterious statement in front of a large audience – his sights were on the FOEP and its limit-value-oriented environmental policy that would be argued using sweet means in today's politics – "if one enthuses on the basis of ingenious ideological but completely nonsensical theories, as to how, in what way and by what means the increase in traffic can be remedied. (...) The limit values are unreasonably high, the methods inefficient, even hot-air. However, air pollution, the waste mountain (...) are realities. It is right for us to work together, not in a Babylonian chaos, but soberly and correctly with an eye to the aims to be achieved." (Quote from speech manuscript).

This relativization of environmental protection policy by "polluting actors" is not restricted to clean air policy but can also be found in water protection in agriculture. It should now be noted that FOEP activities in this context received a far less positive evaluation. Insofar as we can establish it within the context of the discussions which were limited to environmental protection, problems have arisen everywhere in the implementation of rural water protection in agriculture. In contrast to the area of clean-air policy, the FOEP is not seen as helping to speed up cantonal execution. In general, from the perspective of the cantons, the execution of water protection in agriculture mainly fails due to the fact that the agricultural sector and agricultural authorities in the cantons themselves have problems with it. Those cantons surveyed made very clear, however, that the legal regulations and executive aids in the form of recommendations and guidelines etc. provided by the FOEP are not very helpful. The vast majority of the cantonal officials surveyed also said that the federal government and parliament have not made adequate provision to ensure that water protection in agriculture is at one with agricultural policy. The cantons feel that the Confederation is speaking two languages in this instance – that of the FOEP and that of the Federal Office of Agriculture.

We have thus established that numerous problems exist within "transversal" environmental policy which can be traced back in particular to the professionalised sectors. Within the administration this results in strong reproaches being made – at cantonal and federal level. The two blocks stand opposite each other. On the one side there are the federal and cantonal "environmental protectors" and on the other the federal and cantonal "environmental polluters" which include the agriculture, police, highway construction bodies etc. The accusations levelled by these bodies at the "environmental protectors" include:

- their interpretation of the legislative basis is narrow and small-minded;
- they do not show any understanding of other policy areas beyond environmental protection and the situation of the relevant actors;
- they do not provide comprehensive information and fail to issue invitations to important meetings.
- they deliberately put representatives of other policy areas under pressure to act by setting very tight deadlines;

- ultimately, the environmental protectors are not interested in general well-being but pursue only pure and simple environmental policy".

The reproaches levelled by the "environmental protectors" at the "environmental polluters" are the exact inverse of those listed above.

We would now like to take a quick look at the environmental impact assessment policy. None of the cantonal interviewees wanted to fully dispense of the participation of the FOEP in this policy (hearing procedure). The federal ordinance on environmental impact assessment lists a certain number of installations for which the FOEP must be included in the assessment procedure. They saw this procedure together with federal participation as guaranteeing consistency of approach in the cantons. It was possible to establish that in contrast to the general opinion expressed throughout the country and, in particular, by numerous cantonal governments, the relevant specialist offices in the cantons, and also a clear majority of the members of the authorities, welcome environmental impact assessment as a suitable instrument for the co-ordinated implementation of federal and cantonal environmental policy. On the basis of information from another study on the canton of Berne, and statements made by cantonal infrastructure departments, we know that this opinion is not unopposed at the level of cantonal administration. The highway construction offices, hydraulic engineering and energy authorities in many areas are of the opinion that environmental impact assessment is an instrument which obstructs construction. They see it as unnecessary, as prior to its existence, the relevant installations were always tested for environmental effects!

The problems with environmental impact assessment have highlighted many instances of a lack of co-ordination at programme level between different planning policies (particularly environmental protection in landuse planning, forestry policy, highway construction policy, waste and dump planning etc.). A need for material co-ordination has been revealed which must be carefully established prior to the allocation of a permit at project level. This gives rise to real delays. There can be no doubt that environmental impact assessment cannot be held responsible for this function of co-ordination as an "emergency brake" but it is the above-mentioned policies and their extremely complex legal procedural regulations that are to blame! It is not possible to set the fox to keep the geese as some of the interviewees tried to suggest to us. Environmental impact assessment is an essential instrument for establishing agreement between various planning policies which urgently require the integral implementation of environmental policy.

5.5 Some lessons learned

Who should perform the evaluation in such cases? Insiders or newcomers? In the present case, the parliamentary commission opted for the former. Why?

Prior to receiving their appointment by the commission, the evaluators had already studied and evaluated the implementation of federal legislation in Switzerland – in most instances in relation to environmental protection. They belong to a "rare species" of researchers who have traced the development of a topic (environmental protection in agriculture) from federal through cantonal and local-authority levels to individual farmers.

Why did the commission decide to opt for these insiders? The federal auditing commission is under considerable pressure. Our federal parliament is not a professional body (its sessions last together nine to twelve weeks per year); time is a scarce resource and complex evaluations are time-consuming. Secondly, this evaluation provided an opportunity to perform an experiment. At the time of our commission for the evaluation, the federal parliament was discussing the possible creation of a professional, scientific evaluation service; working with a scientific team means gaining experience. This parliamentary service has meanwhile been institutionalised.

The first lesson to be learned from our evaluation is as follows: we suspect that in cases of IGEs commissioned by political bodies, 'absolutely new ideas' are not particularly in demand or, indeed, seen to be of much use – they are, after all, produced in large quantities at the universities. What is of greater interest is the provision of precise information on the complexity of policies and their implementation. This information must be multi-disciplinary, for a parliamentary auditing commission is of necessity interested in several dimensions – legal, procedural, organisational etc. Moreover, it is also interested in insights which can, if necessary, be presented in parliament. From this perspective, thus, only the best is good enough. We also suspect that the members of the auditing commission were keen to use professional experts (insiders) who could be sent down to the next "rung" in the federal system, i.e. the cantons, without shame or loss of face. Thus, it is seen as essential that the evaluators be insiders.

Whether it is actually essential or not, it is impossible for the (professional) experts to avoid appearing as learners who are trying to put together a step-by-step account, and to avoid coming back to their commissioners again and again to readapt the "evaluation structure" they devised together. Thus, the second lesson of our evaluation would indicate that (professional) experts cannot avoid appearing as learners, even if they are insiders:

- irrespective of their previous experience, they must be open to the unexpected;
- they must avoid the application of simple models and the generalised treatment of all actors – such an approach is clearly not suited to a federal system with 26 executive cantons;
- they must be independent and at the same time be aware of the fact that there are only two types of evaluations: politically important and, therefore, quite sensitive evaluations, or merely trifling ones. It is part of the experts' job to work out the political contexts and the consequences of the evaluation.
- openness, being aware of the complexities and interpreting the political contexts are a *conditio sin qua* for Swiss evaluators. Switzerland is a small country where everybody knows everybody else. Gross errors could mean the end of one's activity as an evaluator, or even one's career;
- openness and the willingness to assume a learning attitude is particularly important in the case of an IGE evaluation (as well as in federal systems). The lucky cases whereby an expert is appointed by a government or the federal system as a whole are quite rare. The evaluator, therefore, depends on the good-will and expertise of numerous actors from different executive levels representing different sectoral policies;
- finally, learning is also an essential part of interaction. The evaluator cannot avoid coming back to his commissioners again and again to readapt the "evaluation structure" they previously defined together.

In conclusion, therefore, on the basis of our experience and reasoning, there are many reasons for the acceptance by the evaluator of a learning attitude.

A third lesson: the evaluator is influenced by his clients. Their initial questions must be taken seriously and interactive learning always involves reciprocal influence. The commissioners' questions must be given a more precise formulation, dissected into sub-questions and complemented. Some questions will have to be omitted and in this instance the evaluator must explain to the commissioners the implications of the question in relation to the methodology and the available data. The same applies to the evaluator's questions. Working together to define the important questions serves to stabilise the expectations of all participants and to create something of a common culture. Our third lesson implies that a stable, common culture must be created by means of interaction. This is all the more important in political contexts as politics is by necessity “volatile”.

From a scientific evaluator's point of view, political questions must not be confused with scientific hypotheses. This is our fourth lesson. Hypotheses help to stabilise expectations but, more significantly, they serve in the creation of transparency. The political client and his/her questions do not replace the scientific concept which guides the evaluation. The questions are often linked to specific political contexts and can be subject to constant adjustment by the client in accordance with political popularity and opportunities. Such volatility is not conducive to scientific activity and evaluation which need stability, tranquillity and peaceful seclusion. Volatility may also result in a (partial) change of the system of reference: new notions may be introduced and new questions raised which do not necessarily coincide with those agreed upon. The danger of confusion is a threat to all IGEs. Hypotheses serve scientific purposes and create transparency for all involved, for the choice of concepts is always also a choice of a scientific strategy (i.e. choice for a network or systemic approach etc.). The same holds good for the methods. They create transparency and guarantee “inter-subjectivity”. The selection and mastery of these methods is the job of the evaluator, but it is also his job to explain these and their limitations to the commissioner. This is a complicated process for both parties. As a common “culture of scientific evaluation” has not been established between science and politics in Switzerland, it is not possible to develop and use complex scientific models and methods. The fourth lesson claims, therefore, that political questions do not replace a scientific approach.

Our fifth lesson is simple. Due to the combined effects of the extreme complexity of executive reality arising from the constantly changing conditions and the scientific approach, the evaluation can only be performed in small, iterative steps. This, however, enables the creation of the learning effects which constitute the basis of the evaluation's relevance: it will be taken seriously and be effective. In the course of various evaluations we have also observed that political commissioners do not like surprises: they see them as the product of strategic behaviour. The gradual, iterative approach is, thus, also useful in preventing undesired surprises.

Finally, our sixth lesson concerns responsibility. Even if outsiders evaluate (1st lesson), even if they do it by means of being open and of interactive learning (2nd and 3rd lesson), even if they contribute to transparency by using a scientific approach (4th lesson) and even if they take an iterative approach (5th lesson), it is the (political) commissioner or commissioning body that must take ultimate responsibility for the evaluation. This commissioner is democratically elected and legitimised to draw conclusions, demand change and make decisions. The commissioner must, therefore, be involved in "crucial decisions". In our case these involved the selection of policy areas and cantons, the drawing up of questionnaires etc. Ultimately it makes sense for the experts to work with variants, this guarantees that the politicians must chose and decide. This may be contrary to the experts' preferred approach. Nevertheless, they must accept the paradox of democracy whereby the "people" are always right. Ultimate responsibility must remain with politicians.

It is easy to teach such lessons. How can we guarantee, however, that they are of any value? To answer this question we would like to take a look at what happened to our evaluation.

The federal auditing commission accepted most of our recommendations. In two instances, however, where our recommendations involved other ministries besides the ministry of the interior, of which the FOEP is a part, the commission asked us to work on alternative proposals.

The evaluation has shown that the combination of two federal agencies has not resulted in the establishment of a higher organisational unit but "only" in the FOEP. In this case, the federal government did not exceed its mandate; it had and still has the right to organise the FOEP in the way that it wants to. The status of the auditing commission's findings is merely that of recommendations; in no way do they constitute an imperative for the government. All those involved in the evaluation waited impatiently for the government's response; the government was obliged to inform parliament as to why the recommendations have not been implemented.

The answer was a long time coming, almost two years. The former director of the FOEP has been replaced. He had reached the age of early retirement and accepted to serve his minister as part-time commissioner; his decision to retire was in no way related to the evaluation. The new director had more important things to do than to reorganise his office. He did not avail of offers from the experts to discuss their findings with him. However, the evaluators did manage to obtain some information to the effect that the FOEP had indeed started to work on the development of strategic papers, as had been suggested.

The long-awaited answer from the government finally arrived (1994) and came as something of a surprise. Many of the suggestions made by the evaluators and accepted by the auditing commission have also been accepted by the government:

- reduce the directorate;
- reduce one level of hierarchy;
- introduce a matrix form of organisation:

- introduce task forces within the FOEP and between ministries and offices at national level and enhance their mode of co-operation in order to avoid conflicting outputs. This measure in combination with strategic papers will certainly help the cantons, for they will know what direction to take in accordance with federal strategy and there should be a reduction in contradictory federal statements and advice.

The government did not follow the experts' and the auditing commission's recommendations in the two cases mentioned above which involved other ministries. Here the government argues that switching sections would destroy the existing networks and traditions; this argument is quite valid as the networks referred involve not only the cantons but also the local authorities.

In our study we evaluated the FOEP in the "light of its administrative output" and thus execution. For this reason only certain aspects of environmental policy have been broached. For example, the question of the optimal nature of the policy was not examined. This would possibly have brought different facts to light. We would like to draw attention to the fact that in policy-oriented evaluation, decision-making style should also in principle be taken into account. Under certain circumstances within a federal system, this can lead to what F.W. Scharpf calls the "joint decision trap"[7]. The "historical" tendency of federal systems to reach decisions through the unanimity rule is not exactly optimal in terms of policy outcome. The models developed by the professionalised segments are often unanimously accepted and implemented by the relevant governments. As in the case of the above-mentioned evaluations, the question arises for the governments as to why they should threaten an existing cemented compromise. Such joint programmes are allegedly inefficient, inflexible, unnecessary and somewhat undemocratic!

These programmes are inefficient because, comparatively speaking, they are too expensive due to the participation of several levels; in other words there is a tendency to overspend. They are too inflexible because mostly poor and rich areas (in the case of the Federal Republic of Germany) profit from the programmes, or in other words allocation is not really effective. They are unnecessary in many instances because many important aims have already been attained through federal involvement. Finally, they are often undemocratic because the parliament is often confronted with *faits accomplis* agreed bureaucratically by the two levels. It

[7] Cf. Scharpf, F.W., *The joint decision trap: Lessons from German federalism and European integration*, in: *Public administration*, Vol. 66, Autumn 1988, p. 239–278.

would have been very interesting challenge for us to continue our evaluation along these lines.

References

Kloeti, U. 1987. *Swiss Democracy – Exception or Model?* Zurich: Forschungsstelle für Politische Wissenschaft, University of Zurich

Knapp, B. 1986. Etapes du fédéralisme suisse, in: *Manuel Système politique de la Suisse*, vol. 3, Berne

Scharpf, F.W. 1988. The joint decision trap: Lessons from German federalism and European integration, in: *Public administration*, vol. 66, Autumn 1988: 239–278

6 New Institutional Arrangements for a New Generation of Environmental Policy Instruments: Intra- and Interpolicy Co-operation (1995)

Peter Knoepfel[1]

6.1 Introduction

It is hardly surprising that the European Science Foundation chose environmental policy instruments as the topic of one if its task forces on "environmental policy". This choice testifies the fact that environmental policy lacks the belief that it can achieve its goals with the classical environmental policy instruments. After twenty years of hard work of implementation, many policymakers arrived at the conclusion that its instruments must be supplemented by new, more efficient tools. It seems that after a swift boom in the late 1970s and 1980s, in the light of its current overarching political importance, has already for some time considerably transgressed the simple frame of ordinary policing in the field of health or industrial supervision. More recently, in the search for new instruments, theory and practical experiences of environmental policy has borrowed increasingly from fiscal and tax policy[2]. Moreover, the slowly adopted practice of environmental impact assessments (EIE) had led to first attempts to build a

---

[1] in : Bruno Dente (ed.), Environmental Policy in Search of New Instruments, European Science Foundation, Dordrecht, (Kluwer Academic Publishers), 1995, p. 197–233. Copyright: We thank the editor for authorizing the publication of this article with some minor linguistic adaptations proposed by Susan Cox.

[2] Amongst others D. Helm, Economic Policy Toward the Environment, Oxford (Basil Blackwell Inc.) 1992; Th. Tietenberg, Economic Instruments for Environmental Regulation, Oxford Review of Economic Policy, vol. 6, 1990; for Switzerland: R. Mauch/R. Iten/E.U. von Weizsacker/J. Jesinghaus, Oekologische Steuerreform Europäische Ebene und Fallbeispiele Schweiz, Churand Zurich, 1992 and X. Oberson, Les. taxes d'orientation, diss., Genève 1989.

bridge with industrial, land use and infrastructure policy[3]. The EIS requires the application of an environmental law in the planning and realization of new projects of all governmental policies of environmental relevance. The postulate has always been included in all the relevant laws and at least since the Single European Act of Luxembourg (1987) it figured in the pertinent body of primary law of the European Community (Art, 130 R, Para. 2). In the Maastricht Treaties it received additional normative reinforcement[4].

Everyone who deals with such EIS procedures is well aware of the enormous resistance against these high principles in practice. Neither the EC guideline no. 85/337/EEC of June 27, 1985, nor other known national environmental legislation within and outside of the European Community envisage legally binding, unambiguous regulations stating that facilities cannot be built if the EIS is negative. Nevertheless, the European-wide introduction of the EIS produced the important insight that the most substantial innovation of this instrument is the creation of new administrative structures inside and outside of the traditional environmental bureaucracies, which facilitate co-ordination.

This section will however not focus on the EIS which is still an instrument of classical environmental policy, especially if it is just introduced into already existing permit procedures. Since it focuses on specific projects and operates based on a relatively narrow, traditional understanding of environment, this procedure could be accommodated, at least in the initial phase, within the existing administrative structures. It hardly produced systematic co-operation between environmental policies and the core-actors of infrastructural supply policies on a strategic level. At the centre of this article are the strategies leading to a lasting, systematic and innovative opening of the non-environmental policies towards ecological policies in the comprehensive sense. Among other things, such an extension would improve the access of ecological policy to the instruments of implementation within these non-environmental policies. The spectrum of ecological policy instruments would, therefore, be enlarged not only in the direction of specific economic instruments of fiscal policy, but in the direction of other regulative, incentive or persuasive instruments of potentially all realms of governmental public policies.

---

[3] Cf. P. Knoepfel, Zum Stand des Umweltrechts in der Schweiz/A Survey on Current Environmental Law in Switzerland, Cahiers de l'IDHEAP no 65, octobre 19 89, p. 25f., 57f.

[4] Cf. A. Epiney, H. Furrer, Umweltschutz nach Maastricht – Ein Europa der drei Geschwindigkeiten?, EUR 1992, no. 4

The realization of this ambitious goal presupposes the more or less far-reaching inter-linkage of environmental policy with these policies. Such rearrangements are the main topic of this article. In the first section, the potential scope of this new set of instruments is explored on the basis of three examples (agriculture, transport and land use policies). Some considerations are made regarding the price the environmental policy would have to pay to the non-environmental policies for such a borrowing of instruments (section 6.2). In the second section, the empirical case of Switzerland is briefly discussed, which will serve as an illustration for developing scenarios for restructuring the environmental administration (section 6.3). The centre-piece of the argument is laid down in section 6.4, in which scenarios are presented for reinforcing co-operative structures within an environmental policy, which is expanded in respect of its sphere of application (intra-policy co-operation), and for the extending of instruments respectively, which is based on various concepts of structurally assured co-operative relationships between environmental and non-environmental policies (inter-policy co-operation). This section is entirely drawn from an expert evaluation compiled by Willi Zimmermann and I for the attention of the Swiss Parliamentary Control Commission on the Federal Office of the Environment, Forests and Landscapes (FOEFL) in 1991[5]. In the final section 6.5, these scenarios are briefly referred to with regard to their political and instrumental implications.

6.2 The new potential spectrum of instruments and its price

Most of the instruments of the non-environmental policies, which are of interest to ecological policy, naturally consist of (direct) planning and execution in their implementation phase. At the level of program formulation *per se*, some binding obligations for co-operation of a material or procedural nature can be envisaged. In most cases, these obligations remain however irrelevant for the potential changes of behaviour of the interested target groups as long as there are no (judicial) sanctions in case of non-compliance (for instance an invalidation). It has to be said, however, that

---

[5] See P. Knoepfel, W. Zimmermann, in cooperation with G. Sailer, E. Matafora. Evaluation des BUWAL. Expertenbericht zur Evaluation der Luftreinhaltung, des ländlichen Gewässerschutzes und der UVP des Bundes. EDMZ. Berne. 1991 and Rapport de la Commission de gestion du Conseil national du 18 mai 1992, FF vol. 3, no. 29 du 28 juillet1991, p. 1422ff.

for the extension of the access of the environmental policy to the federal or (above all) regional (= cantonal) political-administrative institutions which implement these instruments some legislative acts are also required. They pertain to the restructuring of existing administrative arrangements at the organizational and procedural level. The most current case is the creation of new regional implementation agencies. They are the key precondition for making accessible the vast field of existing policy instruments to the political-administrative actors of environmental policy with sufficient institutional guarantees. These complementary strategies are the main focus of this chapter.

Before discussing these strategies, a brief description of the problem areas and policy instruments is necessary. Of environmental relevance are all those non-environmental policies, which are objectively in a position to affect substantially the environmental behaviour, defined in terms of space and time, of the target groups of environmental policy. These include, amongst others, all those policies of the federal state or the regions, which can only be realized through the construction of facilities. In the case of Switzerland, this includes the general federal defence buildings' policy (barracks, shooting and training grounds, airports), road construction policy (national highways and main streets), general public transportation policy (federal railways, buses, shipping lines and air traffic), telecommunications policy (construction of transmitters), energy policy (construction of power plants, grids, pipelines), forestry policy (preservation of Swiss forests), environmental policy in the narrow sense (construction of sewage and waste disposal plants), housing policy, regional economic policies (development of mountain areas), tourism policy and, last but not least, the large field of land use policy[6]. In addition, there are policies, which are of indirect relevance for the environment, such as agricultural, research and education policy.

Given this almost unlimited wealth of environmentally relevant, non-environmental policies and the related, individually applied instruments, every strategy of ecologizing non-environmental policies has to set priorities. Concentration on a few policy areas, which will probably vary over time, is mandatory in the interest of practical applicability and political feasibility. Most likely, agricultural, transport and land use policy today form part of this core area in Switzerland and other Central European States:

---

[6] This list is drawn from the Report on the State of Land Use from 1987: Conseil fédéral: Rapport sur l'Etat et l'évolution de l'utilisation du sol et de l'urbanisation en Suisse (Rapport sur l'aménagement du territoire, 1987) du 14 déc. 1987, 87.074, p. 76 ff.

- The tools of agricultural policy comprise regulative instruments such as import restrictions, feed stuff restrictions, production restrictions through the licensing of agricultural building, or the limitation of livestocks through prohibition, as well as the immensely wide field of agricultural product subsidies or product-unrelated direct payments tied to corresponding conditions of extensive production, and the various types of agricultural consulting (persuasive instruments)[7]. Compared with other policy areas, agricultural policy has a wealth of instruments deriving from years of experience, which also reflects the manifold corporatist ties with various groups of agricultural policy (food stuffs industry, producers of agricultural subsidiary products, large, small and mountain farmers, various branches of the food industry). Accordingly, the implantation of concrete ecological elements through an explicit integration of the application of these instruments in the realm of environmental policy would facilitate a multivarious, faceted and differentiated implementation of environmental regulations along the entire spectrum of food production starting from the raw materials, to the farms and to the product sector[8].
- In transport policy, in most countries today, the regulative instruments prevail both with regard to the construction as well as the operation of the transportation network. With regard to the construction of these works, plans and project approval procedures dominate in different phases. In addition, there are of course various budgetary decisions allocating the corresponding investments. With regard to the operation of these systems, the manifold instruments of regulative traffic legislation (for instance, signalization of speed limits, (partial) driving restrictions, neighbourhood streets etc, as well as their control)[9] and also measures of traffic limitation through construction authorization procedures (road blocks, bars, etc.) are of increasing interest to environmental policy: More recently, road traffic policy increasingly uses incentive instruments, which promise fiscal rewards for low exhaust vehicles or reduced driving (and the corresponding transfer to public

---

[7] See Report of the Commission Popp “Les paiements directs dans la politique agricole suisse”, Berne. 1990 and the 7th Report on the State of Agriculture. Beme, 1991.

[8] See P.Knoepfel, L'agriculture et la gestion de l'environnement: Les instruments de gestion en Suisse, Cahiers de l'IDHEAP no. 60, 1989.

[9] See D. Baroni, Brunner, P. Knoepfel, P. Moor, 1992: Strassenverkehrsrecht im Lichte des Umweltrechts – Le droit de la circulation routière face à la législation de l'environnement, Bâle (Helbing & Lichtenhahn), Reihe Oekologie & Gesellschaft, vol. 5.

transportation). Other essential instruments include guidelines for the equipment of vehicles of private and public transportation.

The air hygiene action plans designed by the Swiss cantons in order to reduce air pollution by road traffic provide an impressive illustration of the variety of instruments of regional transport policy which are of interest to environmental policy, and their consequent operationalization by environmental policy. First analyses of the genesis of these plans illustrate that such a building of bridges between environmental and transport policy is only successful if the corresponding organizational and procedural rearrangements at the level of the responsible administrations can be institutionalized permanently[10].

- Surprisingly, in Switzerland and some other European countries, the instruments of land use policy have only been used occasionally for environmental policy. In the case of Switzerland, reasons include the different organization of federal competencies in these two policy areas, the different professional standards, the delegation of these activities to two different federal agencies, and last but not least the quite different nature of the instruments themselves. Nevertheless, the land use policy instruments (such as zoning) would be of substantial interest to environmental policy, since they ultimately decide the nature, the spatial distribution and the density of potential sources of emission. Key works such as "condensed construction", "space saving road planning", "protection through allocating of compensatory areas close to nature", "biotopes" and "nature protection objects", or "definition of livestocks ceilings conducive to the environment in agriculture" are clear illustrations. In Switzerland, there only have been occasional forms of co-operation between these two policy areas so far. Examples include noise protection (identification of noise sensitive areas), environmental impact assessments (definition of the spatial perimeter within which the impacts of new projects have to be examined), action plans for specialized air protection, or (newly required) environmental impact statements for communal land use plans. But environmental policy has not yet ad-

---

[10] See R. Imhof. W. Zimmermann, "Massnahmenplanung" in Switzerland: The strategic air purification plan as an instrument of environmental policy in the field of urban traffic (International Congress: Implementing environmental policies by means of interpolicy co-operation, in Crans-Montana); D. Sperling, Marketable credits for vehicle emissions in California, ibid.; this question is in the centre of the COST-Action no 618 concerning "Institution Building and Information Policies" with regard to air pollution control policies in big cities in Europe (programme "CITAIR"), decided by the COST Council in May 1992.

vanced into the core areas of zoning policy (housing, transport, and industrial zoning policy). A much more comprehensive co-operation and a corresponding utilization of instruments by environmental policy would be possible[11].

It is obvious that the non-environmental policies concerned and the responsible authorities, and also the interest groups standing behind them, will not easily put "their" instruments at the disposal of environmental policy. They would have to concede much more to the environmental policy than they do today in the frame of the ultimately non-binding environmental impact assessments, where they still retain the final decision on the realization or non-realization of a public work. The extension of their instruments to the environmental side would at least imply that environmental policy would not gain a right of initiative, but an (absolute?) veto position. For this purpose, environmental policy would have to pay a price, since the policies concerned would be "slowed down" or even put into question not only at the fringes, but occasionally at the very centre of activities[12].

Because instruments are not only unilateral acts of authority, and because there are often extensive negotiations with the target groups before their application, the non-environmental authorities have also to partially give up their monopolistic control over the various contacts to these target groups in favour of environmental policy.

Stiff resistance by the non-environmental policies can be expected. It can only be broken, if environmental policy in turn can offer advantages to the policies concerned. Such advantages exist indeed: They can be summarized under the title "certificate of ecological compatibility". Such a certificate would provide to the non-environmental authorities and their projects ecological legitimacy, safety and a favourable "climate of allocation". Today, there will hardly be agreement over the political opportunity of issuing such certificates of compatibility. This problem will be addressed again in the context of the presentation of the scenarios in section 6.4. At this point, it only has to be said that with this offer in return environmental

---

[11] See P. Knoepfel, U. Muller, W. Zimmermann, Vorschlag für eine Programmatik sozialwissenschaftlicher Umweltforschung in der Schweiz, BUWAL-Schriftenreihe, Berne, 1994.

[12] See P. Knoepfel, La protection de l'environnement en proie aux problèmes d'acceptation et aux déficits de mise en oeuvre. Mise en oeuvre de la législation environnementale: situation actuelle à la lumière des secteurs Industrie et Artisanat, Agriculture et Politique infrastructurelle de l'Etat, Cahiers de l'IDHEAP no. 70a, novembre 1991 (version française du Cahier de l'IDHEAP no. 70), Lausanne.

policy assumes a high, considering the lack of knowledge in some areas, even very high, liability. Put into question later on, it would not only incur high public legal costs ("protection of trust"), but would also undermine the pact of *"do ut des"* between environmental and non-environmental policies described above. The danger has to be considered as relatively high that given such premises the utilization of the instruments of these other policies would be refused to environmental policy from the beginning.

6.3 The empirical background: analyses and recommendations regarding the restructuring of the Swiss Federal Office of the Environment, Forests and Landscapes (BUWAL/FOEFL)

The above mentioned review of the Federal Office of the Environment, Forest and Landscapes (BUWAL/FOEFL) conducted during 1991 on behalf of the Control Commission of the Swiss National Council constitutes the empirical background of the scenarios presented in this article.

In interpreting the findings the authors followed inductively and deductively the following eleven guidelines in the sense of grounded theories:

1. "The structures of the agency should altogether be simplified and the spectrum of tasks (objectives, functions) be reduced. It should have a basic structure which corresponds better to the requirements of an integrated ecological policy through internal linkages than it does today.
2. Improved internal linkages exist, if the agency is built according to the basic principle of a matrix organization. This form of organization allows to link vertically the sectoral departments with the various environmental media (water, air, noise, nature, landscape, soil) through so-called cross-sectoral department. The cross-sectoral departments regulate issues which – comparable to material circulation (Stoffkreislauf) – circulate through several environmental media and thus pertain to the competencies of several sectoral departments.
3. According to a clear distribution of competencies, the technical responsibilities should be assigned to the departments operating close to the problem. The level of main departments has to be abolished, since these units produce no visible efforts of co-ordination and thus form a redundant intermediate-level at least in the areas under closer empirical scrutiny. These departments should be integrated into the two groups of the

(already existing) sectoral departments and the cross-sectoral departments yet to be created.

4. The sectoral departments have to be structured into emission-oriented sections (following the example of the Clean Air Department). This means that for each major group of polluters (= Emittenten) a section exists (Ansprechpartner). The competencies of these sections for the conduct of selected affairs have to be defined clearly and to be adjusted periodically to changing conditions.
5. In the foreseeable future, the task of the agency should be strictly limited to the performance of environmental protection functions, in order that the agency can encounter the utilization functions administrated by other agencies with the same consequent strategy. A simultaneous representation of protection and utilization functions in one and the same agency creates unnecessary conflicts of interest which are detrimental to the resolute implementation of the agencies' legal mandate.

   Such a solution is likely to be criticized on the grounds that potential conflicts between parts of the federal administration cannot be dealt with inside the agency as in other cases. The Ministry, essentially the already overburdened Head of the Ministry (the Federal Counsellor), would have to decide once more. The integration of different points of view should be achieved at the level of the agencies; it should be part of the quality of an agency head that he can resolve such tensions internally.

   A reply to this frequently aired reservation is that given the consequent implementation of the legal mandate of protection, for which in most cases – namely the definition of norms – scientific and not political arguments are decisive, there is hardly room for compromise at the operational level. Where they are necessary, they should be taken at the strategic-political level of the corresponding Ministry and clearly earmarked as political decisions. Where required by the political implications of such conflicts, regulations should be taken in the framework of environmental delegations by the Federal Council (environment/transportation; environment/agriculture/forest; environment/ energy, etc.). This concentration on the essential mandate of protection corresponds to the existing basic structures of the federal environmental administration. It is mandatory as long as the protection functions cannot be reliably performed by the other federal agencies (for the future, see the scenarios in section 6.4).
6. Strategic leadership of the agency has to be reinforced in a way which results in a reduction of the duties of the Head of the Ministry (Federal

Counsellor). For this purpose, the internal leadership capacities of the Direction (which has been described by all interviewees as too large and too preoccupied with details) should be improved and more operational decision be delegated to the departments.

7. The FOEFL currently employs approximately 320 persons. Such a large agency needs an independent strategic-operative management located outside of the specialized departments which can rely on the functioning of the competent departments. A directive and reductive management is required. Leaders should set priorities and assure their implementation through protective measures against competing demands. For this purpose, procedural rules and guidelines have to be developed.
8. Multifunctional service branches or staff should be avoided. Staff should fulfil either person-, issue- or performance-oriented tasks. Given the size of the agency, a mixture of these functions (in one and the same branch) leads to lack of transparency, inefficiency and frustration on the part of the employees working in the line.
9. The extremely valuable resource of a highly qualified and motivated staff should be used and not abused according to the ecological principle of sustainability.
10. In the cross-sectoral departments in particular, more generalists and social scientists trained in ecology (especially economists, political and administration scientists and psychologists) should be employed. They should not be recruited for particular issues or functions. Rather there has to be more rotation in order to avoid structures being developed according to personality attributes of the existing personnel and not according to issues.
11. Wherever competition allows, where qualified for the performance of the agency's tasks, and available on the open market, such services should be hired from the outside according to market conditions."[13]

In the context of this report, guideline no. 5 in particular requires a brief commentary: proposing a strict concentration of the agencies' tasks on the realm of environmental protection and the related out-placing of the current resource utilization functions in the realm of the forests, the report clearly puts emphasis on improving the co-ordination of the various environmental policies (intra-policy co-operation). Accordingly, in a first step, all environmental protection policies should be concentrated in one agency and implemented according to a unified strategic philosophy of interven-

[13] Report 1991 "Evaluation des BUWAL ...", op. cit., note 5, p. 57 ff. (direct translation)

tion; concerns of resource economy, which potentially weaken this philosophy, should be performed by other agencies.

The report argues that no exception from this principle should be made in the case of the forest which, in regard to protection and utilization, are administered by the FOEFL. While the utilization of the forests has also a "social function", this is the case as well in other policy areas such as energy, agriculture or water economy, which also have to perform a social (or ecological) function, "without policy conceding (them) in a special statute an ecologically legitimized self-regulation ...[14]. As the report continues there could be arguments against this proposal that "the current forest model (against the opinion of the expert) is not a relic of times past, but rather a nucleus for a future new pattern of ecological administration. The point is to integrate the utilization and protection of resources in one and the same administrative structure. In fact this is an interesting scenario for the future ... . For the near future, however, this model is hardly applicable, since there are no obvious reasons to transfer other, ecologically as important or even more important, utilizations of resources such as agriculture or water economy to the FOEFL[15]."

The aforementioned report continues: "For the administrative power of the environmental policy at the federal level demanded by the Control Commission on her behalf, the uniformity of the philosophy of strategic intervention is more important than the special promotion of the ecologization of one particular area which when compared with others (water, agriculture) is ecologically no more relevant. Either the entire agency switches to a strategy of administering ecological resource utilization in general at the level of the philosophy of strategic intervention (...), or it uniformly concentrates on the protection function. The coexistence of both philosophies in one and the same agency is detrimental to the development of a uniform strategy and a shared "corporate identity[16]."

It is no surprise that this proposal has not been spared criticism in the public debate. The Control Commission only partly followed this recommendation[17]. At the centre are different evaluations of the optimal time needed to switch from a protection- to a utilization-oriented environmental policy. The authors of the report, who themselves developed the transitional scenarios presented in section 6.4 of this chapter in the last part of their report, were and are of the opinion that the optimal time for a system-

[14] Ibid. p. 96.
[15] Ibid. p. 96.
[16] Ibid. p. 96 f.
[17] Ibid., p. 1450

atic transition from protection to utilization in Switzerland's environmental policy has not yet come[18]. Over the short to medium term (five to eight years), inter-policy co-operation has to be achieved on the basis of functionally clearly divided responsibilities. Too fast a transition to models which integrate protection and utilization functions in one and the same administrative institution, will result in a preponderance of the utilization and thus in a weakening of the protection functions, as is the case in the view of the authors in the realm of forest administration. The advocates of a quicker pace in turn emphasize the positive ecologization effects of integrating issues of forest economy into the FOEFL, which they claim to have observed since the integration of the forest into this agency in 1989.

The proposed reorganization of the FOEFL concentrates as indicated on short-term improvements in the field of intra-policy co-operation. These proposals are summarized in Figure 1.

These proposals focus on the creation of cross-sectoral departments which will be linked to the concerning sectoral departments according to a matrix organization. The Direction will be streamlined, the intermediate level of the main departments abolished and their operational competencies delegated down to the level of the departments. The table also shows that the utilization-oriented affairs of the forest administration will be transferred to the Federal Office of Agriculture. The protection-oriented affairs of traffic regulation in turn should be newly subordinated to the FOEFL instead of the Federal Justice and Police Department. Finally, the report proposes to place the National Hydrology and Geology of Switzerland outside of the FOEFL and – in accordance with the corresponding concepts in the European Community and other European countries – to transfer it to a newly created Federal Agency for Observation of the Environment. The parliamentary Control Commission has meanwhile adopted most of these proposals (except the proposition to transfer the forest administration)[19]. The proposed structure is presented in Figure 1.

---

[18] Annotation of the author: More than ten years later this situation will change: see chapter 15 of this book.

[19] See ibid., p. 1449.

Figure 1: FAE: Federal Agency of the Environment

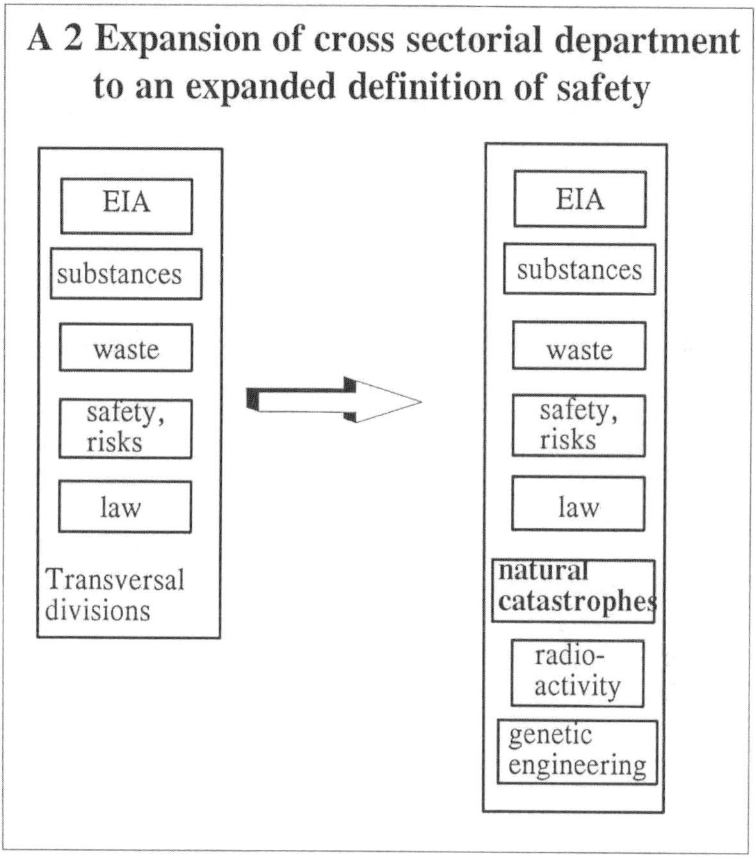

BAP: Federal Office of Police
BLW: Federal Office of Agriculture
EIE: Environmental Impact Assessment
NHG: Law on land and landscape protection

The report includes the following definitions of the sectoral departments, the cross-sectoral departments and the proposed matrix organization which might be interesting for concepts of intra-policy co-operation in reform debates abroad:

- "The six sectoral departments observe and regulate the most important objects of protection according to the classical division of the most important environmental media as they already exist within the FOEFL (nature, landscape, water, soil, air and noise). Their main task consists of protective measure in favour of these media, which they best organize according to the various producing groups (and not according to more specific targets of protection such as ground or surface waters etc.). A second task, which is becoming more important given the discovery of numerous growing interdependencies between the various environmental media, consists of developing concepts of protection of specific ecosystems across the media. This should be achieved through flexible arrangements of regulation among several sectoral departments.

Also increasingly important are co-operation arrangements with particularly relevant "non-environmental" federal policies, which represent utilization interests outside of the agency (e.g. clean air and energy; soil protection and agriculture etc.). Based on existing material, co-operation duties, within the matrix structure (which has yet to be created) necessitate new co-operation agreements which have to be concluded and maintained at the junctions between the cross-sectoral departments concerned and the sections respective departments. The Basic Sections, which should exist in every department (data collection, data analysis etc., in close co-operation with the proposed "Federal Agency for the Observation of the Environment"), should basically be designed according to existing examples (the Clean Air Department, for instance).

- The cross-sectoral departments analyze and regulate those environmental burdens which, through known substances or flows of substances or as a consequence of large projects, concern on a regular basis several environmental media in a way more or less calculable today (waste, substance, EIA). The effective regulation of these environmental burdens requires first a highly specialized, substance- or project-specific technical knowledge, which cannot be accumulated on the media-side. Second, on the side of the media, this regulation requires permanent opposites in the sectoral departments (so called "junctions" in the matrix organization).

These cross-sectoral departments can be structured into sections either according to specific substance- or issue-groups, or (following a more spatial view of regionally specific flows of substances) according to national regions. In both cases, the creation of basic sections is recommended, and these have to deal among others with economic aspects (e.g. international harmonization). The Law Department has a special cross-sectoral function, which will increasingly have to focus on the terminological etc., unity of die development of environmental law given the growing wealth and diversification of partial sectoral regulation by the sectoral and cross-sectoral departments.

- The matrix organization facilitates the coexistence of two departmental groups, whose competencies have been strengthened compared to the present situation, with a minimum of hierarchical guidance and bureaucracy through a concept of pre-structured, horizontal cross-linkages, which in turn are organized under the guidance of a department (and are thus not completely "random"). Experience shows that a matrix structure cannot be "designed exclusively on the drawing-board". In particular, the idea of flexibility facilitating fast cross-linkages should not be turned into the opposite of all-round blockade through perfectionist bureaucratization. Rigid patterns of organization should be avoided and flexible, task-oriented arrangements between individual departments should be promoted outside of the (continuous) cross-sectoral departments. These "task forces" need however a clear assignment and they should be approved by the Direction (without such an approval, there is the – well-known – danger of "flying bureaucratic carpets", which constitute, legitimize, etc. themselves).

  Such a matrix organization is an important organizational instrument for implementing an environmental policy, which is increasingly oriented towards the ecological overall context (and not toward individual laws and directives). It is also likely to form an important precondition for developing a future generation of more planning-oriented executive instruments on behalf of the cantons."[20]

In the following section, we present the two groups of scenarios A (intra-policy co-ordination) and B (inter-policy co-operation) as they appear in the FOEFL report. In this section, we intentionally reproduce the text of the experts in a direct translation without any further comments because this (publicly less debated) piece of our report seems to be of interest to most European countries aiming at restructuring their environmental agencies. They are formulated in a neutral way leaving all political considerations up to the political decision makers. For the purpose of this article, however, we allowed ourselves to add some personal comments on their political implications and a brief analysis of their potential contribution to an enlargement of environmental policy instruments. In the final section, they will be briefly commented on in respect to the problem of environmental policy instruments, which are of a particular interest in this chapter (section 6.5).

---

[20] Report "Evaluation des BUWAL .... ", op. cit. note 4, p. 132 ff. (direct translation).

On behalf of the politicians concerned, the report starts with the following observation: "Current political debates on future environmental policy, as a rule, focus on either the expansion of its protective functions or a general expansion of its scope, namely, a reorientation from traditional environmental protection to environmental uses. There is, for example, a widely shared opinion that the objects to be protected by environmental policy should be defined more precisely, studied more thoroughly and better taken care of with the help of appropriate measures based on the concept of integrated policies. Furthermore, claims are made for an expanded definition of the term "risk" in order to protect the environment from sources of risks that hitherto have received too little attention. A probably equally important category of new political concepts considers the restriction of environmental policy to protective functions per se as inappropriate, because it ultimately hinders environmental policy from taking a direct influence on the cause of a particular environmental use. Environmentally problematic policy areas, by comparison, have more pull due to their longstanding tradition, their good supply of funds and personnel, and their strong support from powerful interest groups. Therefore, it is argued, implementing environmental policy that is restricted to protection is bound to be a Sysyphean task. Consequently, environmental policy should be given better access to non-environmental policy areas wherever these are responsible for the rise of environmental damage.

The expansion of protective functions as well as the integration of environmental uses into the realm of environmental policy not only require numerous modifications with regard to the contents of legislative programs, etc., but also the reorganization of the administrative structure of environmental policy. Since this expert evaluation concentrates on the latter, we shall present in the following a set of scenarios of administrative structures that are based on the various political options outlined above. Needless to say that the implementation of a new environmental political concept cannot start with reorganizing the administration. First, a political decision has to be made on one or the other possible political objective. In the following eight scenarios these objectives can be presented only in rough outlines because a detailed justification and analysis would not be possible within the frame of this expert evaluation."[21]

[21] Ibid., p. 149 f. (direct translation).

6.4 Scenarios for future rearrangements

6.4.1 Two scenarios for strengthening intra-policy co-operation

The first two scenarios deal with the restructuring of the present Federal Office of Environment, Forestry and Landscape (FOEFL/BUWAL) into a new Federal Environment Agency (BFU). Scenario Al illustrates those changes within the agency to be recommended, if an expansion of the protective functions of environmental policy is aimed for.

The present environment agency (BUWAL) consists of sectoral divisions that are each specialized in one environmental medium (air, water, soil, etc.). Scenario Al envisages a better interconnection of these divisions by grouping them around a common core division, thereby bringing them close to the actual objects of protection instead of the environmental media that are now being protected on behalf of the objects. As laid down in the Federal Environmental Protection Act, objects of protection are human beings, animals, plants, and materials. By now, each sectoral division, in performing its tasks of protection, has of course adjusted itself to these objects. Within the administrative structure, however, the objects themselves and their needs are represented only implicitly, and there is as yet no central division that could, so to speak, act as an agent and plead the cause of these objects in their capacity as legal entities. Scenario Al has the additional advantage that synergistic effects (e.g., air and noise; water and biotopes, etc.) could be countered more effectively because their balance sheets of suffering the objects of protection, as a rule, do not differentiate among environmental media.

Figure 2: Scenario AI: Integration of a new sectoral division to strengthen the protective function

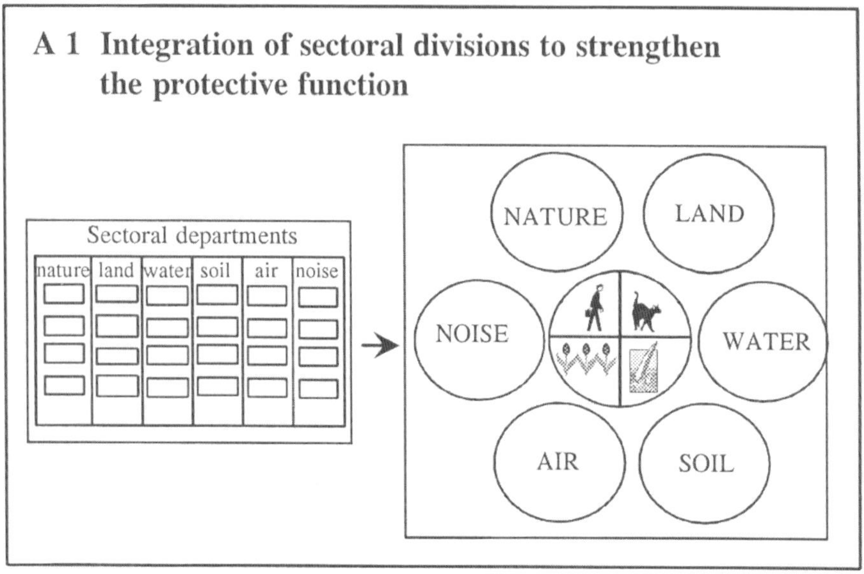

The opponents of such administrative reorganization argue that the interests of protected groups are represented by other large federal agencies. As for human health, reference is made to the Federal Public Health Agency; as for animals, to the Federal Veterinary Agency (located in the Federal Department of Economics; as for materials, to the Federal Institute of Material Testing and Experimentation (EMPA) or to the Federal Agency of Culture (regarding historical monuments). However, a closer look at these agencies reveals that environmental risks and dangers do not play a key role either in the policy of these agencies or in the legislation they administrate. The Federal Public Health Agency, for example, adopted the topic "environment and health" (within its section "medical epidemiology"?) only about two years ago.

Scenario A2 deals with the level of transversal (cross-medial) divisions within the environment agency and is based on an expanded definition of the term "risk". In this sense, it would, for instance, become necessary to consider the integration of radiation protection – today largely performed on the basis of self-control according to, e.g., the German model (Responsibility for radiation protection currently lies with the Federal Energy

Agency and the General Secretary of the Department of the Interior = GSED1)[22].

Figure 3: Scenario A2: New cross sectoral departments according to an expanded definition of safety

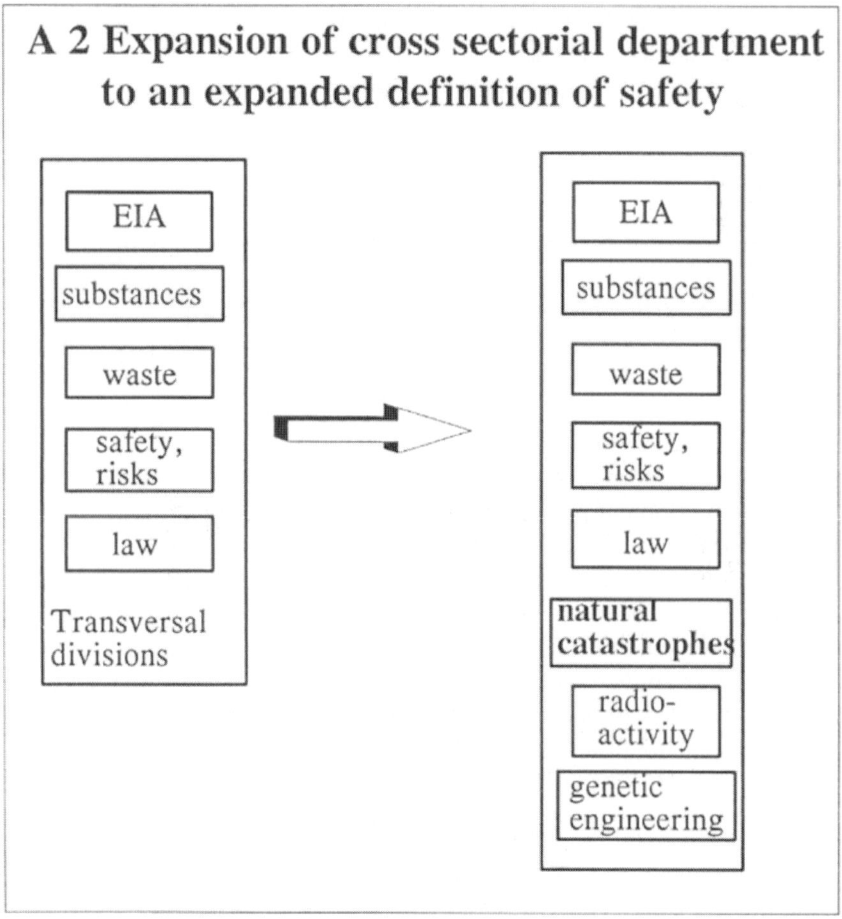

The strict administrative separation of classical immission protection and radiation protection is nowadays hardly justified and can, at best, only be explained historically. Furthermore, risk prevention should be extended

[22] Cf. Ordinance on the Protection from Radioactive Radiation of June 30 1976, SR. 814.50.

to genetic engineering. Considering the growing importance of this field, a new transversal division, “Genetic Engineering”, should be created from the current' section “Paths of substances and Environmentally Hazardous Organisms” which is part of the division “Substances and Soil Protection”.

In the sense of concentrating powers, but also with respect to their relevance for global climatic changes, it should be investigated whether a new division “Natural Catastrophes and Climatic Changes” should also be added. Similar considerations have presumably led to the section “Climatic changes and natural catastrophes” of National Research Program Nr. 31 which was passed by the Federal Executive Council in the summer of 1990. Related functions are currently located in the Federal Water Resources Agency and the Federal Agency of Civil Defence.

6.4.2 Six scenarios for strengthening inter-policy co-operation

Scenario B 1 is the first of those scenarios that take the step from the purely protective functions of environmental policy towards environmental uses. It proposes a bridge between the (restructured) Federal Environment Agency (BFU) and the Federal Agency of Landuse planning (BRP) by introducing a new group, “Region and Environment”.

This scenario corresponds to the political intention to bring the two policy areas environment and landuse planning – often operating disconnectedly – closer together, in the sense of bringing environmental policy back to the region or, vice versa, “ecologizing landuse planning policy”. The need for this has been stressed already in connection with environmental impact assessments (EIA), local and cantonal landuse planning and planning guidelines, or (especially) noise protection. As for the latter, an excellent brochure was published by BUWAL and BRP in March 1988. Furthermore, one should not forget that the “fathers” of landuse planning had originally intended to endow it with a considerably stronger ecological component than it was actually ascribed by the Land Use and Regional Planning Act of June 22, 1979[23]. A “spatialization” of environmental policy would make it possible, in particular, to employ time-tested instruments of landuse planning (e.g. landuse zones) in a “regionally effective” environmental policy. This would be especially favourable in the area environment and agriculture.

[23] 21 SR 700.

Figure 4: Scenario B1: From protection to use: macro-level: integration of landuse planning

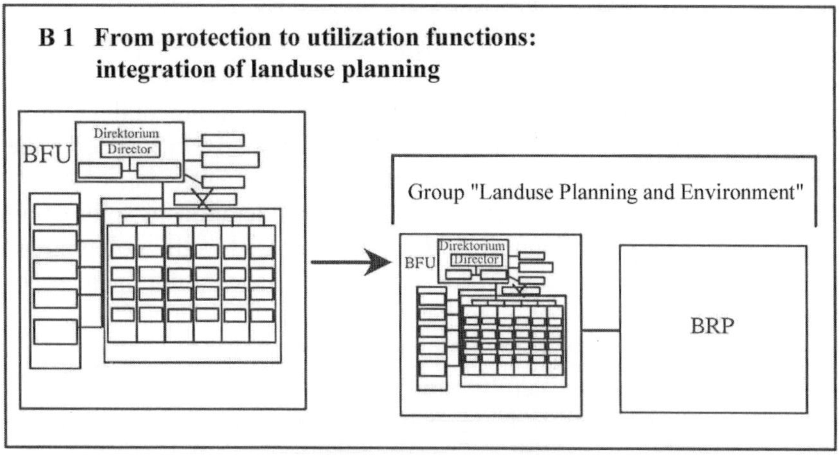

BFU = (a restructured) Federal Environment Agency
BRP = Federal Agency of Landuse Planning

This scenario for a transition from protection to use has, however, one important disadvantage: the protective functions of environmental policy might lose impetus in favour of general interest considerations in landuse planning decisions, as has been the case in EIA procedures. As a matter of fact, the current practice of interest consideration in landuse planning does not at all help to dissipate such apprehensions, all the more so, because landuse planning often fails to set definite priorities among competing user interests. Since environmental policy defines absolute positions that may not be fallen short of (e.g. emission standards), landuse planning, in the case of such a "joint venture", would have to be endowed with clear-cut patterns of privilege. As for the model presented here, it should be pointed out that the relevant federal law allows the formation of groups only if the agencies concerned belong to one and the same department[24]. Therefore, the Federal Executive Council would, first of all, have to amend the Ordinance on the Tasks of Departments, Groups, and Agencies of May 9, 1979[25] and then draft a corresponding bill to be voted on by the Federal

[24] Sec. 46, par. 3, VwOG – Act on the Organization of Federal Administration.
[25] SR 172.010.15.

Assembly[26]. (This vote would be generally binding and not subject to a referendum.) The same applies for scenario B2.2.

The four scenarios subsumed under B2 take yet a further step on the road from protection to use. As everybody knows, landuse planning merely creates the conditions for the implementation of many other policies but plays only a marginal role in formulating these policies. That is why an even greater utilization of landuse planning for environmental political purposes in itself would hardly allow for taking a decisive influence on the other policy areas involved. Infrastructure policies, in particular, will continue to give ecological interests insufficient consideration in spite of an ecologization of landuse planning, although, compared to the policy areas trade and industry or agriculture, they meanwhile have "a dominant share in environmental disruption". Furthermore, there is evidence that, "by comparison, environmental policy in this area has the weakest standing"[27].

According to a political opinion that is increasingly gaining ground effective environmental policy can be realized only if infrastructure policies become "ecologized". Therefore, what is called for is "environmental policy implementation by means of interpolicy co-operation". With respect to administrative structure such an ecologization can be done in various ways. Scenarios B2.1–4 differ in the intensity of co-operation as well as the underlying strategic estimation of the present situation. They can be also seen as subsequent stages of ecological development.

Of the four B2-scenarios, B2.1 is probably the one to interfere least with traditional structures.

This scenario is based on the loose co-operation of a future BFU with agencies that are particularly relevant for the environment, such as the (recommended) Federal Agency of Agriculture and Forestry, the Federal Agency of Transport, the Federal Road Construction Agency, the Federal Energy Agency, etc. Taking the examples of the planned work group on the cantonal plans of measures or the CO_2 group, the Federal Executive Council could[28] establish a permanent interdepartmental coordinative section, "Environmental Policy". But more intensive and politically more influential forms of co-operation are also conceivable: The Federal Executive Council could, for example, appoint a delegation, "Environment", consist-

---

[26] According to sec. 60, par. 2, VwOG.

[27] See P. Knoepfel, La protection de l'environnement en proie aux problèmes d'acceptation et aux déficits de mise en oeuvre. Mise en oeuvre de la législation environnementale: situation actuelle à la lumière des secteurs Industrie et Artisanat, Agriculture et Politique infrastructurelle de l'Etat, Cahier de l'IDHEAP no. 70a, novembre 1991, Lausanne, pp. 12, 20.

[28] On the basis of sec. 27 Institutionalized by sec. 55, VwOG.

ing of the under-secretaries of the federal departments of the interior, traffic, energy and economy or the federal Secretaries Conference"[29] could deal regularly with environmental issues and pass the appropriate coordinative decisions.

Figure 5: Scenario B2.1: From protection to use: Normal Cooperation amongst agencies

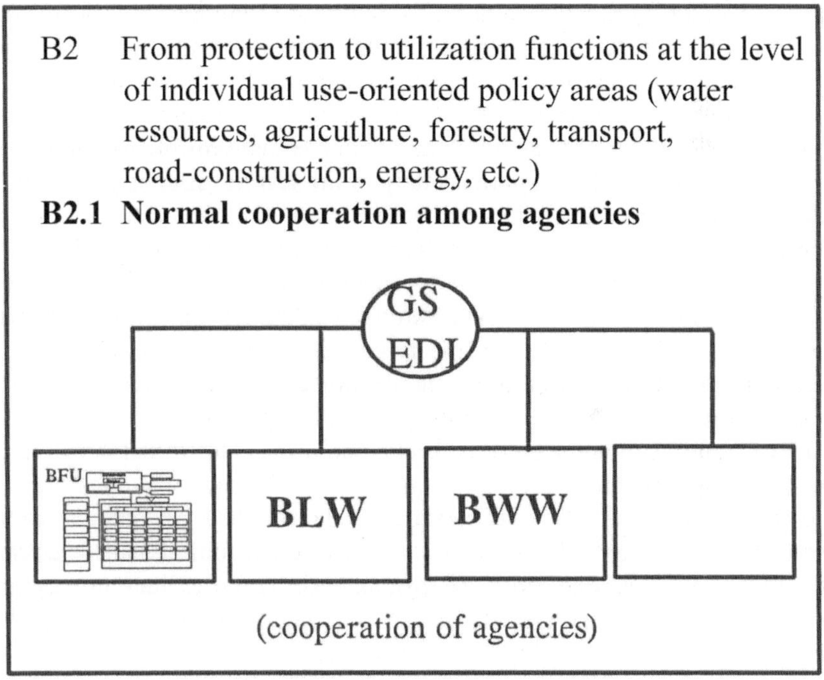

GS EDI = General Secretary of the Department of Interior
BFU = (a restructured) Federal Environment Agency
BLW = Federal Agency of Agriculture
BWW = Federal Water Resources Agency

Scenario B2.2 is based on a conception that is, as yet, quite unusual in administrative practice. It provides for a rotation of co-operating groups that are temporarily formed by different environmentally relevant agencies from various departments and the environment agency.

[29] Institutionalized by sec. 55, VwOG.

Underlying this scenario is the political-strategic consideration that environmentally relevant economic and infrastructure policies at the federal level should be ecologized step by step. A simultaneous ecologization of all agencies involved would be unrealistic not only for technical reasons and reasons of the substance matter, but also for the broad political opposition this would evoke.

With this strategy it would also be possible to annul the recent transfer of forestry to BUWAL as recommended in our proposal for the agency's reorganization. The individual groups, “Environment plus NN”, to be formed should be working for at least four years. During a four-year period it would be possible to develop appropriate means and ways for a consideration of ecological interests both at the level of strategic legislation and at the tactical-operational level. Presumably, this will, in any case, have to be done separately for each policy area. (The principle of rotation, in general, should be applied much more widely at the level of individual federal officials too.).

The realization of the rotation model would of course entail all kinds of activities that would be though of as disturbing the “peaceful” routine within the agencies involved. Therefore, considerable opposition can be expected to grow not only among the civil servants affected by rotation but also among the clientele of these agencies, a clientele that often is politically well-organized (e.g. Swiss Touring Club, Farmers' Association, Swiss water and energy suppliers, etc.). It can be taken for granted that it will not be possible to ecologize the policy areas concerned against the open opposition of these interest groups. Consequently, it would be advisable to create structures for an effective co-operation among the administration, environmentalist organizations and these groups in order to facilitate processes of ecological learning.

Figure 6: Scenario B2.2: From protection to use: Rotation of temporarily formed groups

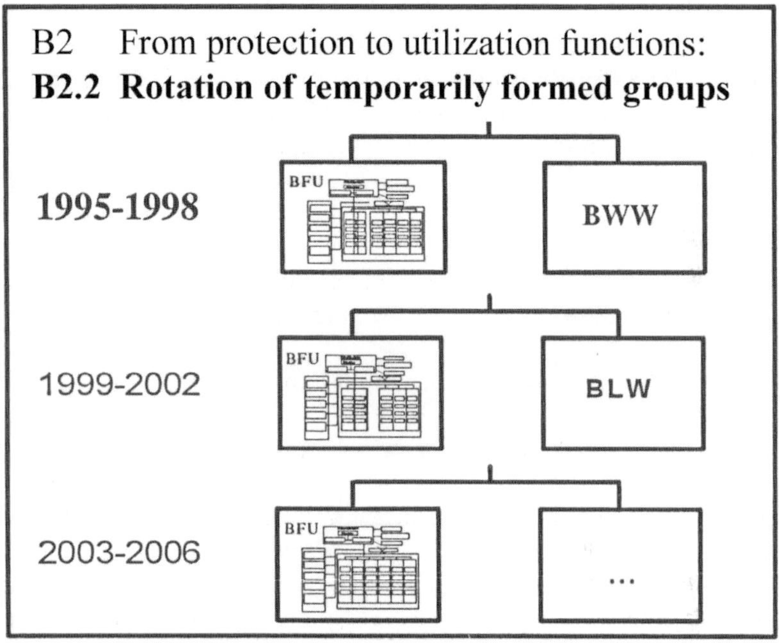

BFU = (a restructured) Federal Environment Agency
BLW = Federal Agency of Agriculture
BWW = Federal Water Resources Agency

The satellite model shown in scenario B2.3 (figure 7) is based on the assumption that ecological strategies should be developed primarily within the affected agencies and policy areas themselves.

Figure 7: Scenario B2.3: From protection to use: the satellite model

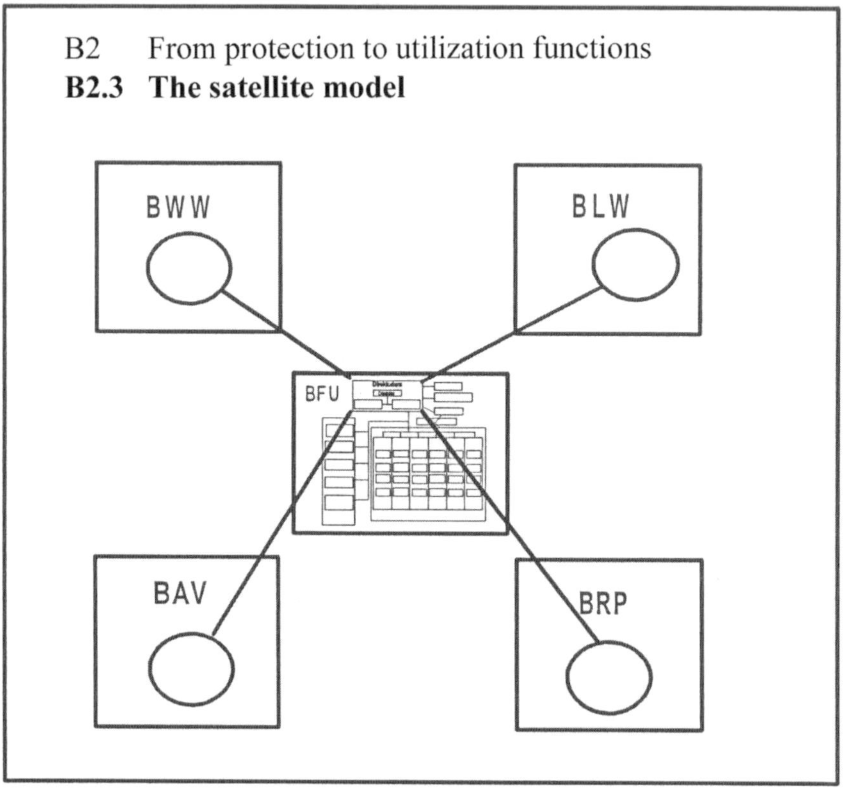

BAV= Federal Agency of Transport
BFU = (a restructured) Federal Environment Agency
BLW = Federal Agency of Agriculture
BRP = Federal Agency of Landuse Planning
BWW = Federal Water Resources Agency

According to this model, and in adopting the principle of "help for (ecological) self-help", the Federal Environment Agency would implant an "ecological cell" in other agencies which would act as a catalyst and accompany the agency's ecological reorientation. Between the first B2 scenarios and this model there is a marked difference in the underlying strategic assessment of the situation (which assessment should actually be applied is, again, a matter of political decision): While the first two B2 scenarios started from the assumption that the affected policy areas themselves are hardly capable of ecological reorientation and must, therefore, be put under more or less massive external pressure, the satellite model is based on the opposite assumption, namely, that all policy areas sooner or later will perform their ecological turn-about in their own way. This judgement is based on the hypothesis that at some point in time each policy area will face a situation of scarcity (scarce funds for tackling surplus production in agriculture; scarcity of land in infrastructure policy; scarcity of consensus in, e.g., transport policy, etc.). Managing decreasing resources, it is argued, forces the setting of priorities and development of plans for necessary cut-backs. Ecological arguments might at first be used only formally to seek acceptance for unpopular measures, but in the long run ecological criteria would become an integral element of planning processes themselves ("economical management of resources").

Whoever sides with this political assessment of the situation will probably be also in favour of the satellite model. And, by the way, he or she will even be able to point out first approaches to its realization within the federal administration (e.g. the "Section Ecology" within the division of agricultural policy in the Federal Agency of Agriculture; the division "Law and Environment" of the Federal Agency of Civil Aviation or the "Environment and Research Office" within the division "Projecting" of the Federal Road Construction Agency). There is, however, reason for doubt that these examples could actually prove the effectiveness of this strategy of ecologization, at least when considering our findings regarding the Federal Road Construction Agency.

As things stand, all these approaches lack one important element: The satellite model requires, at least, the institutionalized regular exchange between the environment agency and the ecological cells in other agencies, unless these cells remain hierarchically subordinate to the environment agency (which is quite unusual in federal administration). It is further required that the "satellite" has sufficient influence on the individual agency's policy. This is generally given when it attains the status of a (though small) division with direct access to the direction.

Figure 8: Scenario B2.4: From protection to use: ecological self-regulation

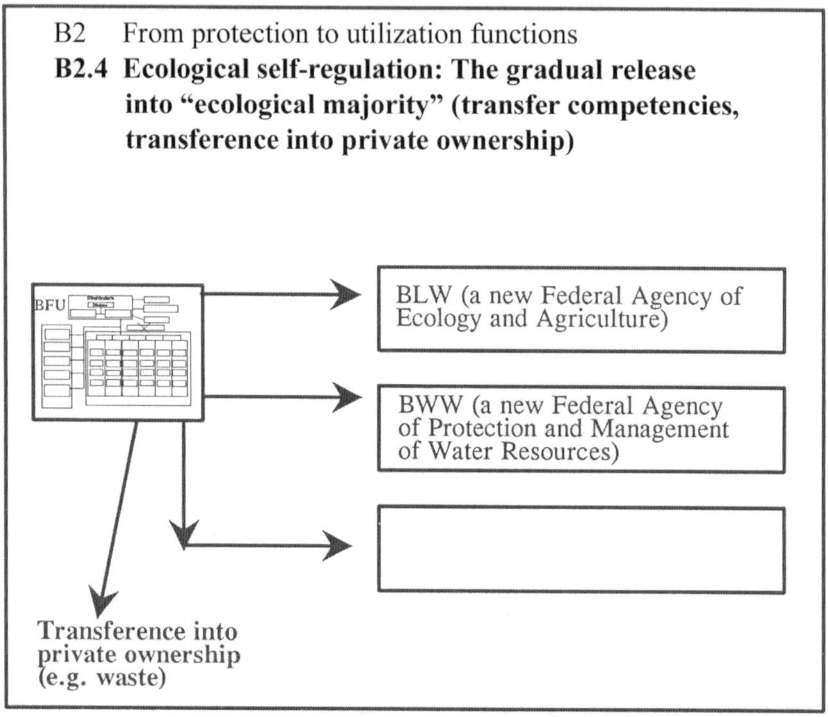

BFU = (a restructured) Federal Environment Agency
BLW = Federal Agency of Agriculture
BWW = Federal Water Resources Agency

The scenario "ecological self-regulation" is a truly futuristic scenario. It is a logical elaboration of the rotation or satellite models in a long-term perspective. It would not make sense to take a binding strategic decision in favour of this scenario today, since the conditions for a gradual discharge of certain emission groups or certain public policies into "ecological majority" are fulfilled neither at present nor in the immediate future. In a certain sense, the scenario represents the inverse image of the current situation. The principle of concentration on protective functions postulated above for the current structure has been dropped altogether. Exactly these protective functions should successively be outplaced from the agency and be transferred to the utilization side.

The ecology would not only be a postulate at the formal or legislation level, but also be secured at the administrative-structural level. It would then fall into the competence of each federal agency respectively become an objective for the entire market economy. The result would be the emergence of a self-sustaining waste economy ("quartiary sector") or the complete integration of environmental costs in the prices at the raw material markets. The same would also apply to the various public policies. The key task of the Federal Office for Agriculture would then be ecological resources management according to clearly defined protection targets. The existing Federal Office of Water Economy would be turned into the Federal Office of Water Economy-and Water Protection, and so on. It has to be observed that with the integration of the Federal Forest Administration into the FOEFL such a step has been taken (hardly deliberately) albeit into the opposite direction. The protection and utilization function have been subordinated (exactly as in the former Federal Office) to the same administrative unit. It is our recommendation that today this fusion should be abandoned. Today, the time for such a construction is not yet ripe neither in forestry policy nor in other realms of environmental policy. Our empirical findings on the (not further scrutinized) forestry policy, which traditionally incorporates strong protective functions, demonstrate, that the later are discounted again and again in favour of the utilization function.

The introduction of economic instruments in environmental policy has been widely discussed for quite some time. Unfortunately, the various proposals hardly ever consider that such a political reform would require a corresponding administrative reorganization. As yet, it would be impossible for both the present BUWAL and a future Federal Environment Agency even to administrate the charges on solvents (VOC), fertilizers and fuel oil to be introduced by the amended Environmental Protection Act. The same applies to the so-called ecobonus, the CO_2 charge and many other environmental-related fees and charges that cantonal governments demand to be introduced.

One way to solve these administrative problems could be to entrust the federal fiscal authorities with the collection of such charges, but as for the ecologically appropriate investment of these funds, the foundation of a "federal ecology bank" would soon become inevitable. This bank would have to ensure that funds are invested in projects that are both ecologically justified and economically profitable. Its institutionalization would become all the more necessary as soon as so-called marketable emission certificates are issued by the national and cantonal governments as a means to finance environmental protection measures that are not required by law. In that case, a federal ecology bank would also have to serve as the central exchange for such certificates.

Figure 9: Scenario B3: Economization

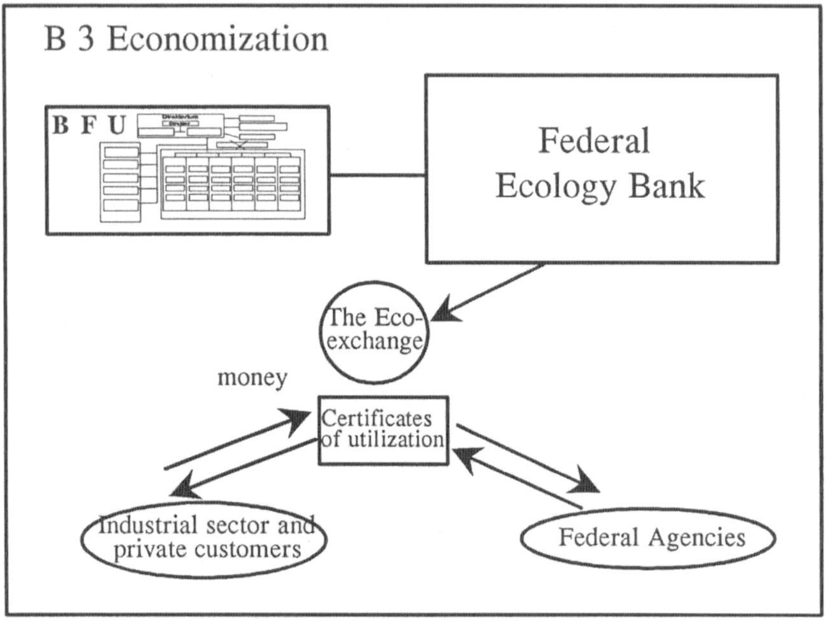

Finally, those who politically side with an even more radical innovation and, in tendency, would like to introduce for all kinds of environmental exploitations prices that would respond to fluctuations in supply and demand, could not do so without a federal eco-bank. The introduction of marketable certificates for air pollution by industries or water pollution by households and water purification plants is frequently discussed. However, it is rarely considered in political debates to extend this concept of infrastructure policies as well, although there seems to be no convincing argument that these policies would not pay for their environmentally detrimental activities. A consistent economization of environmental policy with market-economy means must include this side too.

The introduction of such a system would create an enormous market for certificates which, just as the money market, needs to be supported by a solid administrative structure. This would also involve a certain degree of autonomy (similar to the central bank), not only because governmental departments would also participate in this market but, moreover, in order to ensure that environmental policy can intervene in this market. If, for example, prices for emission certificates threaten to drop unreasonably, the environmental agency must be able to buy certificates in order to stabilize prices just as is routinely done on the money sector. The repurchase of certificates by the state would then not only reduce the total amount of marketable emission certificates but the general pollution load as well.

6.5 Comments on the scenarios regarding their political implications and their contribution to improving the instruments of environmental policy

6.5.1 Preliminary remarks

The eight (A1, A2, B1, B2 1-4; B3) scenarios included in the report on behalf of the Control Commission of the Swiss National Council do certainly not represent all feasible variants. Nevertheless, in political, organizational and instrumental terms, they cover a very broad range of possible options. In the following part, the political options behind each scenario are briefly described – a description which deliberately is not included in the report; in addition, it will be demonstrated for each scenario to what extent it facilitates a possible enlargement of the arsenal of political-administrative instruments of environmental policy. By that we understand, according to the shared definition on which this book is based, instruments with which mainly state agencies can influence the availability of resources to those target groups which are relevant to environmental policy[30]. These resources are "money, goods, legal rights, skilled people, time, information and trust". Since according to our knowledge there are no empirical findings or theoretical designs regarding this topic available in the existing literature, the following arguments are limited to the speculative realm of an essay, whose background consists mainly of Swiss and Central European realities. For this reason, we will not refer to political or administrative science literature which would only provide a very general guideline for our specific topic.

[30] See Klok 1995.

6.5.2 Scenario 1A: new sectoral division: strengthening the protective functions

Without doubt this scenario is based on the intention to anchor environmental policy as an ecocentric policy by extending the object of protection from certain groups of human beings to all ecosystems. The resulting integral point of view is likely to strengthen the classical protection of nature and landscapes and thus the protection of rural areas as the location of a series of protectable ecosystems crucial for the regeneration of resources. The appointment of outspoken advocates of nature, probably in conjunction with the recognition of some fundamental rights for nonhuman, natural beings[31],tendentiously could lead to a radicalization of environmental policy. As known, man is one of the most resistant beings, thus a relatively low level of protection is sufficient. Given the demand for an integral, ecocentric environmental policy, as put forward by the proponents of a strengthened environmental policy, the standards of protection have to be increased at the level of specific norms. One can thus refer to a vertical quantum leap in the sense of raising the standard of protection[32].

This scenario would lead to a less compromising position on the side of the environmental agencies, especially towards the governmental infrastructure authorities, which increasingly operate large systems close to nature in less populated hinterlands of civilization. A rise in conflict and a partial break-down of currently established inter-policy co-operation relationships can be expected. The higher standard of protection could only be achieved at the price of polarization within the administration, but also in politics. In its most extreme version, the scenario would certainly be accused of "ecodictatorship", which, regardless of its justification, would involve societal security risks having repercussions far beyond environmental issues as such.

At the instrumental level, in this scenario, environmental policy would certainly arrive at a stricter level of protection and correspondingly at lower (more rigorous) standards of immission, which in turn would result in a tightening of the corresponding standards of emission. Consequent administrative implementation would lead to a much higher importance of environmental impact assessment than is the case today, since in addition to the impacts on the environmental media air and water quality (which are typical for an anthropocentric point of view), the implications for natural

[31] See P. Saladin, Probleme des langfristigen Umweltschutzes, in: Kritische Vierteljahresschrift für Gesetzgebung und Rechtswissenschaft, 1989, pp. 27–55.

[32] See B. Knoepfel, M. Descloux, Valeurs limites d'im missions: choix politiques ou déterminations scientifiques?, Cahier de I'IDHEAP no. 48, Lausanne, 1988.

ecosystems or landscapes and for established interaction systems between man and environment would also have to be examined[33], which would again require a significant intensification of effort. In such a process, advocates of nature would have to put an (irreversible?) veto on in each case, in which even one – important – species of the ecosystem under protection is endangered. In such a conception, environmental impact assessment would become a more conflictive, institutional mechanism for the accommodation of the interests of "minority" natural beings. The result would be a certain hierarchization of the environmental agencies at the level of central government, but also at the level of the implementing regions. A corresponding politization of science and "nature" could hardly be avoided[34]. Instrumental gains of environmental policy (stricter environmental standards, stronger position of sensitive ecosystems in environmental impact test procedures) could thus turn into a pyrrhic victory, should the baby (stricter environmental requirements) be poured out with the water (environmental policy as such).

6.5.3 Scenario 2A: new cross sectoral departments

Such an extension of the object under protection in the light of newly emerging environmental risks corresponds to the historical development of the European environmental policies, which started with the protection of waters and were extended in the 1960s and 1974s to include clean air, chemicals and protection against technical risks. Politically such an extension means an expansion of the protection policy, which increases the importance of environmental policy, but does not imply a higher standard of protection. This can be described as a horizontal quantum leap, which in the case of nuclear safety and gene technology extends into areas which have traditionally been covered by their own regulating authorities of energy, medical and health policies respectively. The example of Germany, where in 1987 nuclear safety policy was integrated into the newly created

---

[33] Theoretically already required: See Office fédéral pour la protection de l'environnement, Etude de l'impact sur l'environnement, Manuel EIE, Berne 1990, p. 26.

[34] See F. Ramade, L'écologie et le décideur. Les limites de la prise en compte des données scientifiques dans les politiques de l'environnement, in: J. Theys (assisté de V. Liber et M.-P. Palacios) (éd.), Environnement, science et politique, publication réalisée avec le concours financier du Service de la recherche du Ministère de l'environnement (SRETIE), Cahier de GERMES, no 13, Paris, 1991, p. 381ff.

Ministry of the Environment, demonstrates that such an expansion of environmental policy is by all means feasible[35] . The assignment of gene technology regulation to the competency of environmental policy does not necessarily reflect a radical position on ecological issues. In both cases, by the way, the (economic) utilization policies have been allocated to other agencies.

At the instrumental level, through this rearrangement, environmental policy gains the capacity to control new essential risk areas, over which it has exerted only indirect influence so far. Both in nuclear and in gene technology, a broad field of intervention possibilities opens up, through which – not at least in the interest of the respective disposal problems – it can exert influence on emerging risks at an early stage. The Sisyphus work at the waste front (including that of the two technologies) can thus be somewhat relieved. Moreover, both areas provide environmental policy access to the instruments of research and development policy and technology assessment, which goes beyond the normal environmental impact assessment and into which more ecological aspects can be incorporated. The future will show to what extent the price of environmental policy, especially in these two high risk areas, will be under pressure from the enormous economic significance of these two technologies. It will be vital to the protective function and probably issue certificates of ecological compatibility with the new ecological instruments for ecologically highly doubtful activities.

6.5.4 Scenario B1: integration of landuse policy

This "spatialization" scenario implicitly starts from a concept of an increased decentralization and regionalization, because it is mainly concerned about spatially differentiated goals and locally adapted protection means to be, implemented, e.g. by a variety of zoning regulations, etc. Inspired by the principle of subsidiarity, which is currently being increasingly discussed within the EC[36], the scenario assumes that environmental quality cannot be defined by the centre to the last detail, since it has to be

---

[35] See R. Czada, Administrative Interessenvermittlung, am Beispiel der kerntechnischen Sicherheitsregulierung in den Vereinigten Staaten und der Bundesrepublik Deutschland, Habilitationsschrift zur Erlangung einer Venia legendi in Politischer Wissenschaft und Verwaltungswissenschaft, Universität Konstanz, March 1992, p, 120.

[36] See art. 3 B within the Treaty of Maastricht as well as art. 13 0 R, Para 2 of the EC-Treaties. (Sec A. Epiney, H. Furrer, op. cit., N. 4).

elaborated according to local circumstances and with the participation of the population concerned in its entirety. This scenario thus also represents a departure from the schematism of zoning, and it necessarily leads to a regionally differentiated policy of environmental standards. With the recognition of different climatic, topographic, but also cultural and historical parameters of environmental policy, different policy objectives result which vary according to the respective frame of reference. Politically, this scenario corresponds to a position represented in the 1970s and 1980s by the conservative centre. In the 1990s, this view is gaining support both among the former rather centralist left as well as the national-conservative right. In its extreme version, the scenario could result in the abandonment through decentralization of the currently European-wide practised definition of specific standards of environmental quality at the nation-state level. In the case of this extreme version, the problem of potentially high regional disparities regarding guaranteed minimal standards of living quality can be illustrated, and can only be avoided with a probably reduced set of quality standards defined by the central government.

As discussed in section 6.2, instrumental access of environmental policy to land use would be quite remarkable. Along the way, it should be possible to regulate space-critical activities of ail kinds through corresponding space-specific regulations including the aspect of their impact on the environment. According to the varying degree of ecological sensitivity of the soils, various levels of intensity of the agricultural zone (maximum ceilings of fertilizer or livestock utilization), various air or water pollutant volumes per area unit in industrial or settlement zones or different waste regimes could be defined through communal utilization planning. These ceilings could be easier to implement than individual directives regarding all emittents which are active within a given area.

Moreover, such zoning regulations of environmental quality could provide more freedom for agreements and bargaining processes within the concerned collective of emittents in each zone[37]. Finally, such an approach facilitates the recognition of varying sensitivities of the ecosystems concerned through the gradation of zone specific quality goals (probably in the framework of quality targets set by the central government). In contrast to the current widely exercised practice of defining zones which are homogeneous in regard to their utilization (living zone, industrial zone, business zone, etc.), such a regime would also allow the realization of the postulate for more inter-mixing of utilization within the respective zones, which is

---

[37] Definition of emission quotas per zone-unit (ha) as already applied in the field of agriculture (dung units per hectare).

increasingly considered as necessary to achieve the goals of protection defined in qualitative and ecological terms.

Describing the scenario, we have already referred to the potential ecological costs of borrowing instruments from zoning policy. It would have to be taken into account that the environmental aspects also have a procedural priority over other areas of utilization, which would clearly privilege the protective aspects over the economic, infrastructure or consumptive interests of utilization in the balancing process of zone planning[38]. Otherwise there is a danger that they will be caught underneath the wheels of a philosophy of balancing interests in land use planning which treats public protection and utilization interests on an equal basis in solving conflicts over utilization.

6.5.5 Scenario B2.1: intensification of normal co-operation amongst federal offices (traditional pattern)

This scenario is hardly spectacular. Essentially, it is committed to the current basic structure of separating protection and utilization functions based on the division of tasks among various federal offices, which in case of conflict should be accommodated not merely on an administrative, but on an institutionalized, political basis. The scenario is thus suited to render transparent conflicts between protection and utilization, to facilitate their political solution and to avoid covering them up through anticipated administrative compromises. In this respect, it would give environmental policy a politically more visible status. This has been advocated in the light of the growing importance of this policy. The transition to such a scenario would not be without consequences for the functioning of the central and regional governments, since there will be an increase of basic convicts over environmental policy. Along this route, there will be strengthening of the environmental portfolio within governments; at the same time, the number of unanimous government decisions will decrease in favour of majority votes.

Compared with the other scenarios discussed in this section, the co-operation scenario is characterized by its immediate implications for the functioning of the political executive. Inter-policy co-operation is not achieved through changing the contents of the policies concerned, but through (largely policy unrelated) governmental mediation of conflicts

[38] The only case of such an absolute priority in the Swiss environmental law is the (constitutional) obligation to protect moorelands (popular initiative introducing Sec. 24 sexies accepted in the votation of Dec. 6, 1987)

between competing policies, which by themselves will not be resolved systematically. This scenario stands thus in clear contrast to the other versions discussed in category B, which are actually based on such an internal transformation of the essence of the policy areas concerned. It can thus be truly circumscribed as a scenario involving rehabilitating "traditional political parties", whose influence especially in environmental policy is dwindling. Administrations, unions and interest groups of all kinds, which currently profit from a rather "apolitical" functioning of environmental policies, are likely to have difficulties with this scenario.

This scenario cannot be described as providing an added instrumental gain for environmental policy. Replacing administration through political (non-)decisions will change the legitimacy and probably also the quality of the use of a particular environmental policy instrument (for instance, a construction permit for an environmentally disputed project). Yet the set of instruments of environmental policy as such and its relationship with other policy instruments will not be changed in a fundamental way. It will however depend on the political correlation of forces, to what extent the influence of environmental policy on the utilization of these instruments, which according to this scenario will take place in the public political realm, will strengthen their position.

6.5.6 Scenario B2.2: agency groups based on the rotation principle

The rotation model contradicts an important element of bureaucratic tradition according to which once established spheres of activity of bureaucratic units can only be restricted by new political actors (ministers), and not through subministerial co-operation structures. Yet in parliamentary systems of government, shifts of environmental agencies between different ministries occur frequently. These agencies thus fall into varying ministerial contexts, in which co-operative relationships with a changing set of regulatory instances emerge. In most cases, these co-operative relationships are however looser than in those which result in a formation of groups between two or more federal agencies, as in the example of Swiss legislation referred to above[39]. The joint bureaucratic-strategic decision-making centre, which characterizes the agencies linked into groups by

[39] The history of the French Environmental Agency, which changed its status since its creation in 1972 at least six times shows that the policy contents largely influenced by the "Corps des ingénieurs des mines" hardly ever changed.

definition, facilitates a relatively far-reaching and substantial interlocking of the joined public policies, including the use of their instruments.

Rotation is only likely to emerge under strong external (parliamentary and/or governmental) pressure, since it would lead by definition to – disturbing – processes of inter- and delocking within a relatively brief period of time ("docking-on and off" of the respective non-environmental policies and their offices at the level of legislation, administrative structures and decision-making processes). This rotation creates dynamism, openness, substantial potential for mutual learning processes, but also bureaucratic uncertainty, which is tied to the risk that the "wrong" partnership might have been chosen. Similar to the scenario of consequent interagency cooperation, the rotation scenario can only be implemented through political decision. It presupposes the capability to reach a consensus over areas of environmental priority and thus tendentiously contradicts the principle (which, in Switzerland for instance, is widely applied) that in environmental policy all target groups concerned have to make an equal contribution ("symmetry of sacrifices ")[40]. Popular parties which focus on large electoral segments are thus unlikely to support the principle of rotation. The model requires long-term planning periods, which go beyond the four year electoral cycles. This presupposes a certain political stability, which in our view in the 1990s exists less than ever.

The instrumental gains of environmental policy will vary according to the affiliated non-environmental policy. The potentially available instruments have been briefly described in section 6.2 in the areas of land use, transportation and agriculture policy. Beyond these possibilities of formal access to the traditional instruments of the policies integrated in the new group, the rotation model will also fundamentally change the spirit of the administrative units which participate in the application of these instruments. Group building allows the opportunity for rotating officials among the original environmental agency and the non-environmental administrations affiliated during the four year period. In addition, new inter-policy networks will emerge at all levels, which start from shared specific problems and develop into capacities for synthesis, which will persist even after delinkage. Moreover, in a unique way the rotation scenario facilitates the concerted application of several instruments of different control modi (regulative, incentive, persuasive, etc.) in regard to one and the same objective towards a target group common to all inter-linked policy areas. The

---

[40] Widely applied for instance in Switzerland and Germany.

newly created networks form an almost ideal milieu of innovation[41], in which actual explorative laboratories for developing new instruments could emerge.

6.5.7 Scenario B2.3: satellite model

As discussed above, this model starts from the assumption that a stronger presence of environmental policy is necessary in non-environmental policy areas. In contrast to scenario B2.4. ("self-regulation"), this scenario is motivated by the political belief that these policies are (at least at the present moment) not able to be ecologized.

Therefore, environmental policy needs its own representatives within the agencies concerned, who will take a certain position in the decision-making process of these policies. This position varies according to the political point of view of the proponents of this scenario (co-consultation, co-decision, veto). Though hardly openly addressed, in the event of proper blocking power being given to the environmental bureaucracy (veto position) this scenario could lead to a supremacy of the environmental agencies over the non-environmental actors of state policy. In this case, a new superbureaucracy would be created. Instead of a political-decisional, a bureaucratic-administrative supremacy of environmental policy over non-environmental policies would result.

This scenario thus has obvious features of a technocratic-bureaucratic political philosophy which is gaining sympathy among the scientists and scientific-professional civil servants who are becoming increasingly influential. The apparent decrease in the political aspects of politics is met by the growing weight of professionalism, the increasing role of science and bureaucratization. The implementation of this scenario will face substantial resistance from the concerned bureaucracies, and also from interest groups. Each administration has mechanisms to render dislike cells in its interior immune. They range from the structural isolation of satellite branches imposed on by the outside, the keeping away of them from important dossiers to personal pressure upon the disliked "guests". Given this situation, the satellite branch in turn, in the interest of gaining recognition by the host, tends to cut its-ties with the maternal organization and to issue "certificates

[41] See e.g. Glantz (od.), Societal Responses to Regional Climate Change: Forecasting by Analogy, 1988 Boulder (West View Press), 1988 and C. Jaeger, Innovative Milieus and Environmental Awareness. in: Sociologia Inteniationalis, vol. 2 8, 2, pp. 205–2 1 6; D . Nir, Region as Social-Environmental System: An Introduction to a Systemic Regional Geography, Dordecht (Kluwer), 1990.

of environmental compatibility" even for those dossiers which would not have been easily approved by the environmental agency.

At the instrumental level, the potential gain for environmental policy can be similarly assessed as in the land use scenario (zoning) or the rotation scenario (for other policies). Access to the policy instruments of the host policy based on a veto position seems to many committed environmental officials to be the long hoped for, almost unlimited extension of environmental policy instruments; given the lack of political legitimacy among the target groups concerned, it will almost certainly meet opposition or scepticism among the locally active interest groups. Combined with the open resistance of the non-environmental "host" agencies, the environmental satellites would come under strong pressure, which even the most hard-nosed environmental bureaucrat would be unable to resist overtime.

6.5.8 Scenario B2.4: ecological self-regulation

As already discussed, such self-regulating scenarios are proposed as alternatives to all those scenarios which envisage an extension of the influence of environmental agencies in terms of scenarios if intra-policy cooperation (Al and A2), and also in terms of inter-policy scenarios as discussed in Bl (B2.1, B2.2 and B2.3). In the main, representatives of the political centre share the hope or the conviction that environmental policy will sufficiently be taken into constitution today by non-environmental policies. Therefore these components can be integrated into the non-environmental policies with regard to both professional and political standards. Frequently this entails the proposition to reduce the environmental administration in favour of integrating environmental skills into these non-environmental policy areas. Often this concept is seen as a counter-proposal against an alleged quest for hegemony by environmental authorities of all kinds. Where there are no serious efforts by the non-environmental authorities on their own to compensate for the reduction of the influence of the environmental agencies, it can be assumed that the self-regulation scenario is but a pretext to curb the allegedly expansive environmental policy. More serious are proposals which, similar in form to the satellite scenario, attempt to build up substantial environmental capacities in the non-environmental policy areas. Many self-regulating scenarios are based on the philosophical conviction that a rational discourse is the best means to introduce ecological reason into these other policies. The question remains unresolved, while the appropriate moment has come to discharge these policies gradually into ecological independence.

Such ecological independence should be reflected at all levels in the application of the traditional instruments of the policies concerned in political everyday life. In contrast to the satellite scenario, the selfregulation scenario often assumes the opening up of the concerned policies and their implementation procedures "from below" on the occasion of the employment of instruments. The argument is that the employment of ecological instruments can mainly be achieved through procedural guarantees assuring the inclusion of the target groups and the representatives of the environmental media concerned. Tendentiously there is a decentralization and a corresponding augmentation of the policies concerned with the resource "consensus". On the instrumental level, the self-regulation scenario is also characterized by a harmonic conviction that even acute conflicts can be reduced through related negotiations and compromises.

6.5.9 Scenario B3: economization

Economization scenarios as discussed here turn environmental policy largely into a policy of resource exploitation according to market principles. Even in their most extreme versions, these designs envisage relatively strong regulating frameworks through which the marketable environmental utilization rights will be limited in the interest of protecting resource and avoiding cartel-like practices on the various resources market. The environmental agency will continue to play a crucial role in defining these norms. In economic terms, the first condition aims at artificially limiting the availability of existing resources through governmental norms as long as such a shortage has not been achieved. In the framework of traditional political ideologies, these economization scenarios can be attributed to European liberalism as long as they provide no solution to the related social problems. Liberalism and market economy sour the social contradictions between those, which given their economic position can afford environmental utilization rights, and the (probably largely majority) group of persons, who given their social position, is denied access to the newly created markets. In the economization scenario, this contradiction is further accentuated, as the collective consumption of resources based on governmental (infrastructure) policies is also explicitly referred to in such markets.

Instrumentally, these economization scenarios do to provide environmental policy access to the classical economic instruments which are frequently discussed today (taxes, duties, certificates etc.). According to this pure market model, the assigning of resource utilization rights to the various groups of emittents and the transaction of these resource utilization rights would not be handled by public administration, but by the corresponding markets. Environmental policy would not only be deprived of important regulative instruments (substantial dimensions of pollutant limitation through the classical project authorization), but also of essential incentive instruments, which are usually described as "economic instruments of environmental policy". Such scenarios would in turn make accessible to the environmental agency an enormous investment pool, which it would assume as a permanent resource restitution and management activity. It is a political decision whether this newly created "fourth" economic sector[42] should be funded directly by the environmental authorities through the funds of assigning utilization rights, or be privatized in some way. The main instrument of the environmental authorities would be the definition (per unit) of marketable resource utilization quotas (for instance in the form of maximal admissible emission volumes), and the competency probably assigned to the authorities to acquire utilization rights themselves in order to realize specific rehabilitation tasks. The general control function over the resource utilization markets would however be assumed by stock market authority or the cartel control commission respectively. At least for the regulator, die economic instruments environmental policy would not only fail to make a gain, but suffer a substantial decrease in instruments under these economization scenarios. This corresponds to the political will underlying these scenarios to reduce governmental influence in economic and social formative processes. Governmental policy should be replaced by – sufficiently regulated – markets.

6.6 Concluding remarks

In this article, we have deliberately avoided nuances. The eight scenarios are thus not analyzed in regard to their complementarity. In political reality, combinations are however feasible. Nevertheless, on behalf of the political instances concerned, the political and instrumental implications of the various options should be clarified in their most typical form. Moreover, the rapporteur is aware that a typology in the narrow sense is not presented, but rather a list of possible scenarios derived from a political dis-

[42] In the US the waste treatment industry already employs 3 % of the active population. In France, the closing of the borders to solid waste importation, proposed by the Minister of the Environment at the end of August 1992, provoked serious opposition from the Ministry of Economy aiming at protecting the economically increasingly important waste disposal industry.

course. For this reason combinations among the various scenarios are feasible. We hope to have demonstrated that the field of the future generation of environmental instruments has to be looked for not in environmental policy itself, but in the currently non-environmental policy areas. For their development, sufficient structural conditions have to be created first. The key lies in the strategic expansion of intra- and inter-policy co-operation strategies.

References

Baroni, D., Brunner, V., Knoepfel, P., Moor, P. 1991. Strassenverkehrsrechtim Lichte des Umweltrechts – Le droit de la circulation routière face à la legislation de l'environnement. Basel: Helbing & Lichtenhahn (Reihe Oekologie & Gesellschaft, Bd. 5)

Commission Popp 1990. Les paiements directs dans la politique agricole suisse. Report. Berne

Conseil fédéral 1987. Rapport sur l'Etat et l'évolution de l'utilisation du sol et de l'urbanisation en Suisse. Rapport sur l'aménagement du territoire du 14 déc. 1987, 87.074: 76ff.

Czada, R. 1992. Administrative Imeressenvermittlung ... am Beispiel der kerntechnischen Sicherheitsregulierung in den Vereinigten Grosstaaten und der Bundesrepublik Deutschland, Habilitationsschrift zur Erlangung einer Venia legendi in Politischer Wissenschaft und Verwaltungswissenschaft. Konstanz: Universität

Epiney, A., Furrer, H. 1992. Umweltschutz nach Maastricht – Ein Europa der drei Geschwindigkeiten? EUR 1992, no. 4

Glantz, M. (ed.) 1988. Societal Responses to Regional Climatic Change: Forecasting by Analogy. Boulder: West View Press

Helm, D. 1991. Economic Policy Toward the Environment. Oxford: Basil-Blackwell Inc

Imhof, R., Zimmermann, W. 1992. "Massnahmenplanung" in Switzerland: The strategic air purification plan as an instrument of environmental policy in the field of urban traffic, International Congress: Implementing environmental policies by means of interpolicy cooperation, in Crans-Montana

Jaeger, C. 1990. Innovative Milieus and Environmental Awareness, Sociologia Internationalis, vol. 28, 2: 205–216

Klok, P.J. 1995. A Classification of Instruments for Environmental Policy, in: Dente, Bruno (ed.), Environmental Policy in Search of New Instruments. Dordrecht/Boston/London: Kluwer Academic Publishers: 21–36

Knoepfel, P. Descloux, M. 1988. Valeurs limites d'immissions: choix politiques ou déterminations scientifiques?, Cahier de I'IDHEAP no. 48. Chavannes-près-Renens: IDHEAP

Knoepfel, P. 1989. L'agriculture et la gestion de l'environnement: Les instruments de gestion en Suisse, Cahiers de I'IDHEAP no. 60. Chavannes-près-Renens: IDHEAP

Knoepfel, P. 1990. Zum Stand des Umweltrechts in der Schweiz / A Survey on Current Environmental Law in Switzerland, Cahiers de l'IDHEAP no 65. Chavannes-près-Renens: IDHEAP

Knoepfel, P. 1991. La protection de l'environnement en proie aux problèmes d'acceptation et aux déficits de mise en oeuvre. Mise en oeuvre de la législation environnementale: situation actuelle à la lumière des secteurs Industrie et Artisanat, Agriculture et Politique infrastructurelle de l'Etat, Cahiers de l'IDHEAP no. 70a (version frangaise du Cahier de l'IDHEAP no. 70). Chavannes-près-Renens: IDHEAP

Knoepfel, P., Zimmermann, W., in cooperation with Sailer, G., Matafora E. 1991. Evaluation des BUWAL. Expertenbericht zur Evaluation der Luftreinhaltung des ländlichen Gewässerschutzes und der UVP des Bundes. Bern: EDMZ; Rapport de la Commission de gestion du Conseil national du 18 mai 1992, FF vol.3,no.29 du 28 juillet1991: 1422 ff.

Knoepfel, P., Müller, U., Zimmermann, W. 1994. Vorschlag für eine Programmatik sozialwissenschaftlicher Umweltforschung in der Schweiz. Bern: BUWAL-Schriftenreihe

Mauch, R., Iten, R., von Weizsäcker, E.U., Jesinghaus, J. 1992. Oekologische Steuerreform – Europäische Ebene and Fallbeispiele Schweiz. Chur and Zurich

Nir, D. 1990. Region as Social-Environmental System: An Introduction to a Systemic Regional Geography. Dordecht: Kluwer

Oberson, X. 1989. Les taxes d'orientation. Genève (diss.)

Office fédéral pour la protection de l'environnement 1990. Etude de l'impact sur l'environnement. Bern: Manuel EIE: 26

Ramade, F. 1991. L'écologie et le décideur. Les limites de la prise en compte des données scientifiques dans les politiques de 1'environnement, in: Theys, J. (assisté de Liber, V. et Palacios, M.-P.) (ed.), Environnement, science et politique, Cahier de GERMES no 13: 381ff.

Saladin, P. 1989. Probleme des langfristigen Umweltschutzes, Kritische Vierteljahresschrift für Gesetzgebung und Rechtswissenschaft: 27–55

Sperling, D. 1992. Marketable credits for vehicle emissioes in California, ibid.; this question is in the center of the COST-Action no 618 concerning “Institution Building and Information Policies” with regard to air pollution control politics in big cities in Europe, (programme “CI-

TAIR"), decided by the COST-Council in May 1992. International Congress: Implementing environmental policies by means of inter-policy co-operation, in Crans-Montana

Tietenberg, Th. 1990. Economic Instruments for Environmental Regulation, Oxford Review of Economic Policy, vol. 6

7
Regulatory Change and Institutional Rearrangement: Building New Policy Arenas for Ecologization of Agriculture, A Comparative Analysis of Programmes in France and Switzerland (1987)[1]

Peter Knoepfel, Corinne Larrue, Willi Zimmermann[2]

7.1 Raising the question

In the course of general environmental discussion agriculture has come increasingly under fire. Modern agricultural activities in the western industrialized countries have a number of serious ecological drawbacks. These vary in scale and across problem areas, relative to country. Recent political debate within western European countries focuses in particular upon the following aspects[3]:

---

[1] This is a revised version of a paper presented at the ECPR workshop on "Policy Evaluation and the Steering Capacity of Democratic Government", Götenborg, April 1986. It presents some results of an ongoing international comparative project on environmental policies in the field of agriculture, jointly coordinated by the Institute for Advanced Studies in Public Administration (IDHEAP) in Chavannes-près-Renens and the Science Center Berlin. Dagmar Kollander helped us with the translation.

[2] in : Social Science Information, Paris, (SAGE), vol. 26 no 4, 1987, p. 683–731. Copyright: We thank the editor for authorizing the publication of this article in this book.

[3] See C. Conrad, P. Knoepfel, C. Larrue and P. Teherani-Krönner, "Environmental Concerns in Agricultural Policies: Outlines of some Perspectives, Issues and Policy Concepts in a Comparative Perspective", paper presented at the Round Table on Agricultural Policy and Environment, Maastricht, 30–31 May 1985; and J. Conrad, "Regulation of Agriculturally Induced Nitrate Contamination of Water in West European Countries", Science Center Berlin, 1985 (IIUG, pre 85–16). For EEC, see Parlement Européen, "Rapport sur 'agriculture et environnement'", Strasbourg, 3 February 1986 (Doc. a 2–207/85). For France, see Ministères de l'agriculture et de l'environnement, "Rapport du groupe de travail 'Activités agricoles et qualité des eaux'", 2 vols, Paris, October 1980. One of the most in-depth analyses was presented by the German Council for Environ-

1. Agriculture uses much valuable acreage for intensive production contributing to the destruction of biotopes; thus, it leads to a regression in many wild plant and animal populations, ultimately endangering the existence of some species.
2. Excessive and/or incorrect application of fertilizers contributes heavily to the pollution of soil and ground water by dangerous or toxic substances and to eutrophication of surface waters.
3. The excessive chemical treatment of plants results in an otherwise avoidable chemical burden in foodstuffs; it causes damage to many wild plants and animals. Furthermore, it can seriously impair the growth and flourishing of beneficial micro-organisms in soils.
4. Intensive, high-output-oriented livestock production leads to inferior quality of animal products, and it increases environmental pollution through added emissions.

Figures 1 and 2 demonstrate the tremendous increase of fertilizer application in some European countries as well as the recent development of the livestock production. We could add to these figures numerous other more specific data on worsening ground and surface water quality, on tremendously increasing amounts of animal manure, on dangerous nitrate charges in drinking water, and so on. The deterioration of rural ecosystems must be considered as an undeniable fact of the past decade. Whereas in the 1970s environmental politics and policies mainly had to deal with industrial and urban pollution, the challenges of ecological policies of the 1980s are to be found in rural areas as well.

---

mental Affairs: Der Rat von Sachverständigen für Umweltfragen, Umweltprobleme der Landwirtschaft, Sondergutachten, Stuttgart and Maynz, Kohlhammer, March 1985.

Figure 1: Fertilizer application in kg/ha in different countries (100 = 1938)

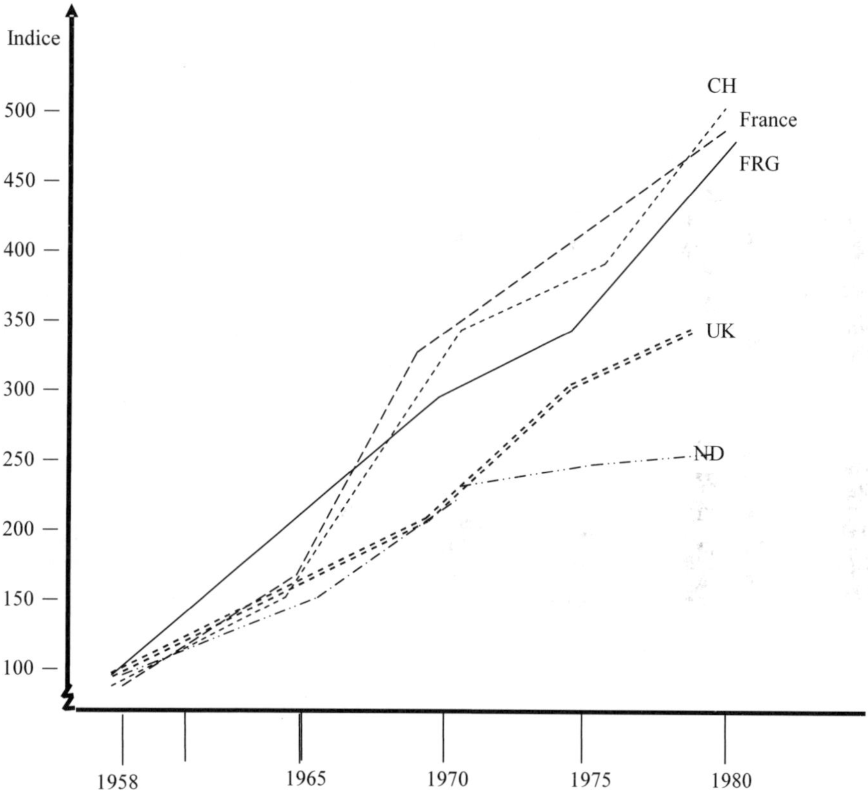

Figure 2: Average number of animals per farm in some countries (cattle and pigs) in 1973 and 1983

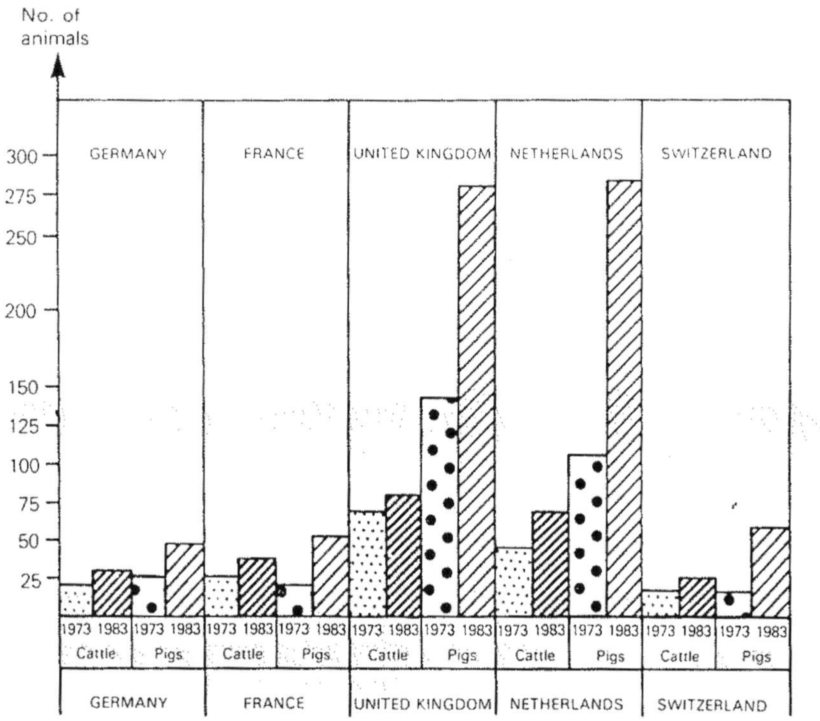

Sources: "La situation de l'agriculture dans la Communauté. Rapport 1984", Brussels, CEC: "Sixième rapport sur l'agriculture suisse 1984", Bern, Eidgenössisches Volkswirtschaftsdepartement, Bundesamt für Landwirtschaft."

In contrast, we know more about how to conceive ecologically sound agricultural activities. Agricultural ecological research, conducted since the beginning of the 1970s, has demonstrated concrete steps towards such new concepts. Within the concerned scientific community there seems to be general agreement on various measures leading to agricultural practices that would meet the following standards[4]:

[4] See for Switzerland, "Wegleitung zum Gewässerschutz in der Landwirtschaft", edited by the Bundesamt für Landwirtschaft and the Bundesamt für Umweltschutz in 1979 (last ed., January 1987). For France, see Ministères de l'agriculture et de l'environnement, "Amélioration des pratiques agricoles pour ré-

1. Plant and animal products free of damaging or toxic residues or other contaminants;
2. Cultivation preserving soil structure, soil biology and soil fertility; it must ensure that the regenerative capacity of soil is effectively maintained on a long-term basis;
3. Agricultural activities preventing or avoiding erosion of the soil and preserving typical features of landscape;
4. No pollution of ground or surface water by agricultural activities;
5. Avoiding noise pollution and pollution through foul smells and pollution of the air with damaging or toxic substances;

Wild plant and animal species, as well as valuable biotopes and elements of the landscape, preserved in accordance with the conditions necessary for species and biotope protection.

We all know that the intensification of agriculture was deliberately pursued – and still is – by national agricultural policies as well as by EEC policy. Thus, the first objective listed in the EEC treaty[5] aims at increasing productivity through technological development. Farmers receive numerous subsidies in order to realize this objective.

In view of the environmental impact of agricultural activities (as presented above), there arises a clear conflict between environmental protection on the one hand and prevailing agricultural policies on the other. Pollution stemming from agricultural activities demands special attention: here, "the polluter pays" principle cannot be implemented as easily as it has been with industrial pollution, mainly because of the well-documented low level of farm income. The agricultural subsidy systems of the EEC and Switzerland – another important objective mentioned in the EEC treaty[6] and in the Swiss Law on Agriculture of 1951[7] – aims at ensuring a minimum income standard for farmers. If "the polluter pays" principle were to be strictly applied, it would mean either additional state subsidies or a radical change of European agricultural policy.

duire les pertes de nitrate vers les eaux", edited by the Groupe de travail "Pratiques agricoles" du COPREN, Paris, October 1986; or Ministères de l'agriculture et de l'environnement, "Les nitrates dans l'eau, un défi relevé", edited by the Mission Eau Nitrates, Paris, 1986.

[5] Art. 39 of the Treaty creating the European Economic Community (EEC), signed in Rome on 25 March 1957.

[6] Also in Art. 39 of the EEC Treaty.

[7] Federal Law of 3 October 1931 (SR 910.1) concerning the promotion of agriculture and the maintenance of the farming population (Landwirtschaftsgesetz).

Looking at the background of this difficult situation, which can be found in most western European countries, it is interesting to examine an individual country's choices of regulatory strategies to control the negative environmental impacts of agricultural activities. On the basis of our preliminary studies on current regulations in different countries[8], we know that neither EEC countries nor Switzerland have, in the last fifteen years, subjected the crucial dimensions of their traditional agricultural policies to ecologically motivated changes. The age of an "ecologization"[9] of traditional agricultural policies has not yet dawned. Nevertheless, most countries have established environmental protection regulations which affect agricultural activities to a greater or lesser extent. These regulations build upon traditional policy areas such as water protection, drinking water regulation, public health or food legislation or even traditional waste disposal Acts. Whereas the concrete ways in which ecological considerations have been built into the regulatory body of different countries vary considerably, there seem to be two common features of these processes throughout western European countries.

1. We can observe the utilization of the results of ecological research and, in parallel, an increasing extension of the scope of pre-existing "older" regulations. This observation leads us to distinguish different phases within such processes. We shall argue in this article that the sequencing of processes of "ecologization" might represent a more general pattern of the ongoing processes aiming at tackling the problem of negative ecological impacts of agricultural activities.
2. The ongoing extension of the scope of such agriculture-related ecological regulations seems to be accompanied by structural changes within the political-administrative system. These changes seem usually to lead towards new regulatory units bringing together actors belonging to different traditional regulatory contexts.

This article, based on studies of Swiss and French activities over the past fifteen years, tries to illuminate these two common features of "ecologization" processes. It offers a comparative analysis of regulatory changes within the areas of water protection and of green vegetables. Whereas the

[8] These reports on regulations in the field of agriculture, rural environment and food in Switzerland, France, Belgium and Italy will be published in French in the "Cahiers de I'IDHEAP" (Chavannes-près-Renens, Institut de Hautes Etudes en Administration Publique) in late 1987.

[9] This somewhat unusual notion is derived from the German word "Oekologisierung" which denotes all attempts to increase environmental concern in different policy areas.

data reported on the Swiss case stem from an in-depth empirical analysis conducted by Knoepfel and Zimmermann during the last year[10], the French case is not yet as completely empirically founded. The French data are the result of a simple investigation into written regulations. Further research must be done in order to complete this presentation[11]. However, we are convinced that the comparative attempt supports an extension and improvement of the original model developed for the Swiss case.

7.2 Conceptual elements for explanatory analysis – a three-phase model

7.2.1 Basic scheme: physical elements and characteristics of three generations of regulations

The main purpose of the proposed conceptual model[12] is to demonstrate the ways in which ecological regulations of agriculture have developed in different countries to take account of ecological systemic cycles of rural biospheres. Furthermore, the model seeks to portray the extension of their scope of protection, starting from the initial object "human health", towards objects within biospheres which are not immediately connected with the health of human beings.

The basic scheme we use is a very rough cause-effect model of the key variables in rural ecosystems which are connected to the anthropospheres on the basis of scientific knowledge up to 1986. These physical elements of the model are depicted in Figure 3. As we shall show, one can detect roughly three generations of agriculture related ecological regulations according to the range of causal factor or protection objects they cover. At the beginning of the 1970s however, not all of the ecological cycles pre-

---

[10] See Knoepfel, P. and Zimmermann, W. 1987. Oekologisierung von Landwirtschaft, fünf Geschichten und eine Analyse. Aarau, Frankfurt, and Salzburg: Sauerländer. Some sections of this article have been translated and adapted from this volume. The research, on which it is based, was conducted in 1984–5 and was financed by the Swiss Science Foundation.

[11] In response to a call for research proposals from the PIREN (Programme interdisciplinaire de recherches en environnement) of the French Ministry of the Environmemt a proposal is presently elaboratedby C. Larrue, which, if accepted, will be carried out in 1988.

[12] This model was first presented by the authors in "Oekologisierung von Landwirtschaft", Schweizerische landwirtschaftliche Forschung, 25(2) 1986: 195–212.

sented in the Figure were well known. There is variance over time in the learning process. Furthermore, the specific objects of protection will vary across countries because of differences between pollution degrees a perceived human health hazards.

Thus, apparently the most significant factor leading to Swiss regulatory activities was the drastically deteriorating quality of surface water (mainly in the lakes in the central part of the country); whereas in France and in West Germany it was the human health issue (drinking water quality stemming from polluted ground waters) which seems to have been the most relevant initial trigger of "ecologization" Nevertheless, the basic causal factors presumably at work within the different ecosystems – at least with regard to the two selected issues "water protection" and "green vegetables" – are similar. This is the first reason why, according to different policy priorities, the strategic choice of countries for approaching ecological problems differs. These choices lead, however, to quite similar final results.

In opposition to alternative substantive policy choices suggested by the model, we are also interested in the different political-administrative structures at work in the initial phases of extending pre-existing regulations. According to what we know about these structures, they restricted the scope of action of each potential actor in such a way as to limit and often to exclude the most relevant causal factors from regulatory activities. The administrative arrangements invariably turned out to be inadequate for extending integrated ecological regulation to agriculture[13]. Our model therefore aims at a combined view of ongoing changes within regulations and corresponding changes within related political-administrative structures. There is a strong argument that the correspondence of regulatory and structural change which we have empirically observed in Switzerland and can make at least plausible in the case of France, might turn out to be a more general pattern of processes of "ecologization" of agricultural activities elsewhere.

[13] This observation holds true for other policy areas too (for example, energy policy, economic development policy, foreign economic policy).

Figure 3: Physical model of rural ecosystems

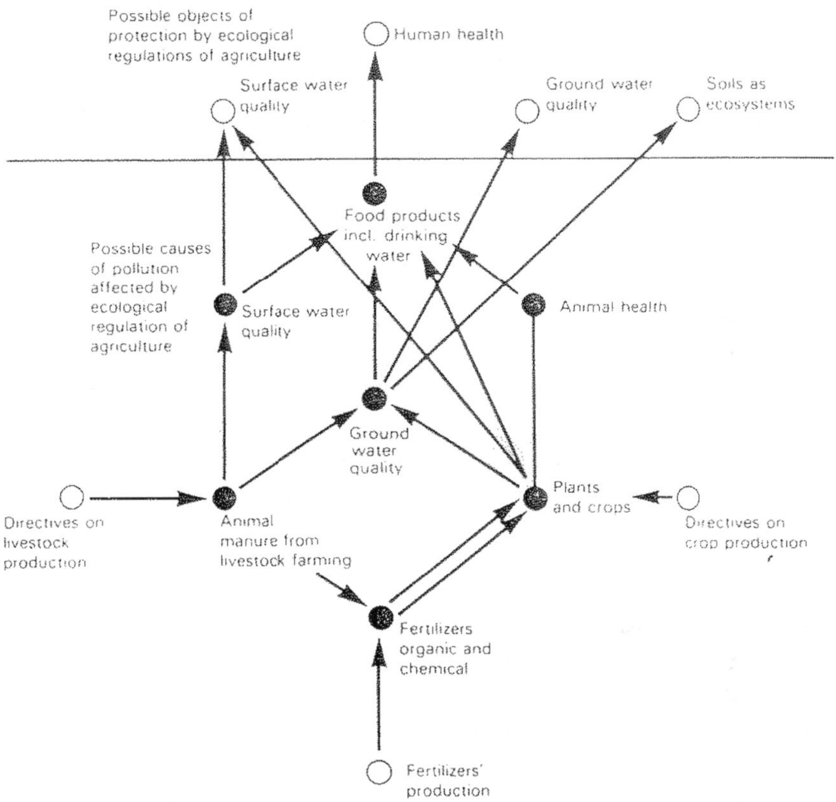

● indicates agricultural elements
○ indicates elements external to agriculture

Our model attempts to determine and explain the conditions under which the processes of ecologization start, develop and finally lead to new, operational regulations. We start from the assumption that the shared desire of various actors to inject more ecological considerations into agricultural policies and practices manifests itself in the form of regulations, guidelines, information brochures, recommended limit values or mandatory limit values, etc. All such manifestations of intent are subsumed under the term "programme". Since our study is restricted to the generation of programmes on the political-administrative level (programme level), we shall also call these "administrative programmes" or "political-administrative programmes"[14]. In-depth implementation research is currently conducted in France and Switzerland[15]. Therefore the model does not yet include relevant implementation processes, such as the inference about potential changes in the actions of regional administrations, agricultural consulting services, or the behaviour of farmers.

A fundamental characteristic of the issues giving rise to ecologization processes is that they mostly were initiated, so to speak, "halfway outside" the framework of traditional agricultural policy; they deal with restrictions imposed on agricultural activities that have been motivated by environmental and human health policy. Nevertheless, such issues and related programmes must also be considered as the results of traditional agricultural policies responsible for rural ecosystems' deterioration. They are somehow connected to those policies, although, in individual cases, it is difficult to define these linkages in a precise way. For this reason, we have chosen the phrase "halfway outside" to express the relationship of issues and programmes to agricultural policy. This phrase is especially appropriate for describing the situation in Switzerland. For example, the Federal Institute for Research on Agriculture, Chemistry and Environmental Hygiene (FAC-L) in Liebefeld, as a strong promoter of the ecologization of agriculture, employs scientific criteria and methods like any other research institute, but it is subordinate to the Federal Office of Agriculture (BLW). This is also true for the French National Institute for Agronomic Research (INRA), though its role as a promoter of ecologization has been less decisive than that of the FAC-L.

Figure 4 depicts the developmental stages of the processes of ecologization in agriculture that underlie our model. Its basic hypothesis states that the generation or modification of governmental programmes to include ecological considerations entails, as a rule, the creation of new, more complex political-administrative structures. In order for ecologization pro-

[14] See Kaufmann, F.-X. and Rosewitz, B. 1983. Typisierung und Klassifikation politischer Massnahmen, in: Mayntz, R. (ed.), *Implementation politischer Programme II. Ansätze zur Theoriebildung.* Opladen: Westdeutscher Verag: 34ff.

[15] The pilot study for Switzerland is conducted in four villages in the Bern area. It is financed by the Federal Ministries of Agriculture and of the Interior as well as by the Cantonal Govemment of Bern. The French study includes local areas in the northern part of France (Manche, Seine and Marne, Pas de Calais and Nord) and is financed by the French Ministry of the Environment. Simliar studies are conducted in Germany (by the Science Center Berlin), in England and in the Netherlands.

grammes to succeed vis-a-vis competing programmes and for a policy of ecologization to be fully implemented, there must be lasting co-operation between these new complex regulatory units ("policy arenas") and various, well-determined traditional policy "homeland" institutions. This co-operation works through the individual actors, who form the new policy unit and who continue to belong to their "homeland institutions". The qualitative change in the nature of regulations may be seen as the successive expansion of areas to be regulated according to the recommendations of new ecological or related social sciences[16]. We distinguish between three empirical types of agriculture-related ecological regulations, characterized by varying degrees of complexity and mutual integration. To simplify, we speak of "small", "medium", and "big" agriculture-related ecological regulations.

[16] Mainly presented by economists analysing different reasons for agricultural production surpluses rooted within related policy decisions.

Figure 4: Three-phase model (overview)

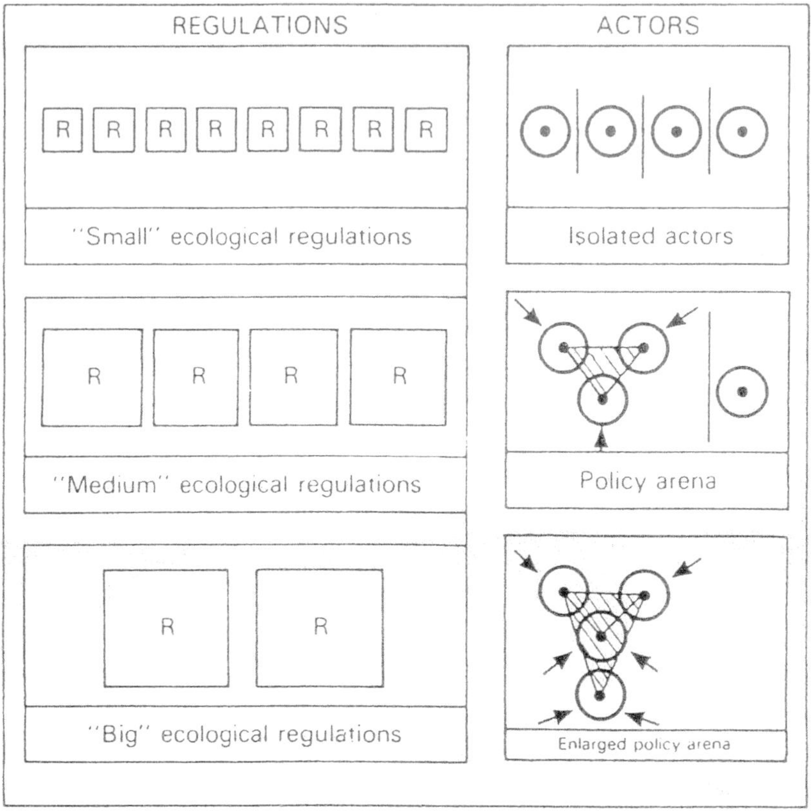

Figure 5 again illustrates the same features of ecologization processes by means of the previous model of rural ecosystems and the relevant factors for their protection. Ecological regulations of agriculture are considered "small" when the ecological context perceived as "relevant" for the effectiveness of corresponding measures has a very simple cause-and-effect structure. Here, the underlying interest in protection refers only to a single object usually a particular kind of food product (vegetables drinking water meat or milk) The reasons for deteriorating quality in the object requiring protection are perceived as simple and easy to overcome. Classic examples of "small" regulations include the old pesticide legislation which focused on toxicological impacts for humans the traditional guidelines on the quality of fertilizers, or older milk-delivery regulations. The most interesting examples of "small" ecological regulations in agriculture are probably the various Food Acts, which contain only a limited instrument of Protection, namely the imposition of bans on consumer products, and provide no opportunity for intervention in the production process[17]. Thus, the basis of "small" regulations is a "small" concept of ecology, limited to the prevention of hazards to human health and directed against only a few simple causes which technically are easy malleable. Surprisingly enough, this also holds good, though to a less significant extent, for the French food control policy in spite of the fact that this body of regulations traditionally belonged UP to 1982, to the Ministry of Agriculture. Thus, even this neighbourhood position at the Core of agriculture policy did not lead to a deeper involvement between the traditional French food control policy and agricultural production control.

Figure 5: Three-phase model based upon the physical model of rural ecosystems: the case of Switzerland

[17] See the foodstuff control policy in Switzerland which was always strictly limited to products and which explicitly exduded production processes within individual farms from the scope of regulation.

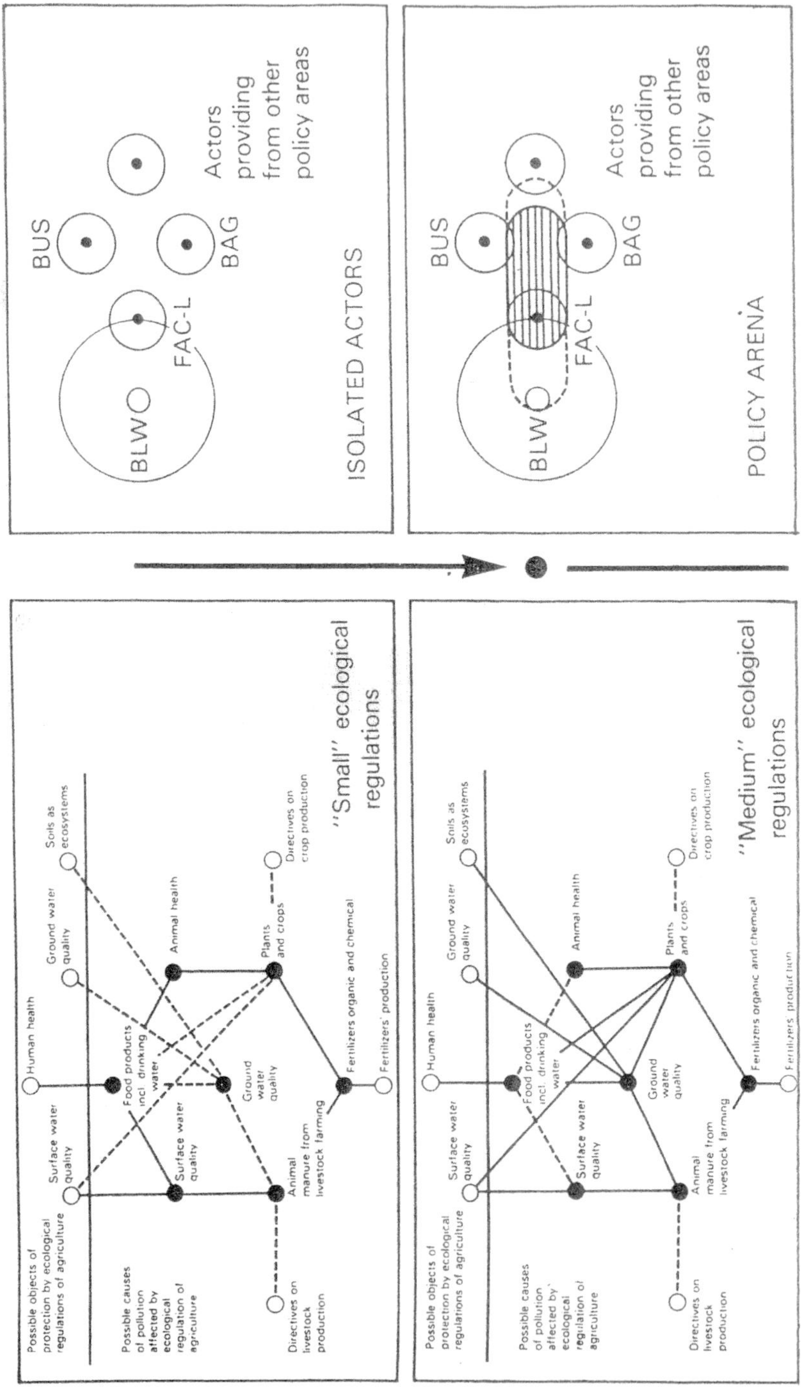
BLW
FAC-L
BUS
BAG
Actors providing from other policy areas
ISOLATED ACTORS
POLICY ARENA
"Small" ecological regulations
"Medium" ecological regulations
Possible objects of protection by ecological regulations of agriculture
Possible causes of pollution affected by ecological regulation of agriculture
Human health
Surface water quality
Ground water quality
Soils as ecosystems
Food products incl. drinking water
Animal health
Plants and crops
Directives on crop production
Directives on livestock production
Animal manure from livestock farming
Fertilizers organic and chemical
Fertilizers' production

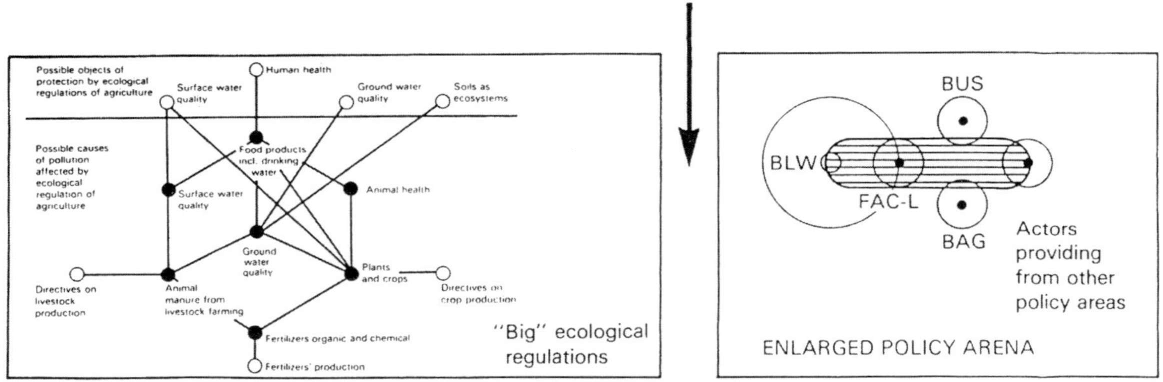

——— indicates connections taken into account in regulations
- - - - indicates basic ecological connections

BLW: Federal Office of Agriculture; BAG: Federal Office of Public Health; BUS: Federal Office of Environmental Protection; FAC-L: Federal Institute for Agronomic Research (environment), Liebefeld (Berne)

These "small" ecological regulations fit neatly into traditional administrative structures of European countries with their typically stable, tightly defined hierarchy of responsibility and areas of competence, and an historical background in the long liberal constitutional tradition Which narrowly defines the possibilities for state intervention. For example, in Switzerland, political-administrative regulation and management of public health is the responsibility of the Federal office of Public Health (BAG) and the Cantonal Food Inspector, While responsibility for the regulation of agrochemicals rests with the Federal Office of Environmental Protection (BUS) and the Cantonal Water Inspector[18]. The trespassing of one administrative agency upon the area of competence of another not only has legal consequences[19]; it can also lead to political tension especially between federal and Cantonal authorities[20]. The most striking situation we found in France, where two independent regulations concerning water pollution in livestock husbandry coexist. The (RSDT)[21] is under the jurisdiction of the Ministry of Health, whereas the other (RIC)[22] is under that of the Ministry of the Environment.

"Small" regulations were the starting point for most of the ecologization processes in agriculture in the early 1970s. Thanks to advances in ecological research, we know today that these regulations were too narrow in scope. In the early 1970s, research in agricultural ecology conducted in Switzerland by the FAC-L and by the universities[23], and in France by the

---

18 The Swiss legislation even makes a difference between general food control, which is attributed to the Federal Office of Public Health (BAG), and meat control, which is under the jurisdiction of the Federal Veterinary Office. Note, however, that this separation has led to conflicts in the past (see the report of the "Kommission Cavelty" of 1981 in Mitteilungen des Bundesamts für Gesundheitswesen, 73, 1982: 229).

19 According to Swiss jurisprudence, the lacking competence of an agency leads automatically to a nullity of the decision.

20 Swiss cantonal governments are very critical of excess power in federal agencies.

21 RSDT – Règlement sanitaire départemental type. See circulars of the Ministry of Health of 9 August 1978, 20 January 1983 and 10 August 1984. In each French department a "Règlement sanitaire départemental" (RSD) is issued on the basis of national norms set up in the RSDT.

22 RIC – Réglementation des installations classées (Law no. 76–663 of July 1976 concerning the installations classified in terms of environmental protection, and Decree no. 77–1133 of 21 September 1977 issued in application of the Law).

23 See FAC-Liebefeld (ed.), "Umweltprobleme und Landwirtschaft. Bericht über das Symposium am 13–14 Oktober 1971 in Bern unter dem Patronat der Abteilung für Landwirtschaft des EVD und der landwirtschaftlichen Forschungs-

National Institute for Agronomic Research (INRA) and by the universities[24], revealed that the causes of ecological problems and human health hazards were far more numerous, had many more facets, and were sometimes more remote from the actual objects of regulation than was initially assumed.

It became clear that ecological problems stemming from agriculture could not be tackled by "small", single-object-directed regulations. Once the larger ecological cycles became the subject of discussion and scientific investigation, members of the academic community urged that regulations be expanded to take into account the newly discovered chains of cause and effect. For example, the co-ordination of soil research with research in chemical and organic fertilizers revealed a chain of complex processes leading to harmful concentrations of nitrates in leafy green vegetables and drinking water such demands for broader ecological regulation still meant nothing more than merely adding supplementary measures to the previous set in order to cover additionally identified sources of ecological problems and better safeguard public health.

In addition, older ecological regulations proved to be too "small" because of their limited scope of protection. The obvious deterioration of environmental quality, especially of lakes and ground water led scientists and, at a later stage, politicians to abandon narrow definitions of objects requiring protection. It was no longer sufficient to confine special political regulatory attention to only a few selected single products. Ground water and soil, as primary supporters of important microbiological processes, had to be reconceived as the basic ecological units which, today, we recognize them to be[25]. Political-administrative regulation had to be readjusted accordingly. Thus the second, expanded version of more recent ecological regulations of agriculture in general are aimed at the protection of ground water and soil.

kommission", Liebefeld Bern, n.d.; and the Symposium on the Protection of our Environment, Federal Institute of Technology, Zürich (H. Leibundgut (ed.), Schutz unseres Lebensraumes. Symposium an der Eidgenössischen Technischen Hochschule Zürich vom 10 bis 13 November 1970. Frauenfeld and Stuttgart: Huber, 1971).

[24] See e.g. F. Ramade, Ecotoxicologie, Paris, Masson, 1977; Annales agronomiques, special issue: "Pollution", 1974; and S. Périgaud, L'agriculture et la pollution des nappes, Orléans, INRA, Collection des Bureaux de recherches géologiques et minières, No. 1, 1977: 155–61.

[25] This extension is mainly due to the inclusion of biological research results showing basic ecological cycles within the soils.

Our results show that "medium" regulations share two features. they were initially postulated by agricultural and non-agricultural ecology researchers, and they were implemented directly through administrative channels. Parliamentary political pressure played significantly smaller role. Not to be underestimated however is the pressure from consumers for improved water and food quality[26], and from farmers in whose interest it was to find ways to cut expenditure on additional inputs such as pesticides and fertilizers. This concern was the point of departure for the French "agronomic relaunch"[27], which will play, as we shall see, an important role in the development of "medium-sized" ecological regulations of agriculture. With these "medium" regulations, we have now arrived at our present-day agriculture-oriented ecological programmes. These latter do not include any of the main dimensions of traditional agricultural policies such as the quantity of livestock or crop production as such. They therefore remain outside the very core of traditional agricultural policies.

7.2.2 Rearrangements of administrative structures: arena building

As our case studies have shown, the step to such broader regulations has generally required some rearrangement of administrative structures because the jurisdiction of the individual administrative agencies involved has frequently been exceeded. Thus, along with the increased scope of regulation, we can also see a sort of "ecologization" of administrative structures (adaptation). Networks of co-operation capable of bridging different, traditional administrative areas of competence have become increasingly more necessary. Once these arrangements have progressed beyond the stage of simple working groups and become new, stable contexts of action, we can then speak of new political-administrative arenas[28].

---

[26] Cf. e.g. in Switzerland, the increased nitrate concentrations in salads and drinking water in 1979–80.

[27] The "agronomic relance" is a policy of the Ministry of Agriculture, launched in 1981–2 with a view to developing the "revival of the agronomy", and based upon the setting up of agricultural advisers, and the agronomical diagnosis of each small region.

[28] See, for the definition of "policy-arena", A. Windhoff-Héritier, Politikanalysen und Policy-Netz. Zum analytischen Nutzen zweier Begriffee, Politische Prozesse und Strukturen, Science Center Berlin, 1985 (IIVG/dp83–212). G Lehmbruch "Parteiensystem und Interessenverbände in der Politikentwicklung", Universität Konstanz, Fachbereich für Politische Wissenschaft. Diskussionsbeitrag No. 6/79, 1979; G. Lehmbruch, 1984. "Interorganisatorische Verflechtung

Experience shows that these new arenas must be capable of creating new initiatives for ecological regulations beyond those issues that initially led to the establishment of new administrative structures. This means that they must be able to attack new problems within the range of issues constituting the new political-administrative arena in question. The new administrative structures must be able to adapt older regulations in accordance with new scientific findings. Good examples of such new arenas are the newly created regulatory body for biological agriculture in Switzerland[29] – jointly developed by the BAG, the (non-governmental) Foundation for Biological Agriculture in Oberwil and the FAC-L – and the new nitrate regulatory unit in France ("Mission Eau Nitrates").

Furthermore, the new arenas must be able to produce programmes which aim at real behavioural changes beyond theoretical preliminaries or strategic declarations of principle. This second prescriptive requirement of an arena means that such bodies must also have at least some influence on the implementation of newly created programmes. It appears as though this condition can be fulfilled only when at least one of the actors involved is willing to open his "Implementation Hinterland" to the enforcement of new, "bigger" regulations, or when the actors involved agree to establish similar administrative arrangements, as achieved for the programme level, also on the level of implementation. According to this concept, the step from "small" to "medium" ecological regulations of agriculture must be accompanied by the generation of a new policy arena capable of structurally supporting the ecologically expanded regulations. Our (admittedly limited) verification of this hypothesis is the case of the failure of an adequate drinking water quality regulation in Switzerland, where there was general agreement among the main actors actually to expand the "small" regulation towards an inclusion of some guidelines concerning farmers' fertilizing practices. Nevertheless, there was no willingness to create an adequate regulatory body to generate and enact such a programme on the federal level. The regulation of nitrates in drinking water, after a short "intermezzo" on the national level, has since been "re-cantonalized" and is far from being solved[30].

---

im Neokorporatismus", in: J. W. Falter et al. (eds), Politische Willensbildung und Interessenvermittlung, Opladen: Westdeutscher Verlag.

[29] See P. Knoepfel and W. Zimmermann, "Oekologisierung von Landwirtschaft", Schweizerische Landwirtschaftliche Forschung 25(2) 1986. 196ff.

[30] See P. Knoepfel and W. Zimmermann, "Nitrat war kein Pfad. Von der ausgebliebenen Oekologisierung der Landwirtschaft auf den Weg über die Debatte zum Nitrat im Trinkwasser in der Schweiz", in: J. Conrad (ed.), Nitrat und Landwirtschaft, Hamburg: Campus, 1987.

Up to now, our model has been confined to the interpretation of processes now completed or at least presently ongoing. With the third type of agriculture-related ecological regulations we shall be treading on the less safe ground of speculation about future processes. Social scientific reasoning[31] and recent developments within the core area of agricultural policy[32] have led us to assume that "medium" regulations represent a transitory state which will be followed by another phase of expansion. As indicated in Figures 4 and 5, the next wave of ecologization will probably hit the centre of traditional agricultural policy, namely the regulation of production volumes.

Towards the end of the 1970s, crop production (in Switzerland) was still considered as a viable alternative to surplus production of milk and meat. By the mid-1980s, however, it turned out that crop production too was no longer immune to problems of overproduction. By means of the well-established system of financial aid for farming in mountainous regions, farmers and administrators have meanwhile amassed enough experience with the instrument of product-independent cash subsidies[33]. Initial mutual distrust on both sides has gradually broken down. In recent years, the view has gained considerable support that agricultural activities, from an ecological perspective, should be treated as public contributions for the benefit of the whole economy. These trends could pave the way, by the second half of this decade, for a cautious expansion of our present ecology based agricultural regulations to include traditional quantity regulation policies.

At present, our information is insufficient to draw conclusions concerning the practical problems that could potentially arise out of the strange "coexistence" between traditional regulations of agricultural production, still partly favouring intensification and overproduction, and more recent ecological regulations. The messages these two bodies of regulations send

---

[31] See, for Switzerland, P. Halbheer and A. Müdesbacher, Agrarpalitik – Interessenpolitik?, Bern and Stuttgart, Haupt, 1985 pp. 87ff. (cash subsidies). and for EEC, see Communauté Européenne, Perspectives de la PAC, le Livre vert de la Commission, Notes rapides de l'Europe verte, No. 33, Brussels, July 1983, p. 62.

[32] Increasing surpluses (no longer limited to livestock-farming).

[33] Such transfers have been paid in Switzerland to farmers in mountain areas since 1980. In France, farmers working in what is called "less favoured areas" receive special subsidies, in application of the EEC Directive no 73/268.

to our farmers are often far from compatible[34]. By and large, we do not know much about the implementation problems of current "medium" ecology-based regulations, where these ambiguities necessarily must lead to critical situations. The fact, however, that regional authorities both in France and in Switzerland have great difficulties enforcing, for example, the establishment of ground water resource protection zones against the declared aims of agricultural crops' production programmes shows that, in practice, the contradiction between measures for increasing farm output and demands for environmentally sound agriculture is increasingly articulated. Such observations confirm the long-held assumptions of scholars.

Accordingly, these factors must be taken into account by "big" ecological regulations of agriculture dealing with the anthropogenic causes of water and soil pollution from agrochemicals. Increasing attention, therefore, must necessarily be paid to ecologically compatible maximum yields from cultivated soils. Whether, at what time and in what concrete ways an expansion of "medium" ecological regulations of agriculture will in fact occur cannot be predicted at present from the limited research results now available to us.

7.3 Ecologization in the field of water protection regulations. The case of Switzerland

Swiss legislation, at the beginning of the 1970s, provided only a few "small ecological regulations which dealt with agriculturally-induced water pollution. Apart from some unimplemented prescriptions prohibiting the disposal of agricultural waste-waters in surface waters (Water Protection Act, 1955, revised in 1971)[35], these regulations consist of some stipulations within the Swiss Milk Delivery Regulation (1971)[36] and the fertilizer guidelines, formally issued in 1972, drawn up by the federal agronomic research institutes of Liebefeld (FAC-L), Reckenholz (FAR), and

[34] Examples are conflicts over the creation of (ground and drinking) water protection zones in fertile animal feed lands (especially subsidized by Swiss agricultural policy).

[35] Federal Law of 8 October 1971 (SR 814.20) on the protection of water against pollution (Gewässerschutzgesetz).

[36] "Schweizerische Milchlieferungsregulativ", of 18 October 1971, edited by the Schweizerische Milchkommission.

Changins (RAC)[37]. Both sets of regulations originated from a purely agronomic context. The basic objective was first to protect livestock and to maintain or improve the quality of milk and milk products. The second objective was to reduce the cost of fertilizing soil.

These regulations did not deal with the enormous quantities of organic fertilizers accumulated from animal husbandry. Nor did they deal with doses of fertilizer and fertilizing methods in crop farming which, at that time, were adjusted according to the needs of individual plants. The quantities and methods of applying fertilizers were essentially considered as a function of sales of agricultural products on a more or less state-regulated market. The problems associated with over-fertilization, particularly with expensive commercial fertilizers, were debated exclusively in terms of soil nutrient loss. Although the licensing procedure for fertilizers was made more stringent in 1972, requiring a toxicological analysis in addition to yield-related quality tests, it was still designed only to prevent negative impacts on livestock, crops and human health.

One step from the "small" regulations for crop farming towards "medium" regulations was made in 1974 when the Water Research Division (EAWAG) of the Federal Institute of Technology (ETH) and the old Agency for Environmental Protection (AfU) issued "Guidelines for an Environmentally Appropriate Fertilization"[38]. At that time the reduced doses of fertilizer recommended in these guidelines were aimed explicitly not only at reducing expensive nutrient loss, but also at preventing ground water contamination by nitrates and surface water contamination by phosphates. This initial step towards expanded regulation is particularly noteworthy because it occurred shortly before public debate opened over the serious euthrophication (mainly by phosphates) of Swiss lakes[39], as a probable result of agricultural practices. There is no empirical evidence

---

[37] Eidgenössische Forschungsanstalten Liebefeld-Bern, Eidgenössische Anstalt für Wasserversorgung, Abwasserreinung und Gewässerschutz (EAWAG), Amt für Umweltschutz (AfU), "Düngungsrichtlinien für den Acker-und Futterbau", Mitteilungen für die Schweizerische Landwirtschaft 2, 1972.

[38] Eidgenössische landwirtschaftliche Forschungsanstalten, EAWAG, AfU "Wegleitung zu einer umweltgerechten Anwendung von Düngemitteln", Mitteilungen füir die Schweizerische Landwirtschafit 7, 1974: 133 ff.

[39] See O. Furrer and R. Gächter, "Der Beitrag der Landwirtschañ zur Eutrophierung der Gewässer, Schweizerische Zeitschrift für Hydrologie 34, 1972; FAC-Liebefeld, "Tätigkeitsbericht über die Jahre 1973–1974", Landwirtschaftliches Jahrbuch der Schweiz, 1975: 315 ff.; Forschungszentrum für Schweizerische Politik, Universität Bern, Schweizerische Politik im Jahre 1974, Bern: Haupt, 1975: 108 ff.

that this first step was initiated by actors from the political arena; our materials rather suggest that the initiative was taken by various research institutions and the Swiss Federal Office of Environmental Protection (BUS). There is no indication of any involvement by the Swiss Federal Government, the ministries, or the Swiss Federal Office of Agriculture (BLW). The official, "authoritative" Fourth Report on Agriculture of 1969 contained no indication of any planned water resource protection measures. The Swiss Parliament's debate[40] over that report focused on the problems of intensive animal husbandry, farmers' incomes and mountain farming. This debate hardly touched the issues of environmental protection and water pollution.

This finding is all the more remarkable since the agricultural aspects of environmental protection had played an important role in scientific discourse since the early 1970s. For example, several papers concerning problems of fertilization and water pollution were Presented at an important symposium organized by the Federal Institute of Technology in Zurich on "The Protection of our Environment" in 1970 and at the "Symposium on Environmental Problems and Agriculture"[41] in 1971 held in Bern under the auspices of the Ministry of Economics' (EDV) Division of Agriculture and the Agricultural Research Commission. This area of research probably supplied the main impetus to the debate on the protection of water resources in agriculture.

Another important ecological impetus probably came from the FAC in Liebefeld, whose research since 1970 had been focused the interrelationships between soil fertilization and inland water bodies[42]. We have seen that the ecological argument for the protection of water resources was repeatedly cited as the justification for the "Guidelines on Environmentally Appropriate Fertilization" of 1974. But the fact that only one year later the Federal Office of Environmental Protection (BUS) found it necessary to install a working group to develop guidelines for water resource protection in agriculture indicated, on the contrary, that the 1974 guidelines were primarily motivated by economic rather than ecological considerations. Although there had been an inclination towards reform in crop farming, at least on the programme level, no obligation was felt to enforce seriously

[40] Bundesrat, "Vierter Bericht des Bundesrates an die Bundesversammlung über die Lage der schweizerischen Landwirtschaft und die Agrarpolitik des Bundes vom 26. Februar 1969", Bundesblatt 1, 1969: 389 ff. The debate in the Lower House (Nationalrat) took place in October 1969.

[41] See note 21 above.

[42] See P. Knoepfel and W. Zimmermann, Oekologisierung von Landwirtschaft, Aarau, Frankfurt and Salzburg, Sauerländer, 1987, Appendix 1.

the restrictions on manure production in livestock farming already imposed by the guidelines.

Ecological regulations were eventually imposed on animal husbandry as well, but this happened only after the above-mentioned public debate on agriculture and the euthrophication of lakes in 1974, the debate over surplus milk production as well as persistent research on constant efforts at persuasion by the BUS and the FAC in Liebefeld. Only in 1979 was the famous "Christmas Compromise" on such directives reached[43]. It is interesting to note that these regulations were enacted by a corporate commission which included representatives of the Federal Office of Agriculture (BLW), of agronomic research institutions, the Association of Swiss Farmers (SBV) as well as the Swiss Environmental Protection Agency. Through this co-operation, a new policy arena emerged: each of the four partners of the "commission" issuing the new regulations contributed to their elaboration. Therefore each will make use of their "Hinterland" in order to ensure implementation.

Thus, with the issuance of the "Guidelines for the Protection of Water Resources in Agriculture", broader regulations, according to our definition, were initiated – namely, regulations aiming at the protection of ground water and, in part, soil. In contrast to 1974, by 1979 it was considered "practicable" to include the production of liquid and solid manures and their application on livestock farms in a political-administrative regulation. Unlike crop-farming regulations, the livestock regulations provided no economic incentives for the farmers. On the contrary, these regulations with their limitations on numbers of cattle, pigs and poultry in fact led to financial loss.

In 1974, it was not possible to enforce such a restrictive regulation against the prevailing growth-oriented agricultural policy in animal husbandry. Only after the brakes were applied to livestock production did expanded regulations for the protection of water resources become implementable. Nevertheless, the new regulation of 1979 still does not correspond to the structure of "big" ecological regulations. For example, although the regulation should control livestock production, it has no sanctioning instrument at its disposal, such as subsidies for improved stables or bonuses for 'reducing herds. Furthermore, it can be shown that the restrictions in the revised Article 19 of the Agriculture Act[44], which contain

---

[43] Bundesamt für Umweltschutz and Bundesamt für Landwirtschaft, "Wegleitung zum Gewässerschutz in der Landwirtschaft", Berne, December 1979, 1987.

[44] The amendment of the Swiss Agriculture Law of 1979 introduced a regulation concerning the construction of new livestock farms (subject to a permit system)

binding limits on the size of herds (passed at the same time, in 1979, as the water protection regulations) were not ecologically motivated by some ideally appropriate ratio between number of animals and acreages. What is certain, however, is that herd sizes frequently do not comply with the standards recommended in the guidelines. Because of this, it is reasonable to assume that the guidelines will have little influence on the core of traditional regulation of agricultural production.

Our present information does not allow us to determine to what extent the guidelines may have resulted in a closer co-operation between the BUS, the research institutes at Liebefeld and Tänikon, and the BLW, whose relationship grew out of the co-operation which had already been established in 1974 to control sewage sludges produced by water treatment equipment under the jurisdiction of the Federal Office of Environmental Protection ("National Sewage Sludges Pact"). However, it seems unlikely that dose co-operation did occur. Many observers have accused the Federal Office of Agriculture of actually adding to the problems in agriculture. It is claimed that the Office's new design for stables was one of the main causes of excessive liquid manure production. We were also unable to detect any intensive co-operation between the FAC in Liebefeld, responsible for fertilizer control, the manure suppliers and the Department of Animal Husbandry within the BLW. The influence of "medium" ecology-oriented agricultural regulations on the actual sources of environmental problems remains therefore limited, even on the policy programme level. Contrary to the impression given by the stipulations for herd sizes in the 1979 guidelines, there has been no progression to "big" ecological regulations.

This fact confirms our hypothesis that the core of traditional agriculture policies is impervious to ecologization even in cases where objectives are almost parallel (reduction of animal livestock farming as a means to tackle overproduction). It also reveals the strong limitations to any expansion of ecology-oriented agricultural regulations going beyond the "medium" ecological regulations of agriculture at that time. Nevertheless, it is remarkable that the co-operation network established between some core actors of the BLW, the various research institutions, the BUS and the SBV for securing "medium" regulations, developed into an important policy arena, planning and issuing agricultural programmes for the protection of water resources. As early as 1980, "Recommendations for the Installation of Liquid Manure Storage Pits"[45] were drawn up, followed by the "Rec-

---

and limitations on the cattle population for big farms (Federal Law of 22 June 1979, enacted on 1 January 1980).

[45] "Hinweise für den Bau von Güllegruben", edited by the Bundesamt für Landwirtschaft and the Bundesamt für Umweltschutz, Berne, 1980.

ommendations for Environmental Protection in Rural Areas"[46] in 1981. Finally, in 1983, the report on manure storage facilities and financial requests for their improvement drawn up through the co-operative efforts of the BLW, BUS and SBV was published by the Commission for the Protection of Water Resources[47].

Contrary to the French situation, the nitrate issue in drinking waters did not play an important role in the recent development of ecologization processes in the Swiss agriculture-related water pollution regulations[48]. As already mentioned, this debate, which carne up in 1978–9, did not lead to a special drinking water nitrate policy arena. And the initial federal regulatory attempts failed because of this shortcoming.

Figure 6 presents a schematic diagram illustrating the step from "small" to "medium" ecological regulations with respect to protection of water resources.

Figure 6: The case of the water resource protection in Switzerland

[46] "Merkblatt: Umweltprobleme auf dem Lande, wie lassen sich Schäden in der ländlichen Umwelt vermeiden ? – Empfehlungen", edited by the Bundesamt für Umweltschutz, the Bundesamt für Gesundheitswesen and the Bundesamt für Landwirtschaft, Berne, March 1981.

[47] See Eidgenössische Gewässerschutzkommission, Bericht über die finanzielle Förderung von Anlagen zur Hofdüngerlagerung. Berne, BUS, Schriftenreihe Umweltschutz no. 11, March 1983.

[48] See P. Knoepfel and W. Zimmermann, "Nitrat war kein Pfad...", op. cit., note 28 above.

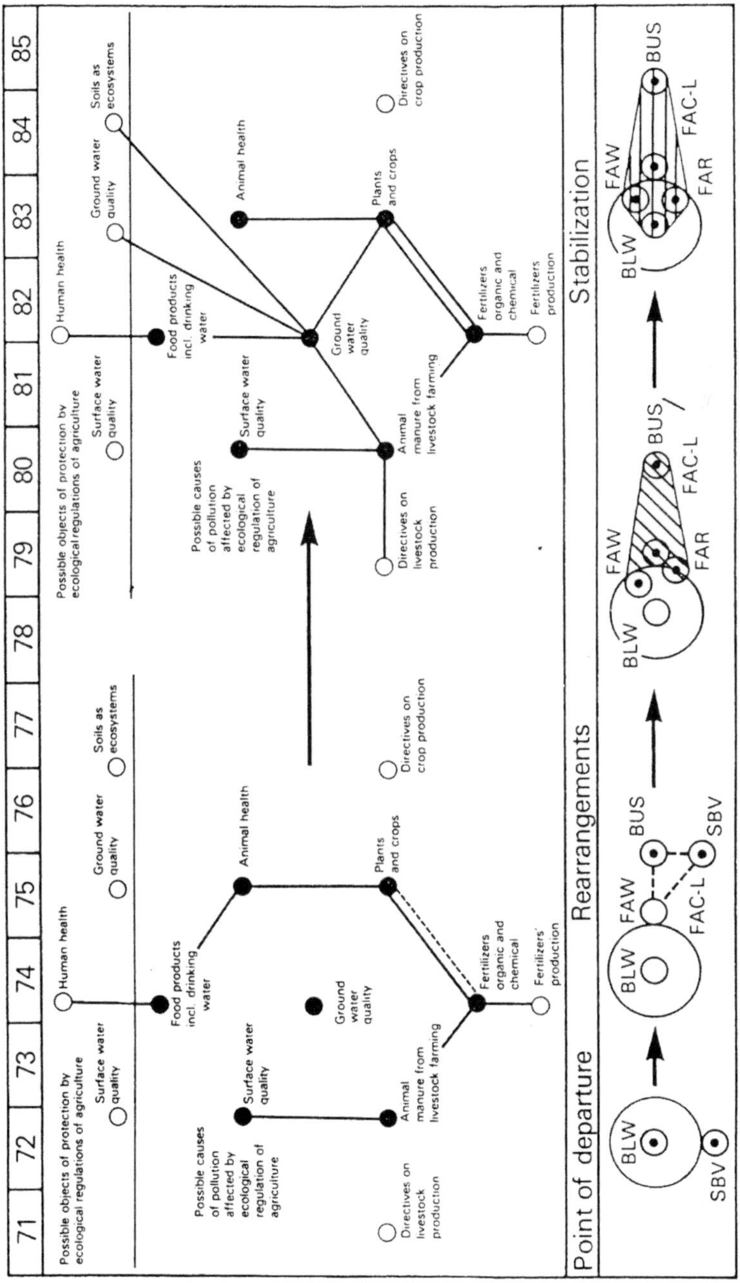

BLW: Federal Office of Agriculture; BUS: Federal Office of Environmental Protection; FAC-L: Federal Institute for Agronomic Research, Liebefeld; SBV: Swiss Farmers' Association; FAW and FAR: Federal Institutes for Agronomic Research (vegetables, arboriculture), Wädenswil and Reckenholz.

7.3.2 Agriculture and water resources – The case of France

At the beginning of the 1970s the French set of "small" ecological regulations of agriculture in the field of water protection mainly consisted of classical health protection regulations, such as drinking water standards[49] and animal health prescriptions. In 1964, these regulations were enlarged by some prescriptions[50] of the General Water Protection Act[51], which mainly prohibited direct agricultural water discharges into surface and ground waters. As in the case of Switzerland, only towards the end of the 1970s could these general prohibitions be made concrete and actually applied. At the beginning of our investigation period, water protection activities were concentrated upon metropolitan areas. Knowledge about methods for protecting ground water in rural areas was still very limited. The only concern at that time was about bacteriological water pollution. Besides these formal classical regulations we found some more informal recommendations[52] dealing with the application of fertilizers which were generated in a purely agronomic context.

These classical drinking water quality standards, sanctioned by bans on consumption, were fixed in 1961[53]. They concerned bacteriological and chemical substances. There were no compulsory standards for nitrates in drinking water at that time; however, standards for bottled drinking water of 44 mg/1 were also strongly recommended for general drinking water[54]. This situation persisted up until August 1985, when the French administration had to implement the 50 mg/1 standard in order to comply with the EEC Directive of 1980[55]. As we shall see, this EEC Directive seems to have played an important role in the ongoing ecologization process in

[49] Set up in the Decree no. 61–859 of i August 1961 and in the "arrêté" of 10 August 1961 concerning drinking water.

[50] Second bock of the "Code rural", dealing with animals and crops.

[51] Law no. 61–1245 of 16 December 1964 concerning the regime and the allocation of water and the struggle against pollution.

[52] See e.g. A. Gros, Guide pratique de la fertilisation, Paris, La Maison rustique, 1962; and Fédération nationale de l'industrie des engrais, Fertilisation, Paris, 1982 (several editions, the 1982 edn is the most recent).

[53] Decree no. 61–859 of 1 August 1961 and "arrété" of 10 August 1961 conceming drinking water.

[54] Circular of the Ministry of Health of 15 March 1962 relating to drinking water.

[55] EEC Directive No. 80–778 of 15 July 1980 relating to the quality of water intended for human consumption.

French agriculture-related ecological programmes. By means of the proposed EEC standards the Ministry of Health made its first attempt actually to limit agriculturally induced pollution. Note that we cannot interpret the absence of a formal nitrate standard in drinking water as a complete lack of knowledge about the health hazards of nitrates, because the old food regulation did in fact contain a nitrate standard for bottled water. But, as in other countries, this standard was established in order to protect babies against methemoglobinaemia[56]. Scientific knowledge about the carcinogenicity of nitrates (in terms of nitrosamines) was not available at that time.

As in Switzerland, regulators were convinced that, by establishing ground water protection zones, the quality of ground water could be guaranteed. At that time, there was no special concern about nitrates and phosphates in ground and surface waters. Not surprisingly, the same Ministry of Health aimed at protecting drinking water quality by the circumscription of such protection zones around drinking water collectors. This instrument had already been introduced by the 1935 Act[57] and reinforced by the already mentioned Water Protection Act of 1964[58]. The latter imposes the creation of these zones for all water collectors created after 1964. Unlike in Switzerland, the French Ministry of Health, by means of its control over the protection zone policy, has an instrument whereby it can deal with some important causal elements for drinking water degradation.

This French instrument goes far beyond the simple possibility of a consumption ban, while it is the normal policy taken for drinking water protection by health authorities in Switzerland[59]. Within protection zones agricultural activities may be limited in the interest of public health. Thus, manure-spreading activities, according to more recent livestock husbandry regulations, may be forbidden within such areas. Although this old instrument of water quality control seems to be one of the first means of a causal drinking water control policy, it turns out to be relatively ineffective due to its limited ability to control long-range movements of ions such as nitrates or sulphates within ground water streams. Furthermore, its application might be highly controversial if it were to cover large and fruitful crop production areas. As they were used at the beginning of the 1970s, these

---

56 Expressly mentioned in the Circular of the Ministry of Health of 15 March 1962 concerning drinking water.

57 "Décret-loi" of 8 August 19 3 5 relating to ground water protection.

58 Art. 7 and 8 of Law no. 64–1245 of 16 December 1964 (see note 49 above).

59 See Knoepfel and Zimmermann, "Nitrat war kein Pfad...", op. cit., note 28 above.

drinking water protection zones. must therefore still be considered as "small" ecological regulations of agriculture.

Another of these "small" regulations in force at the beginning of the 1970s concerned livestock husbandry. It stemmed from the Standard Departmental Health Regulation (RSDT)[60]. Again, the model of this regulation had been proposed by the Ministry of Public Health. In each French "département" such a regulation must be delivered by a prefectoral Decree. These regulations, among others, dealt with possible health effects (mainly bacteriological hazards and smell) of manure-spreading practices as well as with some hygienic aspects of farm construction. Some of these rules, which historically go back to the nineteenth century, seem to have been established in order also to protect surface and ground water from (bacteriological) pollution. In 1974 these regulations were modified in the field of some specific livestock husbandry activities (e.g. for pig farming) because of increasing nuisances stemming from industrialized livestock farming[61]. But again the principal causes for these modifications were the protection of the traditional protection targets – namely human health (especially the health of neighbours). There was no explicit debate on the protection of rural ecosystems as such at that time.

The first step towards a more integrated environmental point of view was undoubtedly the setting up of a new systematic environmental control system for industrial plants by the newly created (1971) Ministry of Environment. The basic legislation for this reform is the Industrial Pollution Act of 1976 (RIC)[62]. This legislation subjected all main industrial activities to systematic environmental control. Among the list of controlled activities we find industrial livestock production, such as poultry, beef-meat and pig farming. We do not know why the list does not contain industrial milk-production farming. Notice that this first attempt of the environmental actor to enter into the agricultural field does not affect thousands of smaller livestock farms which, at that time, still continued to be regulated by traditional small ecological regulation (RSDT). Furthermore, we must stress the fact that this French approach, unlike the Swiss one, utilized the ordinary method of environmental control, aimed mainly at controlling industrial emissions within metropolitan areas. There is no clear-cut distinction between industrial and agricultural activities as far as large livestock produc-

[60] RSDT – Régiement sanitaire départemental type, see note 19 above.

[61] The Circular of the Ministry of Health of 24 May 1963, formal basis of the RSDT at the beginning of the 1960s, was modified by the Circular of 4 March 1974, which included prescriptions on pig-farming activities and environmental protection.

[62] RIC – Réglementation des installations classées, see note 20 above.

tion units are concerned. As we shall see, this control policy opened the way to all kinds of extensions of initially purely health-oriented protection measures, towards ground water and soil protection. Finally, it served as a model for the more recent strategies aimed at controlling the ordinary agricultural activities of smaller farms[63].

As in the case of Switzerland, the presentation of "small" ecological regulations at the beginning of the 1970s would be incomplete if we did not mention a further group of more informal regulations. These concerned efforts to reduce costs of fertilizers. These regulations delivered in the form of recommendations[64], are provided exclusively by agricultural actors, such as the National Institute for Agronomic Research (INRA) and the agricultural technical institutes (semi-public associations playing an intermediate role between INRA and the agricultural consulting activities). The development of more adequate fertilization guidelines for crop farming was one expression of the new form of agronomic research called "agronomic relaunch". As in Switzerland in the early 1970s, concern about water and soil quality played a negligible role in this movement.

In summary, the period of "small" ecological regulations of agriculture in France, which lasted through the 1970s, juxtaposed numerous specific regulations, mostly dealing with the narrow aspects of agriculturally induced water pollution, established in order to protect product quality and human health. This concept of numerous, juxtaposed, individual regulations goes far beyond the situation in Switzerland. It distinguishes not only between the treatment of water pollution in the vegetable and animal sectors but furthermore it treats differently emitters within the area of livestock husbandry, according to the size of the farms and their main products (milk versus meat). This typical juxtaposition also matches the arrangement of the relevant political-administrative actors. Each of the three Ministries (Health, Environment and Agriculture) seems to conduct its own policy and each one is eager to defend its jurisdiction against the other. Contrary to the Swiss situation, there seems, however, in France to be clear-cut domination by the Ministry of the Environment via its dynamic regulation of industrial plants[65]. The ecologization of agriculture-

---

[63] See note 19 above. In the Circulars of the Ministry of Health of 20 January 1983 and of 10 August 1984 (RSDT), prescriptions on livestock activities and environmental protection were set up.

[64] See e.g. Cultivar, special issue : "Fertilisation", 162, June 1983.

[65] See P. Knoepfel and C. Larrue, "Distribution spatiale et mise en oeuvre d'une politique publique: le cas de la pollution atmosphérique", Politique et management public, 3(2), 1985 : 46–69.

related programmes passed by stages through an increasing enlargement of the scope of application of this instrument.

For the first time in France, indeed, this regulation led to an effective control of quantities of manure and the modes of spreading. The prescriptions implemented take into account the amount of fertilizer produced by each animal and the fertilizers needed for several categories of crops. The regulatory application starts with the acreage of a farm and draws from this figure the permitted amount of manure. Thus, for instance, the standard of 40 pigs/ha must not be exceeded. Although this mechanism of animal limitation in terms of the available acreage is quite similar to that of the Swiss regulation of 1979, the French approach again is different because it delivers plant specific regulations. Following the basic philosophy of the Industrial Pollution Control Act, these regulations were enacted for different farming types, even in different years: the first regulation was enacted in 1976 for pig farms[66], the second in 1981 for cattle breeding[67] and the last in 1982 for poultry farms[68]. Note that we still do not have similar regulations for milk production farms.

These regulatory activities of the Ministry of the Environment can be considered as a real attempt by this Ministry, mostly occupied at the beginning of its activities in 1972 with pollution control in metropolitan areas, to enter into the area of rural environmental protection. It is important to note that this new ecologization movement although conducted in cooperation with the Ministry of Agriculture and the National Institute for Agronomic Research (INRA), was considered as a natural extension to rural areas of industrial plant regulations.

For this reason there was no rupture in this political-administrative arrangement and no need for a new policy arena was articulated. But this approach remained partial and the water pollution problem could not be solved. A more comprehensive approach still had to be found. In this sense, the movement from the first generation of smaller ecological regulations concerning agriculture towards medium regulations took place around 1980. At that time the main concern was, again, worsening ground and drinking water quality throughout the country. The nitrate issue was at stake. The more general ecologization process initiated at that time stems from at least two different instances: the first one, as already mentioned, is

[66] Circular of the Ministry of the Environment of 12 August 1976 concerning the reduction of the pollution brought about by pig-farming activities.

[67] Circular of 13 December 1981 concerning cattle-breeding activities (not published).

[68] Circular of the Ministry of the Environment of 1 October 1982 concerning poultry farms and the protection of the environment.

the publication of the EEC Directive in 1980 prescribing, by 1985, the introduction of relatively high nitrate levels for all kinds of drinking water. The second instance was a more internal national event: in 1977 the results of a large surface water quality inventory were published[69]. Figures clearly showed an increase of nitrate and phosphate pollution presumably induced by agricultural activities.

The announcement of the EEC Directive immediately led the Ministry of Health to establish a new inventory of the drinking water quality in 1981[70]. The main concern of this inventory was nitrate concentration. Furthermore, the Ministry made a considerable attempt to enlarge the regulations on livestock husbandry by reforming the model of the Standard Departmental Health Regulation (RSDT) in 1983 and 1984[71]. This reform introduced a "medium" ecological regulation of agriculture in the sense that it contained prescriptions on manure-spreading activities for all farms in order generally to protect drinking water resources. Note that this regulation was not limited to animal manure but included all kinds of fertilizers as well (chemical fertilizers Sewage sludges and compost from waste disposal). The Ministry issued a set of incentives for those farmers willing to adopt fertilizer-spreading plans and mandated local agencies to enforce the regulation.

However, contrary to the Swiss regulation, this new French model does not contain references to the actual quantity of fertilizers to be spread on fields for different cultures. Furthermore the model regulation has to be enacted by a formal departmental decree. It is still uncertain whether the proposed incentive system will actually work. It is true that the Ministry of Health, by means of this recent programme, engaged in some modest co-operation with the Ministry of the Environment in order to co-ordinate their regulations (the model's reform Started in 1978 but was delayed for rural areas until 1983 so as to give enough time to the Ministry of the Environment to deliver its industrial plant regulations). It seems, however, that there has not been substantial co-operation between the two Ministries. As in Switzerland, the Ministry of Health seems to have played the role of a lonesome hero. The chances that its regulatory effort will actually be implemented are low, given the absence of any integration of its programme into a more general ecologization approach supported by both the Ministry of Environment and of Agriculture.

---

69 Ministére de l'environnement: "Inventaire du degré de pollution des eaux superficielles", Paris, 1976.

70 Circular of the Ministry of Health no.1005 of 10 July 1981 concerning nitrates in drinking water.

71 See note 61 above.

It seems that a far more effective strategy was initiated by an internal event (publication of a set of data concerning phosphate and nitrate concentrations in surface water of different areas of France). This startling report provided the basis for systematic co-operation between the Ministry of Agriculture and the Ministry of the Environment since the end of 1979. These two big institutions agreed upon the creation of a joint working group charged with evaluation of the agriculturally induced water pollution as a whole and with the elaboration of proposals for ways to limit this pollution. The working group produced the famous "Hénin report" published in October 1980[72]. The Hénin committee was composed of representatives of the two Ministries, plus members of the National Institute for Agronomic Research (INRA, including the chairman of the working group, Mr Hénin), representatives of agriculture trade unions, agriculture technical institutes and water protection agencies. Again, the near absence of representatives of the Ministry of Health is striking; this Ministry was represented in only one subcommittee.

The Hénin group made numerous substantial recommendations. It also proposed a new permanent regulatory body to be specially charged with the problems of nitrate water pollution. This new administrative unit, called "Nitrate Mission"[73], belongs jointly to the Ministry of Agriculture and the Ministry of the Environment. An advisory board, the CORPEN[74], representing agricultural and environmental actors, is mandated to produce guidelines for the mission's activity.

We can assume that this regulatory body fulfils all of our requirements for an actual new policy arena capable of mobilizing enough regulatory resources actually to tackle the water pollution problem in agriculture. It is now elaborating a complete programme of nitrate pollution reduction, including both main causes of pollution: manure spreading and chemical fertilizers, for which new recommendations have been developed[75]. Note, however, that as far as we know there is no intention to redefine or modify the core of traditional agricultural policy as such. Furthermore, the representatives of the Ministry of Agriculture come from relatively peripheral

---

[72] Ministère de l'agriculture and Ministère de l'environnement, "Rapport du groupe de travail 'Activités agricoles et qualité des eaux'", 2 vols, Paris, October 1980.

[73] The "Mission Eau Nitrates", created in 1984, is presentiy composed of four engineers.

[74] Mission Eau Nitrates", created in 1984, is presentiy composed of four engineers.

[74] CORPEN – Comité d'orientation pour la réduction de la pollution des eaux par les nitrates et les phosphates provenant des activités agricoles.

[75] See "Amélioration des pratiques agricoles...", op. cit., 1982, note 2 above.

units of this Ministry (the water section and the INRA). Again, the similarity with the Swiss arrangement is striking. Nevertheless, we know that, at the level of the EEC, there are attempts to move towards some "big" ecological regulations of agriculture (see the so-called Green Paper issued in 1985)[76]. But again, quite similar to the case of Switzerland (see the 6th Report on Agriculture of 1984)[77], these are declarations of good intentions rather than actual regulatory changes. Figure 7 shows the development of the ecologization of the French regulations of water protection in agriculture over the last fifteen years.

Figure 7: The case of the water resource protection in France

[76] See note 29 above.

[77] Nevertheless, in the EEC Directive no. 797–85, setting the new orientation of the Common Agricultural Policy in the field of agricultural structures, some specific subsidies are now allocated for farmers' investments made in view of environmental protection.

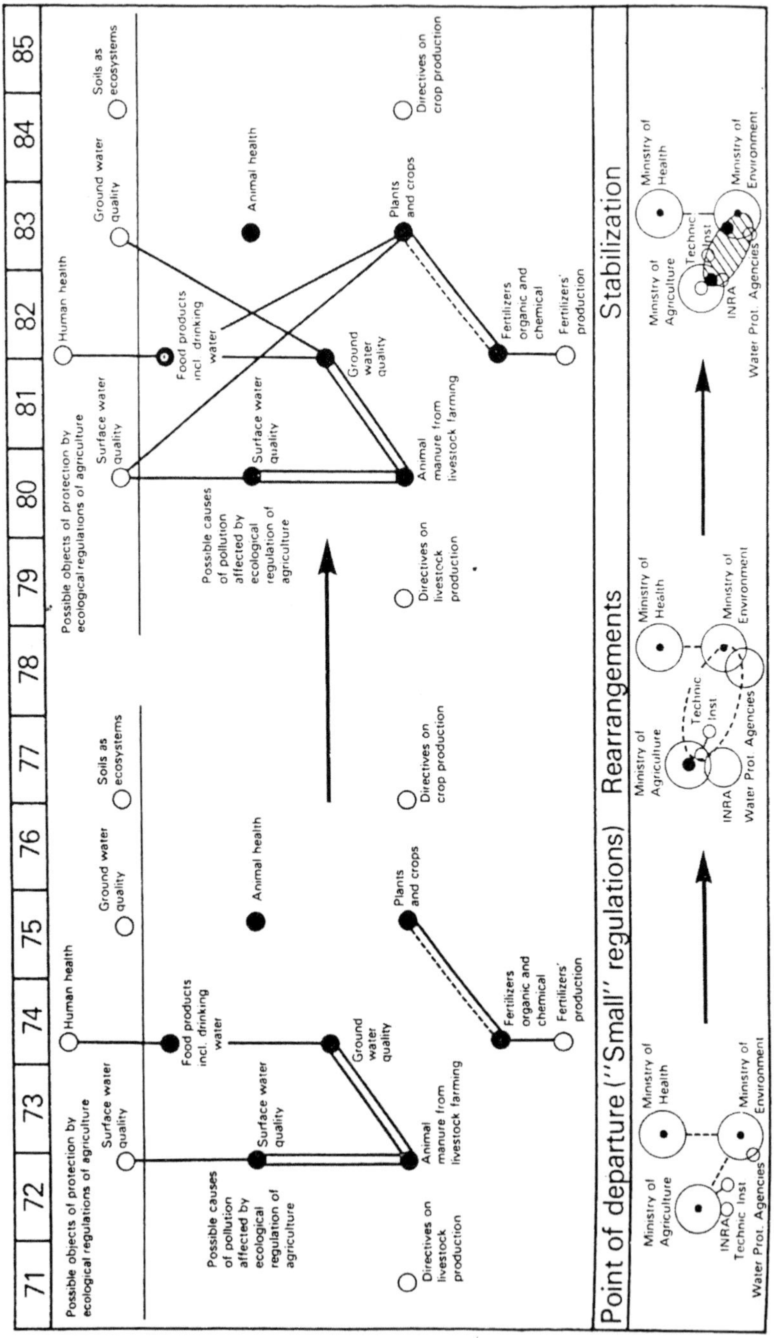
71 72 73 74 75 76 77 78 79 80 81 82 83 84 85
Possible objects of protection by ecological regulations of agriculture
Surface water quality
Human health
Ground water quality
Soils as ecosystems
Food products incl. drinking water
Animal health
Possible causes of pollution affected by ecological regulation of agriculture
Surface water quality
Ground water quality
Animal manure from livestock farming
Plants and crops
Fertilizers organic and chemical
Fertilizers' production
Directives on livestock production
Directives on crop production
Point of departure ("Small" regulations)
Rearrangements
Stabilization
Ministry of Agriculture
Ministry of Health
Ministry of Environment
INRA
Technic Inst
Water Prot. Agencies

7.4 Green vegetables and lettuce: a third comparison

7.4.1 The absent case of France

First of all we must explain the reason for including green vegetable and lettuce regulations in this presentation even though international comparisons in this field will yield poor results, as the French system is still at the stage of small regulations, comparable to the Swiss system at the beginning of the 1970s. The French system stems from a very old French law (1905) against fraud in commerce and trade[78], long used for the purpose of protecting public health. On the basis of this legislation, numerous quality standards for food have been enacted. Its enforcement permits bans on the sale and consumption of unhealthy foods.

Unlike the recent inclusion of production conditions of agricultural products in Switzerland, the French regulation still does not deal with such agricultural production practices. Until 1982, this regulation was implemented under the jurisdiction of the Ministry of Agriculture, because the main products subjected under the Act were agricultural foods. The change of government in 1981 led to the creation of a new Ministry for Consumer Affairs[79], which was charged, in 1982, with the application of this food control regulation. This new Ministry reinforced concerns about health protection in 1985 by amending the older standards on pesticide residues of 1973[80]. The existing nitrate standards, however, are still limited to baby food.

Nevertheless, in 1981, the French Ministries of Agriculture and of the Environment established an inventory on food quality which also pointed out, for example, serious nitrate concentrations in lettuce[81]. On the basis of this inventory, control activities on lettuce have recently been slightly reinforced, but the analyses are still quite unsystematic and usually they are initiated only when there is a request from a particular consumer organization. But concern about nitrate in foodstuffs seems to be an issue

---

[78] Law of 1 August 1905 relating to fraud in commerce and trade.

[79] The Ministère de la consommation was created in 1982. This Ministry disappeared in 1986 (change of government) and the administrative department dealing with consumer affairs presently belongs to the Ministry of Economy.

[80] "Arrêté interministériel" of 7 March 1985.

[81] Ministères de l'environnement, de l'agriculture, de l'industrie, de la recherche, de la santé: "Inventaire national de la qualité alimentaire", Paris, 1982.

limited to medical and health or agronomic research institutes (respectively INSERM and INRA). Contrary to the situation in Switzerland, the actual public debate on nitrates in drinking water does not include the green vegetable and lettuce issue at all.

Among the small ecological regulations concerning vegetable production we find in France – as in other countries – some single prescriptions concerning the prohibition of manure-spreading activities in vegetable gardening. Such regulations have existed since 1937 (Standard Departmental Health Regulation, RSDT)[82]. In 1983, these health regulations were extended to animal health protection (rot) periods and special sewage sludge treatment prescriptions, in the same way as they were introduced into Swiss legislation in 1981. These highly dispersed regulations have been produced and specified by the Ministry of Health. There is little cooperation between this Ministry and other, mainly agricultural actors. However, the standard-setting procedure of foodstuff quality[83], which now is conducted by the Ministry of Consumer Affairs, takes place on a case by case basis with the Ministry of Public Health: it turns out that since the creation of the new Ministry of Consumer Affairs there is no larger and systematic integration of foodstuff control policies into the traditional policies of agriculture promotion. The absence of an inter-organizational network or even of a policy arena, as in the Swiss case, seems to hamper ecologization of the small regulations concerning green vegetables and lettuce in France.

7.4.2 The lonesome case of Switzerland

As mentioned above, Swiss efforts to establish an effective federal drinking water regulation in the early 1970s have failed. The following analysis of the green vegetable- and lettuce-related ecological regulations shows, however, that these ecologization processes might be considered somehow as equivalents to the debates on nitrates in drinking water observed in France[84]. Although the comparison with France justifies little more than

[82] Circular of 1 April 1937.

[83] As far as national standards are concerned. However, standard-setting procedures are mainly conducted on the EEC level.

[84] The same holds true for West Germany, see Der Rat von Sachverständigen für Umweltfragen, op. cit., 1985, note 1, pp. 228 ff; P. Teherani-Krönner, "Tauziehen um eine agrarumweltpolitische Lösung, Die Gülleverordnung von Nordrhein-Westfalen", Science Center Berlin, 1985 (IIUG pre 85–8); J. Conrad and

the simple statement that France is far behind, we shall still present the case because it contains interesting elements of ecologization processes which might, in turn, help reveal some generalizable patterns in other fields of agriculture-related ecological regulations.

The classic "small" regulation governing lettuce and other green vegetables is the oldest of those regulations analysed in our Swiss case studies. The lettuce and green vegetable regulation was issued before 1970. As an integral part of traditional food legislation, its sole objective was the protection of human health. Accordingly, this regulation was confined to prohibiting the consumption of foods containing substances which were deemed by toxicological analysis to be hazardous to health. Its effectiveness was first questioned around 1978. Since the beginning of the 1980s it has successively been replaced by "medium" regulations.

In this case study, our primary interest is the regulation of nitrate concentrations in certain vegetables, green ones in particular. To the best of our knowledge, prior to this regulation neither public policy nor farmers' federations had established any norms for preventing food contamination by nitrates. Even as late as 1977, the Federal Institute for Research on Arboriculture, Viticulture and Horticulture, Wädenswil (FAW), in its "Guidelines for the Use of Fertilizers in Vegetable Gardening"[85], did not consider this problem. This purely human health-oriented regulation was definitely more limited in scope than the older drinking water regulation. It was probably narrower in focus than the special regulations governing pesticides which at least had some influence on the production of vegetables. Not surprisingly, neither the pesticide regulation nor the nitrate regulation under study touches upon any aspects of soil or water resource protection.

The limitations of this regulation would not have become evident quite so soon had it not been for the public debate in 1978 and 1979 over nitrate concentration in drinking water. This debate induced the Swiss government to fix standards for the maximum permissible concentration of nitrates in drinking water[86]. And it was through the issue of drinking water (which, as a target of regulatory measures, soon lost its attraction) that the nitrate concentration in lettuce became a topic of concern. In November 1979, a

M. Gitschel, "Rechtliche Regulierung der Stickstoffdüngung. Nur ein Sturm im Wasserglas ?", Science Center Berlin, 1986, (IIUG pre 86–9).

[85] "Düngungsrichtlinien für den Gemüsebau der Eidgenössischen Forschungsanstalt Wädenswil", 1977.

[86] The present regulation provides a standard of 40 mg/1, see the Ordinance on toxic substances (Stoff-VO) of 9 August 1986. The same standard was fixed within an earlier recommendation of 1969.

report on "Nitrates in Food"[87] unequivocally stated that nitrates pose a serious threat to human health (according to the nitrates-nitrosamines-cancer hypothesis), that a high proportion (70 per cent) of average nitrate intakes by humans is ingested through vegetables, and that there is an obvious cause-effect relationship between agricultural production methods and the contamination of vegetables with nitrates. This investigation into the causes of excessive concentrations of nitrates in vegetables was the basis for new demands that the old Lettuce and Green Vegetables Regulation be expanded to cover actual agricultural practices.

At the beginning of 1980, the Federal Office of Public Health (BAG) and the working group on nitrates in food had adopted a widely acknowledged hypothesis that nitrosamines are carcinogenic. This change in perspective put some pressure on vegetable growers from the end of 1980. Initially, consumer organizations, alarmed by the debate over drinking water and the so-called "hormone scandal" concerning veal[88] called for more effective public health laws. This demand began to focus increasingly on nitrates in lettuce. Public anxiety, however, brought about an immediate response from vegetable growers, made all the more controversial by the fact that consumer protests were strongly supported by "dissident" eco-farmers. The initial strategy of the agricultural federations was simple: build up defences against the outside; persuade the membership to adopt internal self-regulation. At the same time, several ambitious agricultural research projects were underway, and the BAG issued the so-called "action levels"[89] as incentives for farmers to adjust their production methods in accordance with recommendations resulting from the latest research. These efforts by market actors, research institutions, and the BAG were undoubtedly stimulated further by the Parliamentary Control Board's decision to devote a Wee-year period (from 1980 to 1982) of investigation in agricultural policy to the practice and merits of organic agriculture[90].

---

[87] Bundesamt für Gesundheitswesen, "Nitrate in Nahrungsmitteln. Bericht der Arbeitsgruppe 'Nitrate in Nahrungsmitteln' ", Berne, 1979.

[88] Similar "scandals" arose at the same time (end of 1979) in other European countries (lack of control in administering hormones to calves).

[89] See Bundesamt für Gesundheitswesen, "Die Durchführung der Lebensmittelkontrolle in der Schweiz im Jahre 1982. Bericht der Abteilung Lebensmittelkontrolle", Mitteilungen aus dem Gebiete der Lebensmitteluntersuchung und Hygiene (Berne, BAG), 74, 1983: 184.

[90] Geschäftsprüfungskommissionen, "Bericht der Geschäftsprüfungskommissionen an die eidgenössischen Räte über die Inspektionen und Aufsichtseingaben im Jahre 1982 vom 14.4.1983, 'Forschung im biologischen Landbau'", Bundesblatt 11, 1983: 395 ff.

The nitrate regulation was consolidated at the programme level by the end of 1983. The expansion of this regulation occurred stepwise in four directions within different contexts of decision-making. First, with the BAG'S change of perspective, modes of production were explicitly included in the regulations governing conventional vegetable farming. The "action levels", the guidelines and standards set by the BAG which followed, could be adhered to only when certain principles of agricultural practice were applied and the corresponding product controls were carried out. In general, this type of production-oriented regulation was achieved through the professional organizations of the farmers themselves. However, for the formulation of concrete instructions, scientific research was required. This was, in fact, carried out by the FAW. The efforts of agricultural federations to put these instructions into practice (e.g. controlling farm fields, establishing nitrate control rings) were partially subsidized by the Swiss government. In some areas, however, farmers even gave up lettuce cultivation during the winter. This voluntary action contributed to the implementation of the new nitrate regulations as ell as to a better public image of vegetable growers.

Secondly, the detailed regulation envisaged for organic agriculture (drafted in 1985)[91] is to be considered as an important expansion of the old ecological regulation beyond the nitrate issue. This step is quite remarkable because, according to a 1983 report by the Parliamentary Control Board, organic agriculture was practised o only about 1000 farms (out of a total of about 125,000). It signifies a marked reorientation of the political arena, since particularly the Swiss Commission on Nutrition was still campaigning for a ban on the use of the term "biological product" in 1974.

Since 1984, a third direction in the expansion of regulations has become manifest. It is planned to extend nitrate regulation currently applying only to lettuce, to some other leafy green vegetables and beets and maybe even to all vegetables.

In principle, however, these three ecologization steps still aim primarily at improved public health. With the official introduction of the "N_{min}" fertilizer method, recommended in 1982[92], the protection of ground water and soil has become an explicit objective. This fourth step of expansion of regulation can be seen not only as the consequence of new scientific findings, but also as a late outcome of the public debates of the late 1970s on nitrates in drinking water and vegetables. With this last step, the strongly

[91] Bundesamt für Gesundheitswesen, "Änderung der Lebensmittelverordnung Art. 15 a, Entwurf", March 1985.

[92] Cf. C. Gysi, E. Bräm and K. Wixinger, "Überprüfung der Düngungsrichtlinien im Glashaus-Gemüsebau", Der Gemüsebau 13, 1981: 18ff.

co-operation-based Lettuce and Green Vegetables Regulation achieve something which previous drinking water regulations aimed at all o crop farming had failed to do.

It seems that vegetable growers were the only agricultural branch to accept new regulation and were the only ones willing to make a collective contribution to the reduction of nitrates in food, soil and ground water. Some reasons for this may be that vegetable farmers have had a comparatively low status, that their sector has hitherto scarcely been controlled by the state, and that their professional federations were very anxious to improve their ecological image and reputation among the public. With large-scale public relations campaigns they hoped to strengthen their market position. Finally, the high concentrations of nitrates in drinking water which occur mostly in regions with a high proportion of vegetable farming (e.g. Bernese Seeland) may also have provided an additional incentive: vegetable farmers were frequently confronted directly with criticism from alarmed consumers.

Although a uniform, ecological "vegetable production arena" had come into being in 1980, made up of the BAG, the vegetable growers and the FAW, by 1982 a split in this arena ad occurred. The philosophy of peaceful coexistence between conventional farming and organic agriculture was the primary motive behind BAG'S decision, but it was not uninfluenced by the Parliamentary control Board: the new arena thus formed consisted of a stable network of interaction between the BAG, the Research Institution on Biological Husbandry (Oberwil), and the eco-farmers. It developed regulation that extended far beyond the problem of nitrates in lettuce, the issue around which this arena had initially formed. Wit good reason, this regulation can be labelled the "Swiss Charter of Organic Agriculture". The declared willingness to enforce 'his regulation is expressed in the implementation instruments described in the original draft. Furthermore, the network of co-operation between the BAG, several of the research institutes but especially the FAW, the cantonal chemists and the vegetable growers' associations (SGU/VSGP), set up to implement this expanded regulation, is another clear indication that a new policy arena has emerged. The expansion f the nitrate regulation to include other vegetables, the creation of the so-called nitrate control ring, the installation of field test stations, and the sponsorship of several research projects are impressive evidence for the firm commitment to put this regulation into practice. An obvious, positive result of its implementation is the suspension of lettuce cultivation by many vegetable growers during winter months.

Because of the stability of these two arenas, it is possible that advances in the direction of a "big" ecological regulation can occur. But, again, this is largely only speculation. There is, however, some indication that the nitrate regulation will be applied to all vegetables relatively soon. Moreover, our experience leads us to expect that similar production-related regulations will eventually be developed to deal with other harmful residuals in food, particularly hydrocarbons. Finally, it is also conceivable that this legislation will be further expanded to include protection of soil and ground water in other crop farming areas.

Figure 8 shows the development of the ecologization of Swiss regulations concerning green vegetables and lettuce.

Figure 8: The case of green vegetable and lettuce regulations in Switzerland

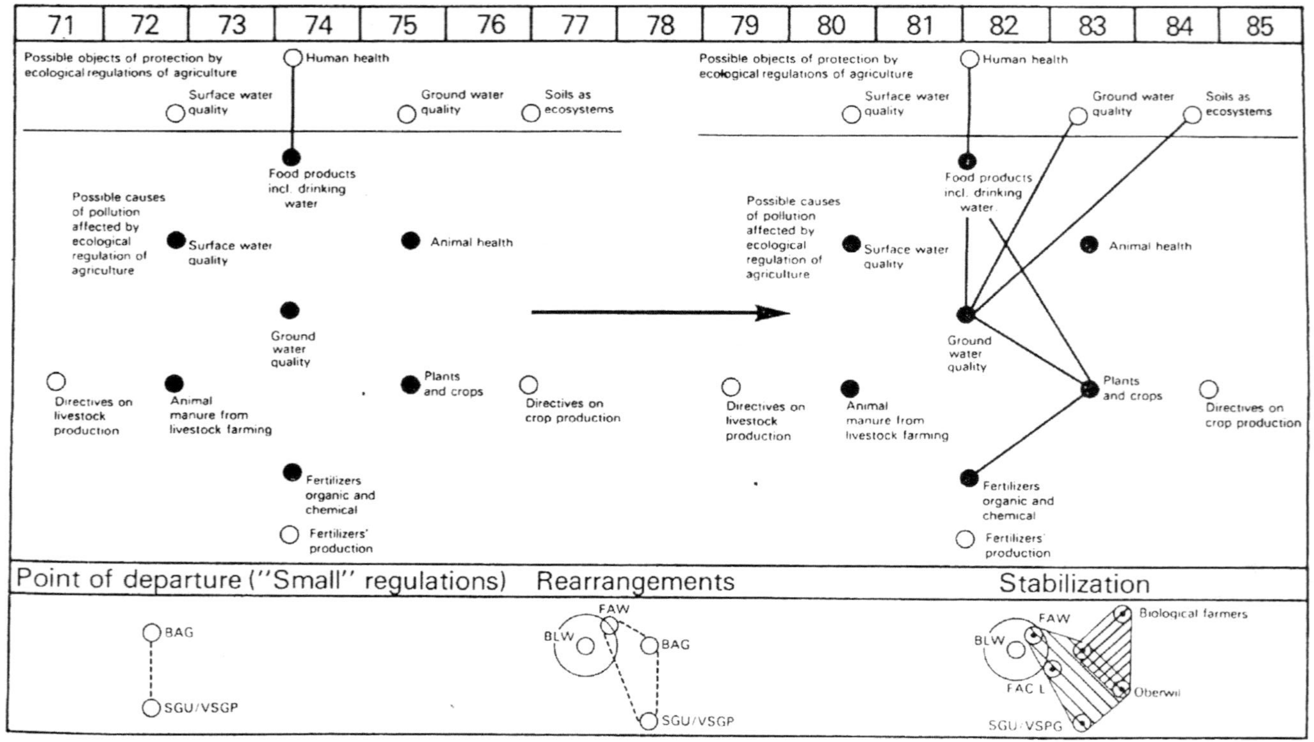

BAG: Federal Office of Public Health; BLW: Federal Office of Agriculture; SGU/VSGP: Association of Swiss Vegetable Growers; FAW: Federal Institute for Agronomic Research (vegetables), Wädenswil (Thurgau); FAC-L: Federal Institute for Agronomic Research (environment), Liebefeld (Bern); Oberwil: Biological Agriculture Institute

7.5 Some comparative hypotheses about the conditions for ecologization processes in agriculture

Given the distribution of power in Switzerland and France as well as in other western European countries, it seems that ecological programmes of action hardly referred to or touched on the core of traditional agricultural policy programmes. This certainly is less surprising for France whose agriculture policy has been transferred to a large extent to the EEC level. The observation, however, holds true also for Switzerland, which has its own agricultural policy. In both cases we observed an interesting coexistence of conventional regulations governing agricultural production and the more recent ecological regulations of agriculture. For the present, however, we have not been able to analyse systematically this (only apparently) peaceful relationship. In our Swiss case studies on the regulations concerning protection of ground and drinking water, we did find some evidence that this coexistence can turn into open confrontation, if the programmes for ecologization of agriculture from "halfway outside" begin to question the central provisions of the conventional regulations of agriculture (case of failed drinking water regulation).

Our results show, contrary to popular opinion that during the last fifteen years, both in France and in Switzerland considerable advances have been made in ecologization, at least at the programme level. This is all the more remarkable since these achievements came about without any major conflicts with conventional agricultural policy. The one exception – the conflict between the French Ministry of Health and the farmers' professional organizations in 1983 about the limitation of some fertilizing practices – rather proves this rule. We shall try to summarize our results in the following six hypotheses on the basic preconditions underlying the development of "medium" ecology-oriented agricultural regulations. As they are empirically based only on our five Swiss case studies and on a more preliminary Study for France, they must be considered tentative rather than conclusive.

1. For the establishment of "medium" regulations it is advantageous to already have "small" regulations in effect because relevant actors will have been institutionalised and can function as competent contact persons or effective promoters of proposed regulatory expansion. Furthermore, "small" regulations provide a foundation for political-administrative responsibilities, and legal competencies (especially at the federal level for Switzerland) upon which to build as well as a minimum of personnel and financial resources for the investigation of ecological

problems. Existing regulatory instruments can provide a suitable basis for expansion, especially "medium" regulations are designed to cover additional sources of ecological problems and if protection of a given object belongs to the obligatory tasks of the responsible agencies under the old regulation. This is particularly applicable to the regulations governing vegetable and drinking water quality in the Swiss case. The assumed preference for initiating ecologization processes on the basis of pre-existing smaller regulations presumably explains the different instrumental approaches of France and Switzerland. Switzerland did not dispose of n Industrial Pollution Act at the time, which could have served as a entrance into the field of rural ecological regulation, as was the case in France. France, in turn, could use this legislation, initially conceived mainly for urban areas, in order to approach rural pollution problems obliquely.

2. The stabilization of "medium" ecology-oriented regulations requires new policy arenas. Thus, the old Swiss federal administrative structure which had drawn up and implemented he "small" regulations was too pigeon-holed. The three agencies basically affected, the Federal Office of Public Health (BAG), of Environmental Protection (BUS) and of Agriculture (BLW), were each commissioned to regulate only one area of ecological problems which, in fact, would not be quite so neatly isolated. Ecologization therefore required the development of new forms of inter-administrative co-operation to initiate the expansion of old regulations and to incorporate new regulatory arrangements into the overall administrative structure. The French "Nitrate Mission" is one of the most interesting r cent creations of such a policy arena.

   In Switzerland, new policy arenas have been established for agricultural regulation to protect water resources and to guarantee the quality of green vegetables. Furthermore, in this country efforts to introduce operable "medium" ecology-oriented regulations governing drinking water quality failed, mainly because the given structures did not allow the establishment of such an arena on the federal level although most actors agreed that such a regulation as needed at that time (1978–79). Since the Ministry of Health is not well integrated into the agricultural policy networks in France, there is no guarantee that such arenas will be established between agricultural actors, the regional water authorities and representatives o the Ministry of Health.

Even if the basic principle of the need for new policy arenas seems not to be falsified by the empirical data, there remains an interesting comparative difference between the two countries under investigation: the arena-building strategy seems to be more common in Switzerland than in France. The example of the presumably successful use of the Industrial Pollution Act by the Ministry of the Environment shows that under some conditions the creation of new policy arenas seems, at first sight, not to be a necessary precondition for an ecological extension of agriculture-related regulations. But this observation is only half the truth: recent developments show that at the moment, when the Ministry aims at an integrated, problem- (and not single plant) oriented approach to rural pollution, it has to pass through the "Nitrate Mission"; the scattered approach through the Industrial Pollution Act – limited to a few huge farms only – will not be sufficient.

New arenas that have prevailed have been able to adjust "medium" regulations on the basis of new scientific information. In that way, they have become relevant promoters of and suitable institutions for developing and initiating "bigger" regulations. Unlike "medium" regulations, "big" regulations must include the classic provisions of agricultural production regulation; an inclusion of the core actors of these policies into the (enlarged) arenas will invariably be needed.

3. For the establishment of "medium" ecology-oriented regulations, it is important to have access to agricultural research with a specific focus on ecology. In this regard, Switzerland seems to be in advance: the research activities of the Federal Institute for Research on Agriculture, Chemistry and Environmental Hygiene in Liebefeld (FAC-L) proved especially helpful in accelerating the ecologization process. First of all, the FAC-L's success was due to the fact that relevant ecological research was concentrated in one institution. Secondly, the FAC-L was an early initiator in combining soil science with fertilizer research. The biological aspects of soil science were considerably enhanced by inclusion of research on soil contamination from heavy metals in sewage sludge. Thus, soil specialists at Liebefeld were equally competent at determining agricultural and non-agricultural causes of soil contamination. This was useful in avoiding either an exclusively "defensive" or an exclusively "offensive" bias. The strategy of balanced research helped the institute to establish a reputation in Switzerland which it had enjoyed for some time in other countries.

   Another factor contributing to the FAC-L's success was its ability to secure a great deal of autonomy in determining its research focuses An important condition for this success was a so its institutional relationship with the BLW. This facilitated access to agricultural interest groups and individual farmers. Finally, the institute was able to combine fruitfully its research activities wit its public service function as a consultant for fertilizer application, sewage sludge and soil. Thus, the FAC-L was able to formulate research projects according to the requirements of agriculture and, at the same time, to disseminate innovative ideas for ecologically sound agricultural practices through its network. This structure was probably responsible for the fact that the ecology-and-agriculture issue was treated as a "techno-scientific problem" without politicizing and creating resistance among Swiss farmers. It also facilitated intra-agricultural assimilation of innovations from the public health and water resources protection authorities.

The French organization of agricultural research has been quite different up to now: here we observe the old, clear-cut distinction between agricultural research and the farmers' consulting institutes. It is interesting to note, however, that this distance between theory and practice has partly been given up, as can be seen 1 the creation of the "Nitrate Mission". This new policy arena (the "Nitrate Mission" and its advisory board the CORPEN), in fact, reassembles representatives of the classical French agricultural research institute (INRA), the consulting institutes ("Instituts techniques agricoles") and civil servants from both the Ministry of the Environment and of Agriculture, and representatives of water agencies, of consumer and environmental organizations. This cooperation will stand a good chance of obtaining an adequate implementation of the measures proposed and a relatively low degree of politization. Ecologization, in this context too, will be dealt with rather as a technical than a political issue. Last but not least, this structure allows ecologization without harming the core dimensions of traditional agricultural policies.

4. It is important that relevant actors are mobilized to initiate and promote the processes of ecologization before institutionalized politicization occurs. In both countries, mobilization as not initiated by national politicians or political parties. There are, however, interesting differences between the two countries: in the Swiss case, "street mobilization", stimulated through mass media, the scientific community, professional organizations and corporate interest groups are important initiating factors. The various issues debated by the public landed directly on the doorstep of the appropriate federal agencies, usually without making detours to political pa ties or Parliament. Federal agencies facing a mobilized public felt committed themselves to take immediate action.

The demand for "better" regulations was frequently reinforced by threats of consumer action such as boycotting lettuce or collectively refusing to pay water bills. Public mobilization was relatively easy to organize because people, as a rule, are highly sensitive when their heath is at stake. Boycotts proved to be an effective and operable strategy, at least in the short run; vegetable growers, for instance were considerably disadvantaged. Swiss federal agencies were, in a way, left holding the bag and had to face public pressure alone. Because these agencies Were often deficient in scientific competence and information, they were more readily inclined to accept external research results and opinions as the bases for developing and supporting new broader ecology-oriented regulations.

In France, we observe an almost complete absence of such "street mobilization". Here, the initiative undoubtedly came from outside. It was the famous EEC directive of 1980, which made things move. At that time, few people thought serious problems would stem from the implementation of this directive, which was mainly perceived as one technical regulation among others. The surprise was considerable When the Ministry of Health discovered that French drinking water quality was far from being in compliance with the requirements of this directive.

Nevertheless, in both countries the dialogue between administrations, researchers, mobilized groups of citizens (Switzerland) or EC bureaucrats was characteristic of the change from the old "Small" regulations to the "medium" ecology-oriented regulations of the 1980s. The absence of traditional political actors (political parties and government) in that process is particularly conspicuous: until shortly before the end of the period under investigation, the official core actors of the traditional agricultural policies were kept out of the line of fire which tended to be directed at the technical responsibility of the FAC-L in Liebefeld in Switzerland and at the INRA or the "Nitrate Mission" in France.

5. Especially under Swiss conditions, "medium" agricultural regulations are a necessary intermediate phase in the ecologization Process. Yet, the longer the time span of change from "small" to "medium" regulations, the greater the regulatory efforts required and the higher the risk of failure The Swiss case of the Lettuce and Vegetables Regulation revealed the longest period of transformation among all cases investigated. It is characterized by many conflicts and significant changes in the position of actors.

   The direct leap from "small" to "big" regulations (skipping the "medium" phase) with compulsory inclusion of relevant parameters of conventional agricultural production regulation did not occur. Even under the condition of seemingly parallel interests (e.g. maintaining livestock production levels and limiting the size of herds) this leap was not possible. Not surprisingly, neither was such a direct transition possible under the condition of openly opposing interests (e.g. drinking water regulations versus increased acreage and crop variety). In the former case, it was not possible to combine directly and compulsorily the ecological and economic reasons for restricting the size of herds. In the latter case, attempts at moderate expansion of the drinking water regulations to include crop regulation failed because of the anticipated opposition of agricultural policy-makers who favoured increased crop farming as one method of reducing livestock production. This observation holds good for the case of France. This is not surprising, because the core dimensions of agricultural policies are decided far away from Paris and blaming "Brussels" for all evils without undertaking unpopular measures at the national level is quite a common pattern in French politics.

6. Finally, it is essential for stabilizing the processes of change in agriculture that the various actors accept "medium" regulations as an important framework, without attacking the main regulatory provisions of conventional agricultural policy. Attempting to "force" the ecologization of conventional agricultural policy would have resulted in vigorous opposition from all the official policy-makers and representatives of agricultural policy. One of the strengths of ecology-minded agricultural researchers in favour of "medium" regulations was that they had gradually gained the confidence of the local farmers. This observation holds good for both France and Switzerland. It is quite likely that in France, when the EEC finally changes its agricultural policies towards ecologically more sound agricultural promotion programmes (mainly for economic reasons), French farmers will be much more reluctant to implement such a reorientation than their Swiss colleagues vis-a-vis similar changes.

7.6 Concluding remarks

The comparison of the three cases of ecologization procedures in France and Switzerland shows – in spite of the differences analysed here – surprising similarities. First of all, the medium-sized ecological regulations for agriculture finally reached have similar extensions with regard to the objects of protection as well as to the recognized causal factors involved in the programmes. Secondly, the contribution shows some empirical evidence for the assumed link between substantial ecologization of agriculture-related programmes and significant changes within the political-administrative structures of the two countries under investigation. Last but not least, there is empirical evidence that such structural changes may often pass through the establishment of new policy arenas rather than through either well established old institutions or newly created bodies exclusively dealing with agriculture, or with public health, or with environmental protection. It is still an open question whether the three-phase model presented here must be limited to the field of ecologization of agriculture or whether it could be applied to the wider field of ecologization of all kinds of policies

References

Annales agronomiques 1974. *Pollution*, special issue

Bundesamt für Gesundheitswesen 1983. *Die Durchführung der Lebensmittelkontrolle in der Schweiz im Jahre 1982*. Bericht der Abteilung Lebensmittelkontrolle, Mitteilungen aus dem Gebiete der Lebensmitteluntersuchung und Hygiene (Berne, BAG) 74: 184

Bundesamt für Gesundheitswesen 1985. *Änderung der Lebensmittelverordnung Art. 15 a*, Entwurf

Bundesamt für Landwirtschaft and the Bundesamt für Umweltschutz 1979. *Wegleitung zum Gewässerschutz in der Landwirtschaft*

Bundesamt für Umweltschutz and Bundesamt für Landwirtschaft 1979/1987. *Wegleitung zum Gewässerschutz in der Landwirtschaft*. Berne, December 1979, 1987

Bundesrat1969. Vierter Bericht des Bundesrates an die Bundesversammlung über die Lage der schweizerischen Landwirtschaft und die Agrarpolitik des Bundes vom 26. Februar 1969, *Bundesblatt 1*: 389 ff.

Communauté Européenne 1983. Perspectives de la PAC, *Livre vert de la Commission*, Notes rapides de l'Europe verte, No. 33, Brussels: 62

Conrad, C., Knoepfel, P., Larrue, C. and Teherani-Krönner, P. 1985. *Environmental Concems in Agricultural Policies: Outlines of some Perspectives, Issues and Policy Concepts in a Comparative Perspective*, paper presented at the Round Table on Agricultural Policy and Environment, Maastricht, 30–31 May 1985

Conrad, J. 1985. *Regulation of Agriculturally Induced Nitrate Contamination of Water in West European Countries.* Berlin: Science Center Berlin (IIUG pre 85–16)

Conrad, J., Gitschel, M. 1986. *Rechtliche Regulierung der Stickstoffdüngung. Nur ein Sturm im Wasserglas*? Berlin: Science Center (IIUG pre 86–9)

FAC-Liebefeld (ed.) 1971. *Umweltprobleme und Landwirtschaft* Bericht über das Symposium am 13–14 Oktober 1971 in Bern unter dem Patronat der Abteilung für Landwirtschaft des EVD und der landwirtschaftlichen Forschungskommission, Liebefeld Bern

FAC-Liebefeld 1975. Tätigkeitsbericht über die Jahre 1973–1974, *Landwirtschaftliches Jahrbuch der Schweiz* 315 ff.

Federal Institute of Technology, Zuich (H. Leibundgut (ed.) 1970. *Schutz unseres Lebensraumes*. Symposium an der Eidgenössischen Technischen Hochschule Zürich vom 10 bis 13 November 1970. Frauenfeld and Stuttgart: Huber, 1971

Forschungszentrum für Schweizerische Politik, Universität Bern 1975. *Schweizerische Politik im Jahre 1974*. Bern: Haupt: 108 ff.

Freiburghaus, D.,Zimmermann, W. 1985. *Wie wird Forschung politisch relevant. Erfahrungen in und mit den Schweizerischen Nationalen Forschungsprogrammen*. Bern: Stuttgart: Haupt

Furrer, O.,Gächter, R. 1972. Der Beitrag der Landwirtschaft zur Eutrophierung der Gewässer, *Schweizerische Zeitschrift für Hydrologie* 34

German Council for Environmental Affairs 1985. *Der Rat von Sachverständigen für Umweltfragen, Umweltprobleme der Landwirtschaft, Sondergutachten*. Stuttgart/Maynz: Kohlhammer

Geschäftsprüfungskommissionen 1983. Forschung im biologischen Landbau. Bericht der Geschäftsprüfungskommissionen an die eidgenössischen Räte über die Inspektionen und Aufsichtseingaben im Jahre 1982 vom 14.4.1983, *Bundesblatt 11*: 395 ff.

Gros, A. 1962. *Guide pratique de la fertilisation*. Paris: La Maison rustique

Gysi, C., Bräm, E.,Wixinger, K. 1981. Überprüfung der Düngungsrichtlinien im Glashaus-Gemüsebau, *Der Gemüsebau* 13: 18ff.

Halbheer, P.,Müdesbacher, A. 1985. *Agrarpalitik – Interessenpolitik?* Bern and Stuttgart: Haupt

Kaufmann, F.-X.,Rosewitz, B. 1983. Typisierung und Klassifikation politischer Massnahmen, in: Mayntz, R. (ed.), *Implementation politischer Programme II. Ansätze zur Theoriebildung*. Opladen: Westdeutscher Verag: 34ff.

Knoepfel, P., arrue, C. 1984. Les politiques publiques comparées: tourisme intelligent ou vrai progrès? le cas des politiques comparées de environnement, *Politiques et management public* 2 (3): 43–63

Knoepfel, P.,Larrue, C. 1985. Distribution spatiale et mise en oeuvre d'une politique publique: le cas de la pollution atmosphérique, *Politique et management public*, 3(2): 46–69

Knoepfel, P.,Weidner, H. 1985. *Luflreinhaltepolitik (stationäre Quellen) im internationalen Vergleich*, 6 vol..Berlin: sigma

Knoepfel, P. 1986. Explaining Differences in the Performance of Clear Air Policies: an International and Interregional Comparative Study, *Policy and Politics* 14 (1): 71–92

Knoepfel, P., Zmmermann, W. 1986. Oekologisierung von Landwirtschaft, *Schweizerische Landwirtschaftliche Forschung* 25(2): 196ff.

Knoepfel, P.,§immermann, W. 1987. *Oekologisierung von Landwirtschaft, fünf Geschichten und eine Analyse.* Aarau, Frankfurt and Salzburg: Sauerländer

Knoepfel, P.,Zimmermann, W. 1987. *Umwelt und Landwirtschaft im politischen Alltag. Drei Fälle für die Schule*, im Auftrag des Bundesamtes für Forstwesen, des Bundesamtes für Umweltschutz, des Bundesamtes für Raumplanung und des Bundesamtes für Landwirtschaft, Bundesamt für Umweltschutz.Bern: EDMZ

Knoepfel, P., Zimmermann, W. 1987. Nitrat war kein Pfad. Von der ausgebliebenen Oekologisierung der Landwirtschaft auf den Weg über die Debatte zum Nitrat im Trinkwasser in der Schweiz, in: Conrad, J. (ed.), *Nitrat und Landwirtschaft.* Hamburg: Campus

Lehmbruch, G. 1979. *Parteiensystem und Interessenverbände in der Politikentwicklung.* Konstanz: Universität Konstanz, Fachbereich für Politische Wissenschaft. Diskussionsbeitrag No. 6/79

Lehmbruch, G. 1984. Interorganisatorische Verflechtung im Neokorporatismus, in: Falter, J. W. et al. (eds), *Politische Willensbildung und Interessenvermittlung.* Opladen: Westdeutscher Verlag

Ministére de l'agriculture and Ministère de l'environnement 1980. *Rapport du groupe de travail "Activités agricoles et qualité des eaux"*, 2 vols. Paris, October 1980

Ministères de l'agriculture et de l'environnement 1980. *Rapport du groupe de travail "Activités agricoles et qualité des eaux"*, 2 vol. Paris, October 1980

Ministères de l'agriculture et de l'environnement 1986. *Amélioration des pratiques agricoles pour réduire les pertes de nitrate vers les eaux*, edited by the Groupe de travail "Pratiques agricoles" du COPREN, Paris.

Ministères de l'agriculture et de l'environnement 1986. *Les nitrates dans l'eau, un défi relevé*, edited by the Mission Eau Nitrates. Paris

Parlement Européen 1986. *Rapport sur "agriculture et environnement"* Strasbourg, 3 February 1986 (Doc. a 2–207/85)

Périgaud, S. 1977. *L'agriculture et la pollution des nappes.* Orléans: INRA (Collection des Bureaux de recherches géologiques et minières, No. 1): 155–161

Ramade, F. 1977. *Ecotoxicologie.*, Paris: Masson

Teherani-Krönner, P. 1985. *Tauziehen um eine agrarumweltpolitische Lösung, Die Gülleverordnung von Nordrhein-Westfalen.* Berlin: Science Center (IIUG pre 85–8)

Windhoff-Héritier, A. 1985. *Politikanalysen und Policy-Netz. Zum analytischen Nutzen zweier Begriffe, Politische Prozesse und Strukturen.* Berlin: Science Center (IIVG/dp83–212)

Annex 1

Abbreviations used in this article

Administrative bodies and terms

 BAG BLW BUS	Bundesamt für: - Gesundheitswesen - Landwirtschaft - Umweltschutz	Federal Office of: - Public Health - Agriculture - Environmental Protection
AfU	Amt für Umwelt-schutz(pre-1978)	Agency for Environ-mental Protection
CORPEN	Comité d'orientation pour la réduction de la pollution des eaux par les nitrates et phosphates provenant des activités agricoles	Guiding Committee for the Reduction of Water Pollution by Nitrates and Phosphates due to Agri-cultural Activities
RIC	Réglementation des in-stallations classées	Rules for Classified In-stallations
RSDT	Règlement sanitaire dé-partemental type	Standard Departmental Health Regulation
RSD	Règlement sanitaire dé-partemental	Departmental Health Regulation

Research and higher education bodies

EAWAG	Eidgenössische Anstalt für Wasserversorgung, Abwasserreinigung und Gewässerschutz, ETH, Zürich	Water Research Division
ETH	Eidgenössische Technische Hochschule, Zürich	Federal Institute of Technology
	Eidgenössische Forschungsanstalt für:	*Federal Institute for Reseqrch on:*
FAC-L	- Agricultur-Chemie und Umwelthygiene, Liebefeld (Berne)	- Agriculture, Chemistry and Environmental Hygiene
FAW	- Obst- Wein- und Gartenbau, Wädenswil (Thurgau)	- Arboriculture, Viticulture and Horticulture
FAR	Eidgenössische Agronomische Forschungsanstalt, Reckenholz (Zürich)	Federal Institute for Agronomic Research
RAC	Station Fédérale de Recherches Agronomiques de Changins, Nyons (Vaud)	Federal Agronomic Research Institute of Changins
INRA	Institut National de la Recherche National Agronomique, Paris	Institute for Agronomic Research
INSERM	Institut national de la Santé et de la Recherche Médicale, Paris	National Institute of Health and Medical Research

Associations

SBV	Schweizerscher Bauern-verband	Association of Swiss Farmers
SGU/VSGP	Schweizer Gemüseunion/ Verband Schweizerischer Gemüse-Produzenten	Association of Swiss Vegetable Growers

Section III:
Emergence of New Analytical Concepts

Policy analysis aims to formulate and empirically test generally valid hypotheses on specific actor constellations characterized by different resource portfolios. These portfolios are attributed to and activated by these groups according to institutional rules allowing or prohibiting them the use of these resources and leading to significantly different identifiable and interlinked products of policy cycles in a given field. Although the results of such research are normally considered valid only for the policy under consideration, scholars have tried since the emergence of policy analysis in the late 1960s to generalize their findings across policy fields. It is still a controversial question whether this generalizing of specific hypotheses is possible and whether policy analysis is therefore capable of building a general theory on the functioning of public policies, at least within the framework of European democracies. Nevertheless, there exists a body of commonly recognized analytical concepts that allows for the answering of relevant questions across policies. Most of these concepts stem from empirical observations made in the field of rather attractive "modern", politically "high ranking" policies such as education, public health, land-use planning, new social policies and environmental protection. This latter policy must be considered especially as one of the main laboratories for the development of such analytical concepts. In this section we present four of these concepts originally developed to better understand the mechanisms of environmental policies that have successfully become general concepts applied in teachings, doctoral theses and practical policy making.

Policy analysts and politicians today still discuss distributional issues usually only in the context of re-distributional policies such as social, fiscal and educational policies. In *Distributional Issues in Regulatory Policy Implementation – the Case of Air Quality Control Policies (1986)*, we demonstrate, on the basis of empirical results of an international comparative study of SO_2 air quality policies, that purely regulative policies working by means of demand and control instruments produce important redistributional effects. Indeed, empirical data show varying models of interventionist priorities in space and time leaving areas controlled or not controlled according to implicit or explicit action plans that favour or disfavour them. The explanation of these discrimination patterns might be political pressure or a specific temporal and spatial distribution of problem pressure. The interesting observation, now generalized for all regulatory policies, is that these two explanatory variables must not coincidence. Hence empirical data may show concentrations of administrative activity in areas where objective problem pressure is low and little activity in highly problematic areas. This distributional element has nothing to do with varying allocation of public money resources. What are at stake the distribution patterns of "public attention", which we can also find in other areas such as public security or public health? Today we use this concept for analyzing social aspects of sustainable development.

A Policy-Monitoring Concept for Promoting the Conservation of Biodiversity (2001) proposes a concept initially developed within the framework of a Swiss National Research Priority Programme dedicated to the environment. It since has been further developed in the direction of systematic computer-based monitoring of outputs and effects of nearly all space-related public policies to be introduced into geographical information systems based on land registries currently under development in Switzerland (registry of public law restrictions of land property according to the project for a new federal law on geoinformation of 2007). Policy monitoring on the basis of systematic observations of outputs and effects of public policies at the level of each parcel is a highly useful complementary element to the normally natural sciences-based concepts developed for monitoring the state of the environment. It is particularly useful for policy evaluation.

Environmental Capacity Building in Switzerland (1997) presents the application of a rather normative analytical concept developed by my colleagues Helmut Weidner and Martin Jänicke who also applied it to many other countries. Today, similar concepts of capacity building are used in other policy fields like health policy, all kinds of social policy as well as security policy. The concept becomes particularly helpful in the field of international comparative policy research and it has the advantage of considering, aside from state agencies, many economic and societal actors contributing significantly to the capacity of the society to move towards sustainable development.

Again with the aim of analyzing the ambivalent role of a specific actor within environmental policies of modern post-industrial societies, *Eco-business and the State – Analyses and Scenarios (1996)* developed a concept that goes far beyond environmental policy. In fact, ecobusiness is nothing more than an actor who profits from promoting public policies without being directly targeted (such as polluters) or considered as an explicit beneficiary group of the policy (such as people suffering from pollution). When looking at the mechanism of most public policies today we always must look at such business actors profiting from effective implementation of enhanced public action such as the building industry (waste treatment plants, noise abatement measures, housing, and infrastructural facility policies), social workers (social policies), increasingly internationally organized accountancy firms (economic regulation), or the safety industry (security policies). We even might suppose that such "policy-business" actors' strategies are more important for explaining policy changes than the strategies of directly implied actor groups.

8
Distributional Issues in Regulatory Policy Implementation – the Case of Air Quality Control Policies (1986)

Peter Knoepfel[1]

8.1 Distributional issues in regulator policies – introductory overview

The clear and widely accepted distinction between regulatory, distributive and redistributive policies with regard to programme structures, policy instruments and administrative arrangements, originally suggested by Lowi (1972)[2] might be the main theoretical reason for the fact that the distribution issue has only exceptionally[3] been analysed explicitly in regulatory policy formation and implementation processes. Without directly raising his issue, a considerable volume of environmental policy research illuminates the existence of selective interest consideration in regulatory programme formation patterns[4]. The very existence or absence of public

---

[1] in : A. Schnaiberg, N. Watts, K. Zimmermann (eds.) : Distributional Conflicts in Environmental Policy, Hants GB, (Gower), 1986: Distributional Issues 363–379 Copyright: We thank the editor for authorizing the publication of this article in this book.

[2] Lowi, T.J. (1972), Four systems of policy, politics and choice, *Public Administration Review*, no. 32, Italy/August: 298–310.

[3] See the following contributions of economists: Jarre J. and Zimmermann, K. Konflikte zwischen Umweltpolitik und Wettbewerbsordnung, papers from the IIES, 81–16; Zimmermann K. Die Inzidenz der Umweltpolitik in theoretischer und empirischer Sicht, IIUG pre 1983–18, as weII as from the same author: Ansatzpunkte einer verteilungsorientierten Umweltpolitik, IIUG – pre 83–2 and: Hartje V.J. Liability versus Regulation, IIES discussion papers no. 81–16. From a more sociological point of view: Schnaiberg, A. 1986. Saving the Environment: From Whom, for Whom and By Whom? A critical American perspective, in: Knoepfel, P. and Watts, N.(eds), Environment Politics an Policies, Harper and Row, London, and Majone, G. Choice among Policy Instruments for Pollution Control, Policy analysis, no. 4 1976, pp. 589–612.

[4] See e.g. White, L.J. U.S. Mobile Source Emissions Regulation: The Problems of Implementation, in: Downing, P. and Hanf, K. (1983), International Compari-

regulation in comparable problem areas can reflect the exercise of selective attention in issue generation processes, resulting in an unequal distribution of public regulation resources amongst competing social groups. Furthermore it is well known that once established, regulatory policies may discriminate against certain enterprises or affected individuals. This again reflects implicit distributional decisions.

Such implicit distributional decisions can be readily found in environmental quality standard and in emission or technological process standards. Whereas the former may even explicitly exclude certain groups of individuals (pregnant women, children, elderly people, the handicapped etc.) from the scope of protection[5], the latter can lead to an attribution of preferred competitive positions to specific producers within politically defined markets. The setting of technical norms and standards for the protection of environmental quality workplace safety or the avoidance of other risks runs the risk of corporative, cartelistic market regulation which will exclude those who cannot meet the standards. Not infrequently such excluded producers are likely to offer more advanced technologies the application of which might lead to more protective effects. By means of capturing state or cartelistic private norm-setting institutions, large and economically influential producers of environmentally relevant products try to block innovations which might lead to more sophisticated but also more expensive standards for technologies. Such corporative standard setting processes of governmental agencies must be considered as economically relevant distributional activities even if they normally lead to regulatory policy programmes.

In addition to these distributional issues in the processes of policy generation and formation, implementation research has raised the following three further important distributional aspect of regulatory policies, of which the first two are immediately related to the position of regulatees. Environmental policy implementation activities are normally accompanied by the delivery of a public service, which consists of technological advice given to the operator of emitting industrial plant. This component might be

---

sons in Implementing Pollution Laws, Boston: Kluwer, Nijhoff, pp. 46–62 and especially Krupnick, A., Magat, W.A. and Harrington, W. Revealed Rules for Regulatory Decisions: An Empirical Analysis of EPA Rule Making Behaviour, in: Downing and Hanf (op. cit.), pp. 63–84.

[5] See Weidner, H. and Knoepfel, P. 1979. Politisierung technischer Werte-Schwierigkeiten des Normbildungsprozesses an einem Beispiel (Luftreinhaltung) der Umweltpolitik, in: Zeitschrift für Parlamentsfragen , p. 160–70. The new Swiss Federal Environmental Protection Act of 7 October 1983 explicitly includes these persons within the scope of protection in art. 13 s. 2.

more (British Alkali Inspectorate)[6] or less (the Italian committees for air pollution control CRIA'S)[7] important; the distributive character of this service delivery becomes evident whenever there is unequal treatment of individual firms or even branches. Furthermore the "delivery" of a legally binding permit can be viewed as an attribution of specific legal positions allowing polluting activities whenever emissions remain below the threshold fixed in the related permit conditions. Permit delivery based upon regulatory programmes can therefore also be considered as a distribution of rights of access to environmental resources (air, water, soil) amongst different economic subjects[8]. Whoever operates his plants in compliance with permit conditions is legally assumed to be protected against legal action for damages. The advantages of the resulting reversal of burden of proof are attributed by governmental regulation. This distributional issue contained in implementation of regulatory policies becomes evident if we compare the legal position of emitters subjected or not subjected to regulatory activities. In those cases, at least, where the threshold of permitted emissions is set at a relatively high level, the position of regulated emitters is much better than that of emitters whose activities are outside of governmental control. This comparison can be made in the case of Italy where the anti-smog legislation only applies to the highly polluted metropolitan areas included into the control zones A and B whereas emitters located outside of these zones are not subjected to this regulation. Thus the latter do not have access to individual permits guaranteeing their protection against claims for damages, the distribution these permits being limited to industries located within zones A and B[9].

If we look more closely at the regulation of individual industries we can again observe considerable differences with regard to the preciseness of the content of permits. The better pollution can be measured, the more large emitters in particular will try to achieve precisely defined permit

[6] See Hill, M. The Role of the British Alkali and Clean Air Inspectorate in Air Pollution Control, in: Downing and Hanf (op. cit.) fn. 3, pp. 87–106.

[7] Only in the case of the Lombardy CRIA does there exist enough technical skill to actually consult emitters. See: Dente, B. and Lewanski, R. Implementing Air Pollution Control in Italy: The Importance of the Political and Administrative Structure, in Downing and Hanf (op. cit.) fn. 3, pp. 107–28.

[8] See e.g. Hartje (op. cit.) fn. 2, pp. 5–10.

[9] Such plants are only subjected to the old sanitary regulation (T. U. July 27, 1934 – N. 1265); there do exist a lot of old plants which are in operation without any permit. In these cases especially penal court suits according to arts 636, 674, 734 and 650 of the Italian Penal Code might lead to drastic interqentions. See: Knoepfel, P. and Weidner, H. 1980. Handbuch der SO_2-Luftreinhaltepolitik, Teil II: Länderberichte, Berlin: E. Schmidt: 576f.

conditions, which allow them to take long range planning decisions and thus provide investment security. Again it seems that some strong and important individual emitters or branch corporations achieve more in bargaining with regulatory agencies than others. Preciseness of permits thus becomes an important criterion, still applied too infrequently in the evaluation of distributional effects in permit granting by public agencies after bargaining processes.

The last and by far the most interesting distributional aspect of regulatory policies is (suggested by the observation) that administrative attention for environmental concerns of different local areas may vary considerably. In our comparative research in the field of air pollution we could determine specific patterns of distribution of administrative attention within regional implementation systems. Such unequal distribution of regulatory attention as presented in the next paragraphs will normally also lead to concomitant unequal treatment of emitters. The main interest of the analyses, however, will be in the differences with regard to varying life quality of inhabitants of different local areas due to different levels of pollution control activities by one and the same administrative agency. Conceptually and empirically the following presentation bases on the comparative project Helmut Weidner and I conducted together with national research teams in the field of SO_2 air quality control policies in selected European countries. The complete data as well as the comparative interpretation are presented in the final report to the German National Science Foundation[10]. To save space the following presentation is limited to examples from the Italian[11] and the French[12] findings.

8.2 Distribution through implementation of air quality control policies in Italy and France

This research analysed implementation activities of three regional agencies within two or three local implementation areas (LIAS). The results showed

[10] Knoepfel, P. and Weidner H. 1984. Schlussbericht zum Projekt: International vergleichende Analyse von Implementationsprozessen in der SO_2-Luftreinhaltepolitik in ausgewählten EG-Staaten. Berlin. (mimeo).

[11] See: Dente, B., Knoepfel, P., Lewanski, R., Mannozzi, S. and Tozzi, S. 1985. Analisi di una politica regolativa: Politica contro l'inquinamento atmosferico. Roma: Ufficina. (forthcoming)

[12] See: Darbera, R., Larrue, C. and de Magalhaes, H. 1982. The Implementation of SO_2 Air Quality Control Policy in France, final research report, Paris/Berlin. (mimeo).

that quality and quantity of regional administrative outputs (permits, clean-up orders and control activities) affecting comparable samples of emitter within the selected local areas of the same region did vary considerably.

Table 1: Regional implementation systems of France: variables affecting spatial distribution of regional activities 1970–80

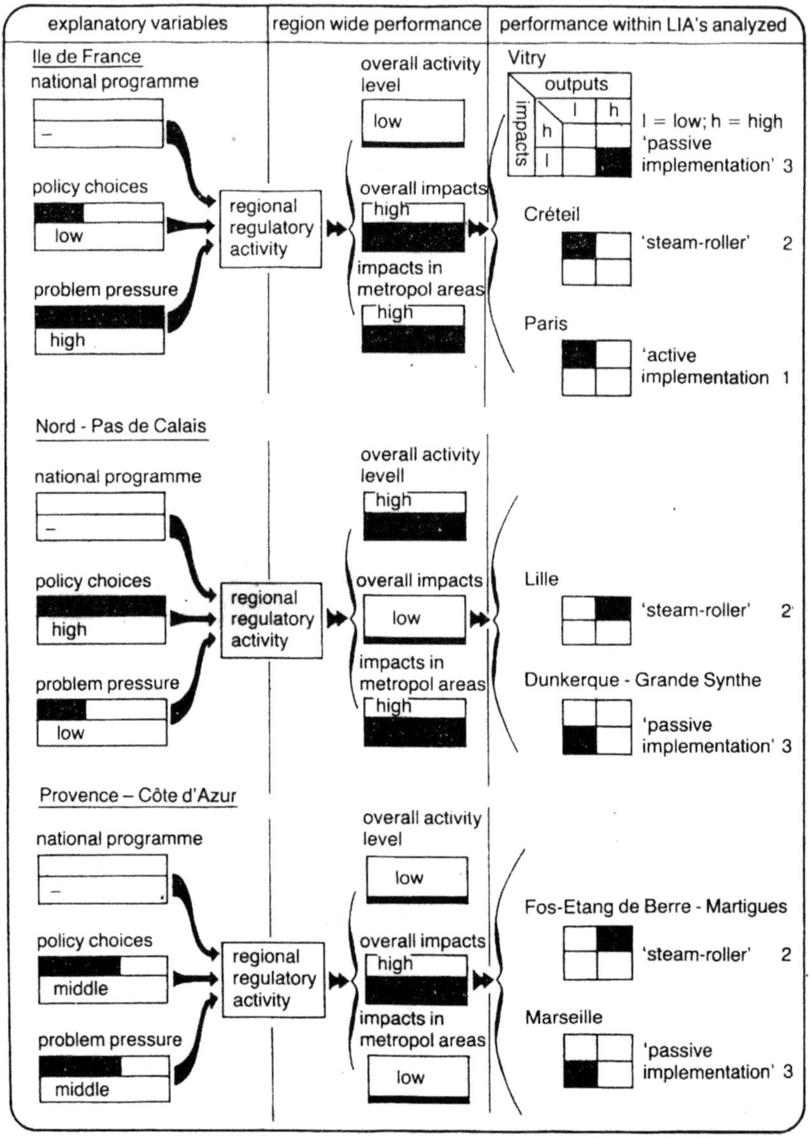

Source: P. Knoepfel, H. Weidner: Final report ... 1984 (fn. 9)

Table 1: Regional implementation systems of Italy: variables affecting spatial distribution of regional activities 1970–80

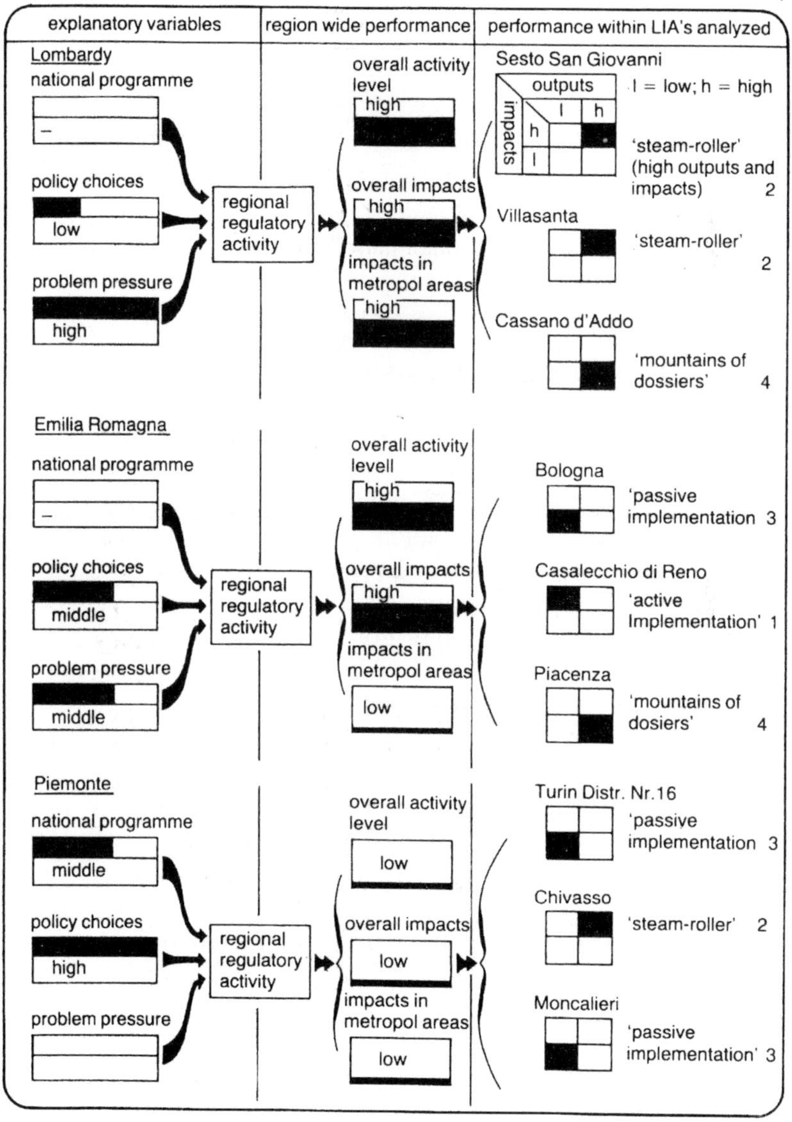

Source: P. Knoepfel, H. Weidner: Final report ... 1984 (fn. 9)

The same is true for the impacts of these outputs. Relatively independent of the composition of the emitters within these LIAs we could attribute the implementation activities observed in each LIA to one of the following four types. (see tables 1 and 2)

1. active implementation policy (low number of outputs/high impacts);
2. implementation through the steam roller (high number of outputs/high impacts);
3. passive implementation policy (low number of outputs/low impacts);
4. implementation through mountains of dossiers (high number of outputs/low impacts).

We tried to explain these variances amongst LIAS of the same regions, reflecting a specific distributional pattern of administrative attention, by the following variables:

8.2.1 Varying immission situations

French and Italian air quality control policies are strongly immission oriented; according to this philosophy it would be reasonable to concentrate regulatory attention primarily on highly polluted and densely populated areas. Spatially selective distribution of regulatory attention could therefore reasonably be expected in response to different degrees of problem pressure as indicated by pollution levels. But this expectation is only fulfilled in two out of our six regions. For the distributional patterns found in these two regions (Ile de France in France and Lombardy in Italy) variations in problem pressure (expressed in different local immission data trends) did have a strong explanatory capacity. Surprisingly, this variable was much less important than initially expected for the explanation of distributional patterns in the other four regions. Here, we found either very limited regulatory activities in typical “problem areas” and/or allocation of considerable administrative resources in areas with relatively low pollution levels. The explanation of such discrepancies between the measurable distribution of problem pressure and the distribution of regulatory activities had to be explained by one of the two following variables – the programme and implementation policy.

8.2.2 The programme

Confirming results of other implementation studies[13] the explanatory capacity of national pollution control programmes for actual regional implementation activities as a whole and – a fortiori – for their spatial distribution turned out to be relatively low. As indicated by tables 1 and 2, amongst the six regions the national programme played only a limited role in the case of the one Italian region of Piemont. For all other regions this programme only accounted for general initiation of, and legal support for implementation policies as a whole; they never systematically guided policy choices by setting specific distributional priorities. However, if we were to include in the comparison the Western German or Dutch findings the relative position of the national programmes would turn out to be much stronger. For the purposes of this brief theoretical presentation such an inclusion is not necessary. In the one case of the region of Piemont, "missallocation" (viewed from a problem pressure perspective) of administrative resources was directly attribuable to a highly legalistic and literal application of the Italian pollution control programme itself by an understaffed regional agency. The actual pattern of distribution of the scarce administrative capacities is an immediate consequence of the agency's passive and waiting role. Permits and especially clean-up orders are thus treated incrementally and decided upon according to a schedule determined by the time of arrival of emitters' demands for permits etc. This result is programmed by the Italian anti-smog legislation itself, which commits every industrial emitter of the whole region located in zone A or B to ask for a permit at the same point of time. Thus in 1972/73 the agency was overwhelmed by mountains of demands which according to the legalistic rationality of the Piemont agency had to be answered and therefore treated without substantial problem-related investigations. No specific priorities have been set for this initial phase of implementation such as there were in the case in the other two Italian regions exposed to the same legal requirements. Thus, the programme's procedural arrangements hampered the application of its own philosophy favouring immission-oriented strategies. (Similar restrictions to active implementation policies could be found especially in some Western German regions where an extremely literal "execution" of the Federal Air Quality Control Act also led to considerable spatial misallocations of regulatory resources by the Bundesländer. Here again, the reason was a highly

[13] See the excellent presentation of the discussion by Barret, S. and Fudge, C. 1982. Examining the policy-action relationship, in: Barret, S. and Fudge, C. (eds), Policy and Action, Essays on the implementation of Public Policy, Methuen, London and New York: pp. 3–42. See also chapter 4 of this book.

developed legalistic rationality finally leading to a formal equality of the treatment of emitters, independent of their substantial contribution to local immission situations[14]. Nevertheless, this spatial misallocation of regional regulatory attention did not depend essentially on this programme dimension. As the tables indicate, by far the most important variable is the following one.

For all regional implementation systems analysed in the two countries, implementation policy choices do have at least a low explanatory capacity for patterns of spatial distribution of regulatory attention They may (Lombardy, Ile de France but to a lesser extent also Emilia Romagna and Provence-Côte d'Azur) or may not (Nord-Pas de Calais, Piemont) enhance problem-related distributional patterns, in either case, policy choices resulting from bargaining processes affect the way in which regional interventions actually regulate local industries. The next paragraph will focus upon some specific aspects of such policy generated distributional patterns in air quality control policies. These implementation policy choices actually present the best example to illuminate the importance of the distribution issue in regulatory policies.

8.3 Implementation policy as a quasi distributional policy

Regional implementation policy includes all decisions on individual permits, clean-up orders, control activities or enforcement orders as well as eventual collective outputs such as the creation of special protection zones or collective alarm systems. Related policy choices result out of negotiation processes between representatives of the (formally competent) regional air quality control offices, the local authorities (mainly representatives of local public health agencies and/or of local political systems) and the individual emitters or their interest organisations. In Italy and France especially, other actors might enter into these processes: environmental protection organisations, individuals affected by pollution, courts, other state agencies and – in the case of France – the central government.

The interaction network is prestructured by procedural and organisational arrangements already partially laid down in the national pro-

[14] See: Knoepfel, P. and Weidner, H. 1983. Die Durchsetzbarkeit planerischer Ziele auf dem Gebiet der Luftreinhaltung aus der Sicht der Politikwissenschaft. Ergebnisse aus einer intemationalen Vergleichsuntersuchung, in: Zeitschrift für Umweltpolitik, No. 2: 87–117.

grammes. But in both countries we find modifications of or even alternatives to these nationally defined arrangements which either simply reflect practical requirements (Italy)[15] or specific local emitter and power structures (France)[16]. Furthermore, the national programme partly defines those actors who are to initiate interaction processes. Thus, the Italian legislation – as mentioned – attributes an essential initiating role to the emitters by its requirements that each operator of an industrial plant should ask for a new permit independent of whether or not the firm was already in operation. On the other hand, the same legislation obliges regional agencies to intervene by opening real clean-up procedures whenever the "industrial immission standards" are exceeded[17]. The French legislation on the contrary does not provide for such a systematic initial action to be taken by each firm. Here, the discretion to initiate clean-up activities lies much more with the competent regional agency (Direction Interdépartementale de l'Industrie – DII).

Independent of the different legal definitions (which do not attribute formal decision making power to local government in either country) variances in the actual role of local authorities in initiating and influencing these interaction processes are considerable. The behaviour of local governments within the interaction networks analysed turned out to be by far the most important factor affecting regional patterns of distribution of regulatory attention. As we will demonstrate below, this observation holds for both problem and non-problem-oriented distribution patterns as well as for regions with some local areas relatively "overregulated" or, on the contrary, neglected.

The position of the regional environmental agencies (which formally have jurisdiction over industrial environmental regulation) varies considerably in both countries within interaction networks. Again, those variations are relatively independent of formal prescriptions in the related legislations. The strongest positions within these networks are those held by the regional agencies of Ile de France and Lombardy. Both administrative bodies are comparably well staffed and relatively resistant to political pressures challenging their own (problem-oriented) priorities. On the other hand these agencies are flexible enough to make productive use of local governments' or other actors' policy initiatives to the benefit of their own

[15] The clean-up procedure as provided by the national anti-smog law of 1966 is too complex and therefore not operational. See: Dente and Lewanski (op. cit.) ln. 6, p. 108.

[16] See for the cases of Fos-Etang de Berre-Martigues and Dunkerque: Darbera, Larrue and Magalhaes (op. cit.) fn. 11.

[17] This 'industrial immission standard' is again a very strange and hardly manageable instrument. See: Dente and Lewanski (op. cit.) fn. 6, p. 108.

regulatory priorities. Thus it must be stressed that the relative success of regional activities within the two systems mentioned does not rely merely on the existence of a strong regional agency.

The decisive element for the Italian region is the good functioning of an integrated network between local and regional agencies which are not completely locally dominated. As Dente and Lewanski point out, in the reverse cases ("inequality of power ... to the advantage of the local offices") only strengthening of the position of local authorities runs the risk of destroying these interaction networks: "Once it becomes clear that the regional administration is unable to solve all the problems raised locally, the municipal administration will give up the idea of a joint effort and will concentrate its own effort either on the most important polluters or on the problems it feels able to solve alone (...). The consequences (...) for (...) control processes are quite clear. A generally weak network will not only prove highly ineffective in terms of the quality of its outputs but will also tend to adopt a formal/authoritarian pattern of operation and a case by case approach. At the other end of the continuum a strong network will generate processes characterised by a systematic pattern and strong bargaining with the industrial plants. In this case both the municipality and the regional offices will have direct access to the controlled firms, while in the case of a weak network virtually all contacts will be made through letters and official documents"[18].

Even if for constitutional and political reasons the weight of local governments within these regional networks in France is generally much more limited than in Italy, we nevertheless find comparable variances in power distribution between local and regional actors within our three French regions. The role of local governments in both the northern (Nord-Pas de Calais) and the southern region (Provence-Côte d'Azur) is considerably stronger than in the top-heavy and relatively centralised central region containing the city of Paris and the surrounding departments (Ile de France). The resulting patterns of distribution of regulatory attention amongst different local areas turned out to be the same: a relatively weak position of the regional agency led to an increasing importance of political choices in implementation policies compared to the problem pressure and/or the programmed dimension.

In both countries environmental regulation by weak regional agencies tends to depend largely on whether local governments ask for such activities. Problem related priorities, developed for the whole territory of the regions, which might guarantee a certain equivalence of environmental

[18] Dente and Lewanski (op. cit.) fn. 7, pp. 124f.

quality amongst different local areas within the region were basically missing during the 1970s. The precondition for a balanced distribution of regional regulatory attention amongst local areas thus consists of a balanced interaction network which avoids discrepancies between spatial distribution of regulatory attention and the actual spatial distribution of problem pressure whether based in policy choice or allegedly legitimized by the programme. Such conditions have been met in the cases of Lombardy and the Ile de France, although the two systems represent different types of interaction networks shaping implementation policy choices.

8.3.1 The Lombardy-type

To exemplify the operation of this type we can take the case of the LIA Sesto San Giovanni. This municipality is located in the heavily polluted metropolitan area of Milan. Owing to its highly active local health office, which gained strong support from political representatives of the municipality, local government was able to establish a complete emitter inventory, to identify problematic emitters and to attract regional regulatory attention by transmitting this information to the regional agency at the beginning of regional implementation activities. This agency was also autonomous enough to decide not to intervene in all cases but to concentrate its activities on the most heavily polluting plants (iron and steel industry) which in their turn should have been cleaned up all over the metropolitan area of Milan. In this way technological skills of the regional agency which had to be exercised in one industrial branch of Sesto San Giovanni could also be utilised for other highly polluted areas. By setting this priority the regional agency (CRIAL=Comitato Regionale Contro l'Inquinamento Atmosferico Lombardo) did not wholly fulfil local governments' requests for intervention but the success of these demands set a precedent when entering into negotiations with iron and steel industries within the municipality and elsewhere in the metropolitan area of Milan. A complete clean-up activity for the single LIA of Sesto San Giovanni would have absorbed too much of the capacities of the regional agency, to the disadvantage of other equally polluted local areas. One important feature of this "Lombardy-type" is the fact the Lombardian Regional Agency created and maintained relatively constant interaction networks with each local government of the most polluted areas. These networks became manageable due especially to their separation from each other and the quasi monopolistic position of the strong and well-equipped regional agency, simultaneously present within all of them. Within this system local policy demands served as an excellent information basis and as political support for selective problem-oriented interventions of regional agencies within local areas. This well integrated network by the way was the precondition for the partial reorientation of the agency's strategy from an immission towards an emission control philosophy[19].

8.3.2 The Ile de France-type

This top-heavy system succeeded in reconciling distributional implementation policy choices with distribution of problem pressure by means of the creation of a regionwide interaction network. Within this arrangement local government as well as other local actors play only a marginal role. The network is characterised by an institutionalised co-operation of the competent regional agency with corporate interest groups (especially the association of industries' representing the four important power plants of Eléctricité de France (EDF) as well as about 15 other large emitters) and the national Ministry of the Environment. This arrangement was due to the ministry's consideration of the metropolitan area of the French capital at that time (1974–81) as a pilot region for a highly flexible implementation of the national air pollution law (1961) and the related decrees (especially that of 1974). This corporative and region wide approach (including 8 departments) enabled the regional agency to maintain a powerful position within the newly created network and to depoliticise air pollution control policy by introducing a technological air quality management for the whole region, for which the agency finally achieved agreement from the main actors. The result was the creation of a regional alarm system and the extension and improvement of the old special protection zone of the city of Paris to the four surrounding departments (petite couronne). The introduction of this policy technology by means of superimposing a collective regionwide control system did not leave local or departmental governments important leeway for their own policy decisions concerning individual plants, although according to the industrial environmental regulation law departmental prefects would have been competent. Thus, policy demands for regional intervention to be subject to the consent of all important actors[20] have been avoided by means of the establishment of a technocratical

---

[19] See: Dente B., and Lewanski R. (op. cit.) fn. 7, p. 115.

[20] Darbera, Larrue and Magalhaes (fn. 12) point out that there has been a conflict between the regional and the departmental level in the case of the police prefecture of Paris (pp. 155ff.); the provision for inclusion of (weak) local governments into the networks could not be realised at that time because of the different political colour of national government and the majority of municipalities.

managerial system which allows a problem-oriented distribution of administrative attention amongst local areas. Therefore, implementation policy choices lead to decisions partly contrasting with the problem distribution. Such "misallocation" of regional regulatory attention is primarily due to the weakness of regional agencies within the relevant interaction networks and/or due to concomitant strong positions of local actors. We can distinguish between the following two constellations.

8.3.3 The powerful local brakemen-type

Such situations are well known by scholars of implementation research[21]. The brakeman role in our case is played by industrial associations especially formed to deal with air pollution control policy (Marseille and Dunkerque) or by single large emitters (the ENEL power plant in Piacenza). The success of such brakemen, in the case of Marseille, particularly can only be explained by their support by local government itself. Against the powerful mayor of Marseille (Deferre), who held and continues to hold an important position in French national politics, neither the competent prefecture nor the politically weak technical regional agency (DII) were successful in requesting the creation of a special protection zone in the area of Marseille. Although in second place amongst French cities with regard to air pollution, the economically weak and normally old-fashioned emitters of this city hardly became regulated during the seventies. Local government's protection of the mainly small and medium-sized firms (iron and steel), metal, chemical and food industries) could only be overcome in 1981 with the reaction of a – relatively weak – special protection zone accompanied by an alarm system for large emitters. Instead of intervening in the politically difficult Marseille area the regional agency preferred to allocate the bulk of its resources to the much less populated area of Fos-Etang de Berre-Martigues. A similar though less extreme pattern could be found in the case of the privileged treatment of the well organised large emitters of Dunkerque compared to the generally smaller industrial plants of the city of Lille. The Dunkerque brakemen succeeded in their opposition to the creation of a special protection zone at the beginning of the seventies due to their strong relations with national government. Contrary to the smaller firms of Lille the old industrial pool of Dunkerque is considered as of national interest. The strong brakeman position of the Turin automobile industry (FIAT) against regional activities in

[21] See: Bardach, E. 1977. The Implementation Game: What happens after a Bill Becomes a Law? Cambridge, Mass., London: MIT-Press.

this most polluted northern Italian city must be viewed against the background of attempts by his industry to promote self-regulation partly in coalition with the Turin health office, even though the relationship with the regional regulatory agency at that time was relatively conflictual.

8.3.4 The powerful local mobiliser-type

This type is also discussed in implementation theory[22]. It is especially well known in the field of distributional public policies[23]. Highly active local mobilisers which succeeded in attracting regulatory attention of regional agencies for emitters operating in their localities could be found in the case of LIAS Bologna, Lille, Fos-Etang de Berre-Martigues and Chivasso. As already mentioned, such local activities are not only characterised by an improvement of the situation of these LIAS but they simultaneously lead to a concomitant neglect of other areas which are equally or even more polluted. This "zero-sum-constellation" reflects the weak position of the competent regional agencies condemned to simple react to local initiatives. Probably the most extreme case is that of the LIA Chivasso which, in spite of its relatively favourable immissions situation (compared to the LIA Moncalieri for example), was able owing to a political mobilisation campaign, to attract a considerably higher portion of regulatory attention for its emitters than were the more polluted metropolitan areas, "district Nr. 16 of Turin" and Moncalieri. The same is true for the political mobilisation campaign of the three southern French municipalities Fos-Etang de Berre-Martigues against large industrial emitters of national interest located in their industrial zones to the disadvantage of the inhabitants of Marseille.

8.4 Conclusions

On the basis of selected empirical data from two out of five comparatively analysed countries we can briefly summarise our findings concerning distributional issues in regulatory policy implementation in the case of air quality control policies in the following four points:

1. Besides regulatory programme formation, regulatory programme implementation processes especially actually deal with distribution issues.

[22] See: Bardach (op. cit., fn. 21).

[23] This is the basic issue examined in a large-scale international comparative project in the field of social policy, directed by D. Ashford, in which the author is participating with an analysis of Switzerland.

In our case the most important distributional issue lies in the way scarce regional regulatory resources are distributed among different local areas. The cross-national comparative approach proves especially useful in detecting similar patterns of distribution of such resources throughout regions belonging to different countries. Furthermore the comparison is able to provide similar explanations for such patterns. This demonstrates the relatively marginal role of national programme in these distributional patterns[24].

2. Surprisingly enough, even in countries with a strongly immission oriented air quality control philosophy, the spatial allocation of administrative regulatory resources coincides only in a minority of analysed regions wholly with the measurable variations in local immission situations, which are a relatively reliable indicator of problem pressure. Such a coincidence could be found in two regions with strong regional agencies and interaction networks capable of reconciling distributional implementation policy choices with distribution of problem pressure either through strong co-operation with active local governments or with strong regionally organised corporate interest groups. A precondition for the functioning of such regional systems seems to be a highly integrated co-operation between two equally relevant political power centres independent of which centre has formal regulatory jurisdiction.
3. In the one case where the national programme did have an explanatory capacity for the distribution of administrative attention amongst local areas, the distributional effect turned out to be opposite to the problem pressure dimension. This result is likely to indicate that regional distributional policy choices are difficult to programme ex ante. It is probably more effective to define the implementation distribution issue as an explicit matter for regional jurisdiction.

4. Distribution of regional regulatory attention which has obviously not coincided with problem pressure distribution (misallocation) could be found in those regions where there were weak regional agencies and locally dominated interaction networks. In such regions we can find either the local brakemen type leading to the successful protection of local emitters against regional regulatory attempts or the probably more interesting type of local mobilisers. The latter puts the local area in a

[24] See: Knoepfel, P.and Larrue, C. 1984. Les politiques publiques comparées: tourisme intelligent ou vrai progrès? Le cas des politiques comparées de l'environnement, in: Cahiers de I'IDHEAP no 18, March, pp. 17ff.

privileged position, again not justified by problem pressure criteria, to the disadvantage of other areas equally or even more polluted.

If this analysis is correct, then the most surprising observation is that in spite of such presumably quite important distributional effects of implementation policy choices, this distribution issue has not been articulated or politicised in any of the six examined regions. Supposing that our empirical observations are not actually analytical phantoms but reflect reality, we need to answer the question why there is no conflict about this distribution issue as is normal in the case of distributive policies. I hope that after publication of our empirical data, political circumstances will raise this issue and thus render this academic question superfluous. Or is the public more "lowian" than Lowi himself, thinking that the distribution issue is reserved to distributive and redistributive policies?

References

Bardach, E. 1977. *The Implementation Game: What happens after a Bill Becomes a Law?* Cambridge, Mass., London: MIT-Press

Barret, S., Fudge, C. 1982. Examining the policy-action relationship, in: Barret, S., Fudge, C. (eds), *Policy and Action, Essays on the implementation of Public Policy*. London and New York: Methuen: 3–42

Darbera, R., Larrue, C., de Magalhaes, H. 1982. *The Implementation of SO_2 Air Quality Control Policy in France*, final research report, Paris/Berlin. (mimeo)

Dente, B., Lewanski, R. 1983. Implementing Air Pollution Control in Italy: The Importance of the Political and Administrative Structure, in: Downing, P., Hanf, K., *International Comparisons in Implementing Pollution Laws.* Boston: Kluwer, Nijhoff: 107–28

Dente, B., Knoepfel, P., Lewanski, R., Mannozzi, S., Tozzi, S. 1985. *Analisi di una politica regolativa: Politica contro l'inquinamento atmosferico.* Roma: Ufficina

Hartje, V.J. 1983. *Liability versus Regulation*, IIES discussion papers no. 81–16

Hartje, V.J. 1983. *Liability versus Regulation*, IIES discussion papers no. 5–10

Hill, M. 1983. The Role of the British Alkali and Clean Air Inspectorate in Air Pollution Control, in: Downing, P., Hanf, K., *International Comparisons in Implementing Pollution Laws*. Boston: Kluwer, Nijhoff: 87–106

Jarre, J., Zimmermann, K. 1983. *Konflikte zwischen Umweltpolitik und Wettbewerbsordnung*, Papers from the IIES no. 81–16

Knoepfel, P., Weidner, H. 1980. *Handbuch der SO_2-Luftreinhaltepolitik, Teil II: Länderberichte*. Berlin: E. Schmidt: 576f.

Knoepfel, P., Weidner, H. 1983. Die Durchsetzbarkeit planerischer Ziele auf dem Gebiet der Luftreinhaltung aus der Sicht der Politikwissenschaft. Ergebnisse aus einer intemationalen Vergleichsuntersuchung, *Zeitschrift für Umweltpolitik* 2: 87–117

Knoepfel, P., Weidner, H. 1984. *Schlussbericht zum Projekt: International vergleichende Analyse von Implementationsprozessen in der SO_2-Luftreinhaltepolitik in ausgewählten EG-Staaten*. Berlin. (mimeo)

Knoepfel, P., Larrue, C. 1984. *Les politiques publiques comparées: tourisme intelligent ou vrai progrès? Le cas des politiques comparées de l'environnement*. Cahiers de I'IDHEAP no 18. Chavannes-près-Renens: IDHEAP

Krupnick, A., Magat, W.A., Harrington, W. 1983. Revealed Rules for Regulatory Decisions: An Empirical Analysis of EPA Rule Making Behaviour, in: Downing, P., Hanf, K., *International Comparisons in Implementing Pollution Laws*. Boston: Kluwer, Nijhoff: 63–84

Lowi, T.J. 1972. Four systems of policy, politics and choice, *Public Administration Review* 32, Italy/August: 298–310

Majone, G. 1976. Choice among Policy Instruments for Pollution Control, *Policy analysis* 4: 589–612

Schnaiberg, A. 1986. Saving the Environment: From Whom, for Whom and By Whom? A critical American perspective, in: Knoepfel, P., Watts, N. (eds), *Environment Politics and Policies*. London: Harper and Row

Weidner, H., Knoepfel, P. 1979. Politisierung technischer Werte-Schwierigkeiten des Normbildungsprozesses an einem Beispiel (Luftreinhaltung) der Umweltpolitik, *Zeitschrift für Parlamentsfragen*: 160–70

White, L.J. 1983. U.S. Mobile Source Emissions Regulation: The Problems of Implementation, in: Downing, P., Hanf, K., I*nternational Comparisons in Implementing Pollution Laws*. Boston: Kluwer, Nijhoff: 46–62

Zimmermann, K. 1983. *Ansatzpunkte einer verteilungsorientierten Umweltpolitik*, IIUG pre 83–2

Zimmermann, K. 1983. *Die Inzidenz der Umweltpolitik in theoretischer und empirischer Sicht,* IIUG pre 1983–18

9
A Policy Monitoring Concept for Promoting the Conservation of Biodiversity (2001)

Christoph Bättig, Peter Knoepfel, Katrin Peter, Franziska Teuscher[1]

Summary

Since the late 1980s, the natural and social sciences, environmental administrations and statistics offices in many European countries have been trying to record "policy" in addition to the traditional data on the state of the environment and changing socio-economic and socio-cultural factors. Up to now, efforts to document political-administrative decisions as definitive dimensions for changes in environmental quality in a sufficiently operational form and in this way relate them to environmental data have not had much success. This contribution argues that the reason for these difficulties lies in the previous lack of co-operation between environmental monitoring and traditional policy evaluation. It presents a concept, which has been tested in practice and links the dimensions of traditional scientific environmental monitoring with those of policy analysis. Hence it enables the documentation of "policy" in spatial and temporal terms with the help of a geographical information system (GIS) which means it can be used as an operational explicative dimension in the natural environment.

9.1 Introduction

Compared with other countries, Switzerland has a relatively substantial body of environmental legislation. This legislation has been very successful in achieving its objectives in certain sectors (e.g. protection of waters, waste, (industrial and commercial) air quality control, hazardous incidents, environmental chemicals)[2]. In other respects, however, little progress has been made in the protection of the natural environment. In the latest OECD

---

[1] German version in: Zeitschrift für Umweltpolitik und Umweltrecht, 2001, no 1: 21–60. Copyright: We thank the editor for authorizing the publication of this article in this book.

[2] Knoepfel 1994.

report[3], Switzerland receives rather low grades when it comes to the protection of semi-natural habitats, the landscape and biodiversity. It is widely believed that the failures in these areas are not, for the most part, due to inadequacies in the legal basis but to deficits in the implementation of existing legislation. This is particularly true in the case of rural areas where a barely perceptible trickle of officially sanctioned interventions are implemented with gradual accumulative effects. No state body has a general overview of the uses of land which are authorised, prohibited, promoted or co-designed by the state, local authorities and cantons.

It is against this background that increasing demands are being made for the adoption of suitable instruments for monitoring the success of environmental policy on a continual basis. This does not merely involve the very expensive and labour-intensive (ecological) monitoring of the state of the environment. What is actually needed is a proper policy-monitoring system which would not only examine the development of the natural environment itself but also the activities of the authorities with respect to the formulation and implementation of environmental policy.

For theoretical and practical reasons, this kind of monitoring system should fulfil the following five requirements.

(1) It should record all federal, cantonal and local-authority policies involving activities that affect the environment. The confinement of the monitoring system to traditional environmental and nature conservation policies, as is the case today in the environmental status reports, is insufficient as this gives rise to either "doctored" or even "unjust" results. These falsifications can be explained by the fact that their ecologically positive objectives and efforts cannot be counteracted by the activities arising from non-environment policies. Also, a restriction to state or cantonal policies would not give a fully integrated picture; the inclusion of all state levels, which exercise a positive or negative influence on environment-relevant action in a given region, is essential.

(2) It should record environment-relevant data on public policies, action and environmental states with the help of a geographical information system (GIS), which as a relational database enables the establishment of links between different sets of data. The data must be retrievable at all times along certain dimensions as non-causal links and according to different starting queries.

(3) It should record the data on the basis of a suitable policy-analysis model which simplifies the extreme complexity of the processes involved in response to practical requirements by means of standardisation so that

[3] Cf. OCDE 1998.

non-expert officials will be able to record data with the help of instructions.

(4) As a monitoring system it should be capable of being linked with traditional policy evaluations. The basic policy-analysis model must correspond to those used in standard policy evaluation: this is not the case with most systems for monitoring environmental policy, which (at least partly) take into account the policy dimensions[4].

(5) Its spatial resolution should be plot-based. The basic units used in regional or nature monitoring (e.g. hectare networks (Hektarnetze)), surface of protected object etc.) are not suitable for policy-oriented monitoring because they do not enable the assignation of observations to the responsible actors (owners, authorised users). From the perspective of policy implementation, a precise knowledge of the policy target groups responsible for the environment-relevant action is essential. In good European tradition, these are usually the authorised plot users whose claim is based on property rights or their tenants.

The monitoring system described in this article fulfils these requirements in that it integrates observations on changes in the state of the environment and observations on policy activities. The system presented here was developed as part of a research project funded jointly by the Swiss National Research Fund "Environment" programme and various administrative bodies. It makes policy accessible to an extensive standardised monitoring system in the form of a relational database in that it differentiates between the various policy products which emerge in the course of the policy cycle.

This article starts by explaining the basic concept which is founded on a tried and trusted policy cycle model from policy analysis and policy evaluation. The concept essentially starts from the notion that the frequently confusing process of policy implementation can be recorded on an ongoing basis by means of the generated policy products. This represents a significant extension of the scope of standard monitoring systems which are limited to the long-term monitoring of potential policy effects ("environmental monitoring").

Finally, in the last chapter we explore the potential applications of the proposed monitoring system and briefly describe some of the options for practical application and some theoretical insights for the further development of policy analysis arising from the initial test runs.

---

[4] For example, the environmental reports produced by some cantons, the OECD ("Evaluations") and the federal authorities (Swiss Federal Office for the Environment, Forests and Landscape and Swiss Federal Office for Statistics).

9.2. Policy products and environmental states: the basic elements of an integrated policy and environment monitoring system

The traditional environmental policy monitoring systems usually concentrate on recording environmental states which environmental policy programmes aim to change.[5] Based on the empirically widely substantiated knowledge that the efficacy of environmental policy programmes often suffers as a result of flagrant implementation deficits and, less commonly, inadequate policy programmes, the proposed integral monitoring system examines both the implementation criteria and the policy programmes. To this end, in addition to the state of the environment, the system also monitors the policy products created by the authorities. In accordance with the first of the above-listed requirements, the monitoring system includes all policy products whose effects are significant for the monitored environmental states and the actions which are assumed responsible for it. These include various use policies (construction policy, infrastructure policy, agriculture and forest management etc.) in addition to traditional nature conservation and environmental policies.

The phase model of policy analysis[6] which was developed by Knoepfel et al. (1997), mainly for the purpose of empirical policy evaluation, provides a suitable basis for this kind of comprehensive monitoring system. Figure 1 shows a version of this model which has been slightly adapted for the present purpose. In this model, the formulation, implementation and effects of policy are presented as a multi-phase process. In the diagram, the arrows stand for the processes taking place during the different phases. They point to certain (intermediate) results which are symbolised in the diagram by boxes. In political reality, these phases seldom occur in sequence; they tend instead to take place simultaneously. Corrective retrospective links to previous phases also occur at each stage.

The basic idea behind the monitoring system proposed here rests on the notion that a policy process can be documented and made accessible for evaluation through the recording of such (intermediate) results. The allocation of the individual products to phases in the above-described model

[5] Swiss environmental policy today has data at its disposal from long-term monitoring systems in a wide range of environmental sectors (cf. Peter/Grolimund). A similar situation exists in the Federal Republic of Germany and other European countries (cf. Weidner, Zieschank, Knoepfel 1992).

[6] For a more detailed description, cf. Knoepfel/Bussmann 1997; Knoepfel/Varone/Bussmann/Mader 1997.

represents the main ordering principle for the recording of the products in the relational database.

The individual phases and the products which arise within them are explained in the following chapter. We follow the ideal model from "top to bottom" and discuss programming first before moving on to implementation and policy effects. By using the phase model of policy analysis, which is widely used in policy evaluation in Switzerland by the administration, research and consultancies, the proposed monitoring system fulfils the requirement that the monitoring be capable of being linked to evaluation studies (which build on it).

Figure 1: The phases of policy generation and implementation

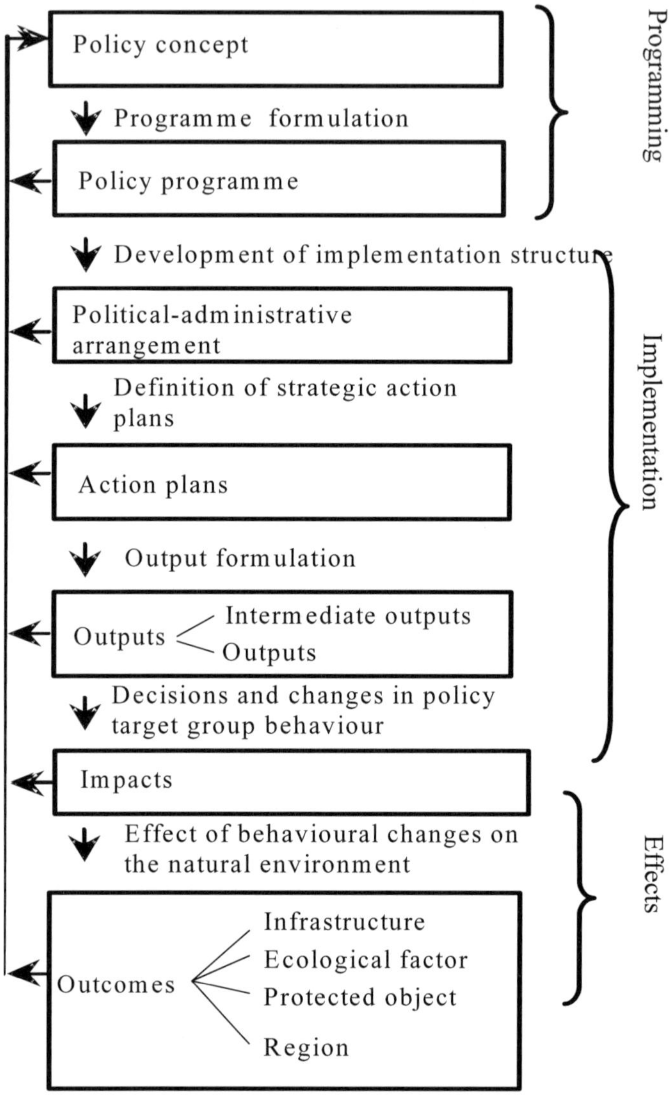

Modified version based on Knoepfel/Bussmann 1997: 70

9.2.1 Programming

9.2.1.1 The policy concept

The first phase of *programming* includes decision-making by the relevant political-administrative actors leading to the definition of the collectively acknowledged policy problem. The resulting product is known as the "policy concept" which, in addition to an approximate identification of the problem to be solved, contains two basic elements (and the actors responsible for them): [7]

- one or more causal hypotheses which identify the assumed causes of the collective problem to be solved;
- one or more intervention hypotheses which describe the assumed way the selected mode of intervention works and its expected effects in the intended field of action.

[7] The question of the "right" policy concept is currently the subject of controversial discussions in Swiss national and cantonal nature conservation policy. The main consideration is whether a nature conservation policy which concentrates on the placing of species and individual biotopes under protection represents the correct approach for a policy for the preservation of biodiversity. Other issues, for example, the question as to why dry meadows and pastures are disappearing are also the subject of controversial discussions and this is reflected in the diversity of the measures implemented for their preservation. Discussions focus, for example, on the influence of non-economic value positions of farmers in the farming of dry meadows and the contribution of airborne nutrient input to the development of vegetation in dry meadows. More recent scientific studies in the context of the Environment Focus Programme of the Swiss National Science Foundation (Körner Group) claim that CO2 concentration has an influence on biodiversity (for an initial overview see Körner 1996). For similar discussions in agriculture, see Knoepfel 1998; BLW 1998a und 1998b. The assumed causal hypothesis is reflected in the corresponding selection of target groups. Depending, for example, on whether air pollution or farming is identified as the cause, those responsible for the policy will consider different measures with different target groups. In one case, they will try to bring about a change in the behaviour of motor vehicle drivers, industrial and commercial concerns and households, in the other of farmers. The intervention hypothesis defines the instruments used in intervention. The fact that the farming contract is the most commonly used instrument in Switzerland suggest, for example, that farming practices are strongly dictated by economic concerns and can hence be controlled through economic incentives in the form of contractually based payments.

It would be beyond the capacities of policy monitoring to comprehensively record this often implicit, complex and generally controversial policy concepts. Moreover, controversial policy concepts are at least in part reflected in (inconsistent) administrative programmes which often comprise measures with very varied conceptional backgrounds.[8] Thus, the system proposed here does not include the relevant policy concepts. It starts at the second phase of policy programming, the product of which is the administrative programme.

9.2.1.2 The administrative programme

The *administrative programme* is located on the interface between policy formulation and policy implementation: it operationalizes the policy concept with respect to its administrative implementation. The administrative programme creates a legally binding definition of the (political) objectives to be implemented and the means to be used to achieve them. According to Knoepfel/Bussmann's definition (1997: 71f), the administrative programme is "all the legislation and concretizing instructions issued by the national and cantonal authorities and deemed necessary by governments and parliaments for the implementation of the policy concept in way that is beyond legal reproach and in the form of administrative action plans and, in particular, legally valid administrative action *vis à vis* the policy target groups."

Some definitional clarification is necessary with respect to the recording of the policy products "administrative programmes" in the standardised monitoring programme. It is particularly important to clarify the criterion, on the basis of which different laws and instructions from different legal sources should be empirically assigned to one or other administrative programme. An explanation of the horizontal and vertical demarca-

---

[8] The causal and intervention hypotheses of the policy concept are also expressed in the instrumental presentation of an administrative programme. Depending on whether air pollution or incorrect forest management is identified as the cause of the Waldsterben, those responsible focus on different measures and different target groups. In the former case, efforts will be made to bring about a change of behaviour among the users of motor vehicles, industrial and commercial operations and households while in the latter, the forest owners and forestry operations will be targeted. The selection of the instrument largely depends on the intervention hypothesis (regulatory levy on fossil fuels, mandatory use of catalytic converter, speed limits, management contributions, subsidising of forest roads etc.).

tion of administrative programmes is essential in this context.[9] The horizontal demarcation refers to the demarcation from other programmes in similar policy areas and the vertical demarcation refers to the different state levels.

The horizontal demarcation of different administrative programmes acts as a means of distinguishing policies from each other. As accepted at the outset, the proposed system should record all environment-relevant public policies. Thus, it must provide a satisfactory answer to the question of how they are distinguished from each other. There is no simple universal criterion for the demarcation of different policy fields. Policies, and the corresponding administrative programmes, can basically be analysed on a variety of levels. Thus, it is equally valid to speak of an environmental protection administrative programme or noise abatement, clean-air, water-protection or water-management programmes. It not really possible to establish on a theoretically deductive basis whether the activities undertaken in connection with the environmental impact test should be classified as part of the "regional development", "road construction" or "environmental protection" administrative programme. What is needed here instead is an empirical-inductive criterion. In consideration of the background of our research interests, we have selected as the constituent element for the classification of individual legal rules to an administrative programme, the circumstance that the corresponding decision figures in one and the same ordinance (Sachverordnung).[10] Thus, we record general legal rules which were assigned to a specific ordinance in the course of the process of legal enactment as belonging to a uniform administrative programme.

Moreover, we work on the basic assumption here that, as a rule, all of the relevant provisions of superordinate laws and constitutional provisions are "consumed" in the relevant directive, i.e. integrated in a form that renders them concrete. Cases in which it will be necessary to identify "non-consumed" elements from legally superordinate levels as additional components of administrative programmes should prove exceptional. The Rothenturm Article is an example of such a case, whereby very concrete regulations concerning the protection of marshlands are found at constitu-

---

[9] Chapter 9.4 contains a more detailed account of the content-based structure of administrative programmes (and the subordinate policy products) referred to above.

[10] Unlike procedural or competency decrees, this criterion is not always satisfactory because the assignation of a topic to one or other decree (particularly in time series) depends inter alia on coincidences, political opportunity considerations ("packaging") etc. In terms of routine use, the selected definition has the significant advantage of clarity and clear (political) assignability.

tional level but are not explicitly mentioned or included in their full scope in the corresponding directive.[11] In such cases, it will be necessary to process regulations from legal sources that are superordinate to the directive when recording the administrative programme. For reasons of clarity, such additions must be restricted to obvious cases because the purpose of the monitoring is not to verify whether the directive is an adequate concretization of superordinate provisions or whether it is unlawful or unconstitutional.[12]

The instructions, ordinances and regulations based on decrees must also be included as components of the administrative programmes (e.g. implementation aid for the Alluvial Sites Decree). Guidelines (*Wegleitungen*) are particularly important in this context. In practice, guidelines often operationalize the legal requirements to an extent that they can act as a direct lever for administrative action. Guidelines of this kind often even attain a "quasi legal" status.[13] We have only recorded such guidelines, instructions, regulations and other subordinate ordinances in those administrative programmes whose objective is to trigger priority effects in the (physical) environment. The "consumption principle" is also applied here: requirements from ordinances which are superordinate to the instructions shall only be adopted if they are not included in the actual instruction.

Vertical demarcation is significantly easier to implement than horizontal demarcation. The problem of vertical demarcation arises from the multi-level structure of the federal Swiss state. The Swiss federation only implements its administrative programmes directly in exceptional cases

---

[11] Constitutional article (Art. 24 seties Paragraph 5): "Moors and moor landscapes of significant beauty and national import are protected objects. They must not be built on or their soil changed in any way. Facilities which serve the preservation of the aim of protection and previous agricultural use are exceptions to this." ("Moore und Moorlandschaften von besonderer Schönheit und von nationaler Bedeutung sind Schutzobjekte. Es dürfen darin weder Anlagen gebaut noch Bodenveränderungen irgendwelcher Art vorgenommen werden. Ausgenommen sind Einrichtungen, die der Aufrechterhaltung des Schutzzweckes und der bisherigen landwirtschaftlichen Nutzung dienen.")

[12] The corresponding passage of the – concretizing and, in our opinion, unconstitutional – Upland Moor Decree (SR 451.32, Art. 5 Para.1c) determines that the cantons ensure that "buildings and facilities shall be only erected and changes made to soil for the preservation of existing agricultural uses which do not contradict the aim of protection. ("zur Aufrechterhaltung der bisherigen landwirtschaftlichen Nutzung nur solche Bauten und Anlagen errichtet und nur solche Bodenveränderungen vorgenommen werden, die dem Schutzziel nicht widersprechen".)

[13] These are often quoted by the Swiss Federal Court in its judgements.

and usually commissions the cantons with their implementation. The cantons integrate the federation's requirements into its own administrative programmes (which sometimes have to be created)[14]. In extreme cases, a single federal programme can give rise to 265 cantonal (part) programmes. For example, in the canton of Berne, the regulations contained in the Federal Law on the Conservation of Nature and Habitats (*NHG* of July 1st 1967, SR 451) are included in a cantonal nature conservation law (and the corresponding directive). On the other hand, in the canton of Zurich, it was decided not to pass a cantonal nature conservation act. In this case, the federal requirements were integrated into the cantonal planning and construction legislation (and the corresponding directive). Given that the discrepancies in the implementation of federal programmes is controversial issue – a fact that should be demonstrated by a policy monitoring system – we are recording federal and cantonal programmes as independent units which can be linked electronically. As opposed to this, it has been decided not to include local authority administrative programmes in the proposed system (for the time being).

9.2.2 Policy implementation products

9.2.2.1 The political-administrative arrangement

The first policy implementation phase consists in the assembly of a suitable political-administrative arrangement.[15] Metaphorically speaking, the political-administrative arrangement corresponds to the machine which is intended to make the policy products defined in the administrative programme. This administrative implementation structure determines which tasks, rights, competencies, financial and personnel resources are assigned to which actors in the implementation process (for a more detailed account of this process cf. Knoepfel/Bussmann 1997: 71). We will not record this arrangement in our monitoring system as it is strongly resistant to standardisation. However, for each monitored policy implementation product, we record those actors who played a key role in its generation and those who participated in terms of expressing an opinion. Thus, each policy product corresponds to an empirically recordable actor structure which

[14] Cf., for example, Kissling-Näf, Wälti 1999.

[15] This phase can also be conceived as the termination of policy programming as it only contains implicit substantial policy concretization: cf. Knoepfel, Larrue, Varone 1999: 200 ff., vol. 1.

despite the formal similarities of the policy products can vary from case to case.[16]

9.2.2.2 Action plans

Today, all public policies do not necessarily provide such action plans as an automatic phase in the policy implementation process. Due to the increasing lack of public resources, however, they are now becoming more common (and often also informal). The explicit formulation of action plans has increased significantly in recent years, particularly in the area of environmental policy. Action plans define spatial-temporal priorities for the generation of certain policy products and the corresponding allocation of resources. An action plan can, therefore, be based simultaneously on several administrative programmes[17]. The purpose of action plans is to create concretizing production planning for the action of persons responsible for policy implementation in the individual administrative units. In our policy monitoring system, we make a distinction between four categories of such plans:

- (Multi-) annual plans: these consist of an internal plan by an official instance for activities and the targets to be reached as part of "traditional" budget planning. This instrument is used, for example, in the area of noise abatement and water protection (priorities for the cleanup of waste water treatment plants).
- Performance mandates (Leistungsaufträge): this instrument, which is an increasingly common feature of public administration, consists in a formally defined mandate to an official instance with specific defined performance objectives (plans for noise abatement measures extending over several years). A performance mandate was issued, for example, by the canton of Aargau to the cantonal office for nature conservation.[18]
- Measurement plans: these consist of formal resolutions concerning a measure to be carried out as part of the implementation of administra-

---

[16] E.g. according to the authorities requested to comment.

[17] Example: Nature conservation programme which implements both the protection of biotopes as prescribed in the nature conservation legislation and agricultural direct payments for special ecological services.

[18] Performance mandates or contracts are the central instrument used in Switzerland in the implementation of new public management. They are often combined with what is known as a global budget (cf. federal, cantonal and local authority administrations) (cf. Hofmeister, Buschor 1999).

tive programmes, which the cantons are obliged to implement, for example under the Swiss Clean-Air Decree of 16 December 1985 (SR 814.318.142.1) (Art. 31ff.).

- Protection plans: this instrument, which is widely used in the area of nature conservation, summarises in a single document all of the measures which the relevant authorities are obliged to implement for a specific area. Such measures include all kinds of maintenance work, fertiliser restrictions, access restrictions and regional development measures (e.g. exclusion of protection zones). The (often highly controversial) resolution as such is at most binding at official level; there are no legal implications until the subsequent definition of implementation measures.

9.2.2.3 Outputs

We define outputs as all of the intermediate and end products generated in the course of the implementation of a particular public policy which do not fall under the category of action plans or political-administrative arrangements. Outputs are direct or indirect services provided to policy target groups by the administration and other (private and para-state) instances which are charged with the implementation of state-ordained tasks. Examples include all kinds of decrees (administrative acts), payments, contracts and state consultancy activities. A distinction must be made between intermediate outputs and outputs in the strict sense.

Intermediary Intermediate outputs
Intermediate outputs are preliminary products required by the administration for the generation of outputs. Such intermediate outputs often become components of the end products (e.g. use of an inventory mapping as a basis for farming contracts). Unlike outputs in the strict sense, intermediate outputs are not aimed at policy target groups. Also make a distinction here between (a) comments made in the context of hearings and consultation processes (*Mitberichtsverfahren*)[19], (b) reporting within the administration, (c) basic tasks (inventories, measurement, survey campaigns) and (c) directional and subject/content planning (*Sachplanungen)*.

Outputs in the strict sense
We understand outputs as direct services to policy target groups provided by the administration and other (private and para-state) instances which are

[19] This (typically Swiss) term refers to all kinds of file circulation processes among different authorities.

charged with the implementation of state-ordained tasks. As a rule therefore, outputs are the end product of public policies.[20]

Two main dimensions stand to the fore in the characterisation of outputs: the steering mode and the nature of the target group reference. In a simplified adaptation of policy analysis, we distinguish between regulatory, persuasive and distributive (financial) outputs on the basis of the nature of the selected steering principle. In the case of *regulatory* measures, certain modes of behaviour or target values are prescribed to the target groups and sanctions are threatened for failure to adhere to these regulations. In our monitoring system, we make a distinction between orders (licenses/permits, bans), local authority and cantonal use plans and concessions.

Distributive instruments (subsidies/tax relief, payment/compensation, taxes/duties, fees) aim to control the behaviour of policy target groups by means of financial incentives. This category also includes farming contracts, which are widely used in the area of nature conservation. Unlike the other financial instruments, the policy target group in this case has no legal right to payment from the state; this does not exist until a contract is completed. For this reason, the monitoring system records contracts as a separate category.

In the case of *persuasive* instruments are used in an attempt to influence the knowledge and motivation of the policy target groups and to prompt them to change their behaviour. Here, we make a distinction between information campaigns, information events, officially organised educational and further training events and specific individual consultancy services.

Direct services, whereby a state authority achieves the target status through the direct provision of its own services, and thus without the involvement of social policy target groups, constitute a special case.[21] For example, maintenance measures in conservation areas are partly implemented by employees of the cantonal authorities. These kinds of direct services also represent a separate output category for our policy monitoring system.

[20] This applies at least for individual concrete outputs while general concrete outputs often come second last in policy production and still involve an individual concrete output. (For an explanation of the difference between individual and general concrete outputs, see following chapter). "Evaluative statements" which are not recorded as independent policy products are excluded in our monitoring system (cf. Knoepfel, Larrue, Varone, 1999: 257 ff. vol. 2).

[21] Cf. Klöti 1997: 41f.

In the case of regulatory and financial instruments, the supervision of compliance with rules and agreements is an important criterion for efficacy. *Control outputs* are hence recorded for each of these outputs.

The target group reference of the different output categories is concrete in either individual or general terms. Individual concrete outputs are directed at specific target groups whose behaviour shall be influenced by this output. As opposed to this, general concrete outputs are directed at everyone who is actually present in the situation envisaged by the output. Typical examples of general-concrete outputs are traffic signals (local authority and cantonal), use landuse plans and access bans (cf. Moor 1991: 117f.; 290). Unlike mere intermediate outputs, the consequence of general concrete outputs is a binding legal relationship between the administration and the policy target groups.

By combining the "steering mode" and "target group reference" criteria, it is possible to categorise the different outputs as shown in Table 1:

Table 1: Categorisation of authorities' outputs

Mode of Control	Target Group Reference	
	individual-concrete	general-concrete
Regulatory	License, individual ban, concession, control, sanction	Landuse plans, general plans
Distributive	Tax relief/ subsidy order, tax/duties order, fee, contract	
Persuasive	Individual consultancy	Offer of training, information, campaigns

Provision of direct services is not included in the table because it does not have a target group and is not, therefore, subject to any specific steering mode.

9.2.3 Policy effects

In its assessment of policy effects, policy evaluation draws a distinction between impacts and outcomes. We explain below why it is impossible for us to take impacts into account systematically as part of a routine monitoring project. Our method of recording outcomes is explained in a later chapter.

9.2.3.1 Impacts

According to Knoepfel/Bussmann (1997: 73), the impacts of a policy consist in all changes in the behaviour of the policy target groups which can be directly related to the policy outputs generated in the course of the implementation process. It is less common for changes in behaviour to be directly based on the coming into force of administrative programmes.[22]

To estimate the influence of measures, sufficient insight is needed into the actors‘ motive for their (non-) policy-conforming actions. If real and detailed policy evaluations frequently encounter problems with this dimension, monitoring based on a significantly simpler design must omit it as far as possible. Valid and recurring actor surveys would be simply too expensive.

For practical reasons, therefore, the monitoring system presented here will be restricted to the observation of real actions. To be precise, we actually only document those actions which can be directly or indirectly observed in the test areas.[23] We have to rely on the direct observations of third parties (local residents, gamekeepers, nature conservation supervisors etc.) for the recording of such actions and indirectly surmise the actions from "action tracks". Thus, hoof marks would suggest riding activities, insurance numbers indicate the existence of home insurance contracts and burnt patches in the ground suggest activities surrounding leisure outings. For survey technical reasons alone, it is not feasible to aim for comprehensiveness in this area.

Thus, what will be recorded as part of the proposed policy monitoring is not impacts in the real sense but mere (assumed policy-relevant) actions, which, in turn, are only recorded on the basis of their visible traces. For the purpose of categorising these directly or indirectly monitored actions, for practical reasons we have construed certain action areas which correspond to specific policy fields (e.g. forestry, transport, leisure/tourism, hunting/fishing, military, energy and raw-material extraction).

9.2.3.2 Outcomes

Knoepfel/Bussmann (1997: 73) define outcomes as the "totality of intended or unintended effects arising from changes in the behaviour of pol-

---

[22] E.g. tax legislation, future prohibition legislation.

[23] An at least partial extension of the monitoring to actions which take place outside the test area is basically possible but it will not be at all possible to achieve comprehensiveness here (cf. chapter 9.4.2).

icy target groups on the constitution of the problem, which the policy set out to solve". It is completely possible for a policy to produce sufficient outputs and also bring about the desired changes in the behaviour of the policy target groups without alleviating the problem it set out to resolve.

Just as it is impossible to establish a causal link between the changes in the behaviour monitored at the level of the impacts and policy outputs, it is also impossible to directly monitor the contribution of altered behaviour to changes in the state of the environment in the context of a basic monitoring project. However, in order to demonstrate correlations between policy objectives and the actual state of the environment, the system proposed here tries to measure the outcome in terms of the dimensions of the relevant administrative programme target definitions. This is difficult in the context of long-term monitoring because the objectives of the environmental or nature conservation policy or relative use policies can change. The ongoing tracking of policy changes requires a finer resolution and the allocation of policy targets and monitored environmental states. Thus, the proposed system makes a distinction between four different outcome levels. These four levels correspond to the levels of the targets, on which the environment-related policies can (but do not have to) be based.

The four outcome levels used were formulated from the perspective of nature conservation and (the more recent) biodiversity policy. We are working on the assumption here that the preservation of biodiversity is one of the main objectives of environmental policy or that it even represents a kind of "sustainability policy". Against this background, the proposed system should be able to provide a good gauge for the evaluation of the (direct) environmental effects of all policies. Moreover, its internal differentiation facilitates a certain degree of transferability to systems for monitoring the success of more strictly defined region-related policies. For example, the monitoring of the location factor "surface water" is also of interest from the narrower perspective of protection of waters.

We draw a distinction between outcomes at the levels of use / infrastructure, location factors, protected objects and regional biosystems. Within this model, we are working on the assumption that the different levels constitute links in a causal chain. Thus, for example, the use of hydro-electric power affects the location factor "water", which in turn affects the quality of nature conservation protected objects (such as alluvial sites) in the relevant locations. This allocation cannot be applied *a priori* for the outcome levels "location factor" and "protected object". The material pollution of the soil can, for example, result from excessive fertiliser use or emissions from motor traffic. The complete reconstruction of these causalities cannot, therefore, be adopted as the purpose of the monitoring. Instead, the status parameters, which generally change in spatial-temporal terms, shall be differentiated and recorded on the basis of these four levels. Their causal linking will have to be implemented at a later stage with the help of expert knowledge or specific additional studies. The four outcome levels are defined and described in greater detail below. We define categories and indicators for each outcome level which are based on prescriptions in the federal environmental legislation.

Outcome level 1: Infrastructures/use traces
The first outcome level involves the recording of persistent physical action tracks over a minimum time span (months, years). These comprise the consequences of action and conditions for action, or both. We refrained from summarising the infrastructures in hierarchical categories. Instead, we work with an open list of individual categories which is created on the basis of observations in the field.[24]

Outcome level 2: Location factors
The status of location factors is recorded in the second outcome level. The descriptive dimensions used for this purpose are based on instructions defined in the relevant Swiss decrees. We draw a distinction between the pollution of the location factors of water, ground water, soil and air.[25] For each location factor, we will record the existence of sensorially perceptible (foam on water, smell etc.), material and radioactive pollution as well as changes to physical properties (e.g. surface tension of water). Thus, for example, the discharge regime/volume will be described for the location factor “water”, and “noise” will also be recorded under the location factor “air”.

The quality of location factors is not systematically recorded as part of the pilot system. Data from existing surveys and inventories will be taken into account. However, it is not possible to represent data on location factors outside the test areas in the GIS (the cost of digitalization costs are too high). Any information available about the quality of location factors *within* the test areas is included in the GIS.

---

24 The list for our test areas includes the following units: railway line, building site, secured footpath/road/car part, artificial snow spreader, rubbish dump, dumping ground, barbecue/picnic area, forest area, high voltage line, agricultural area, planning, allotment, sports centre, transport system, footprint/track, unsecured path, shooting/training/practice range, hydraulic engineering structures/bank stabilisation, water discharge, water pumping plant/volume

25 The location factor “climate” is not included because it cannot be influenced in the context of the policies being monitored or the relevant influences cannot be proven.

Outcome level 3: Protected Object

Protected objects (such as alluvial sites and dry sites) are located on the plots whose location factors are being recorded. Overall, this outcome level represents a gauge for the monitoring of the success of "traditional" nature conservation policy which focuses on conservation areas. This differs in that it is mostly aimed at the maximum conservation of objects which are worthy of protection. The intended protected objects include habitats as well as individual species of flora and fauna. In accordance with existing Swiss legislation, we make a distinction here between 13 categories of target objects.[26] Outcome level 4: Region.

The fourth outcome level describes biodiversity at a regional level. This level represents probably the most important gauge for the success of all modern nature conservation and environmental policies but it is also the most difficult to operationalize. Despite numerous efforts it has not yet been possible to establish either generally accepted concepts for the definition of biodiversity or the corresponding indicators. The relevant data is not available for our test areas. For this reason, we have refrained from categorizing this outcome level.

9.3 Integrated policy and environment monitoring versus success monitoring/evaluation

As mentioned at the outset, the proposed system records policy within the framework of a relational database. Up to now, we have explained how the numerous policy products created in the context of public policies and environmental states can be categorised on the basis of an idealised phase model. Furthermore, we outlined the extent to which the recording of such data is feasible within a permanent monitoring system. In the following sections, we respond to the question as to how the recorded policy products and environmental observations are linked in a relational database. For without intelligent linking, the system will only provide a collection of data about existing policy products and environmental observations which is ordered on the basis of the logic of the policy cycle. With respect to its suitable application in the context of success monitoring, it should be able

[26] In terms of "target objects" to be protected we include the following individual categories: alluvial sites, upland moors, lowland moors, other wetlands, dry and infertile pastures, hedges and copses, forests and forest perimeters, other biotopes, water and bank areas, ecological compensation areas, individual flora objects, individual fauna objects.

to make visible the potential correlations (of effects) between the monitored policy products and environmental observations. This requirement is met by the principle of non-causal linking of target formulations which are found in the various policy products (which are connected in series) with real monitored policy products and environmental states.

9.3.1 The phase model as a basis for the non-causal linking of policy products and environmental observations

Each policy product and each observation of actions or environmental states is initially recorded by the proposed monitoring system as an isolated piece of data. For example, the system processes a farming contract for a certain dry meadow at a specific point in time in exactly the same way as it processes individual policy products such as the cantonal "dry meadows" administrative programme. It does not initially link the two products. The concept of non-causal linking serves in bringing the different products together in a systematic context. It assumes that the monitored intermediate results (policy products and environmental observations) can be linked in a certain way.

The system records the policy products by means of double assignation. Firstly, it registers the phase in the policy cycle to which they belong (administrative programme, action plan, output etc.). Secondly, it records which substantial objectives the product contains for which of the subsequent product phases. For example, environment policy administrative programmes contain information about the action plans and/or outputs to be produced and the environmental state is to be achieved as a result. Thus, for example, the Swiss clean-air decree states that the cantons must develop action plans in the form of measurement plans; the nature conservation legislation of the canton of Berne states that valuable areas shall basically be safeguarded through the awarding of contracts.

The policy products from all phases also require, however, direct targets for the modes of action and environmental states. For example, in its administrative programme for the protection of alluvial sites of national importance, the Swiss state sets itself the aim of "preserving (the alluvial sites) in undiminished form". Other administrative programmes contain very concrete quantified and quantifiable target values, like that for the protection of waters with its detailed regulations for minimal residual water volumes.[27] The spatial dimension is fundamental to the proposed policy monitoring. Policy products and observed environmental states are linked through their spatial dimension (see also Chapter 2).

As the system is designed as a relational database, thus objectives with respect to the outputs formulated during the administrative programme phase can be compared with the outputs actually produced. Similarly, the objectives formulated in the outputs with respect to changes in the behaviour of the policy target groups can be compared with observations of actual action.

The causal link between the objectives formulated in policy products and corresponding observations at the subsequent monitoring level suggested in the model is not, however, tenable within the actual monitoring carried out in the context of our system. Nonetheless, it is possible to estimate policy influence on the basis of the comparison of the action intended in the policy outputs and the action actually observed in the monitoring. To be precise, however, it is only possible to prove the inefficacy of policy products. This is the case, for example, if none of the actions intended in specific policy outputs can be observed. Due to the frequent presence of accompanying effects, it is impossible to provide clear proof of positive efficacy. In such cases, all we can do is monitor policy-conforming actions which cannot or cannot not only be explained in terms of the incentive created by the (monitored)[28] policy outputs. This phenomenon is widely observed in general policy evaluation.[29]

Although the monitored changes in behaviour can be traced non-causally to policy outputs within our system, the contribution of outputs and resulting behaviour modification to changes in the state of the environment is not directly demonstrable. The same applies to the linking of policy products from different phases: it is far from being the case that each observed output can be linked back to an action plan which intended to achieve the corresponding output.

The principle of non-causal linking does not help here. It enables, for example, the monitoring of the fulfilment of targets in relation to implementation and the state of the environment (implementation and target fulfilment monitoring). It refuses to specify, however, the extent to which the attainment of the target is due to the influence of political programmes (ef-

---

[27] Swiss water protection decree (Gewässerschutzverordnung) of 28 October 1998, SR 814.201. However, such concrete and quantified objectives are largely unknown in nature conservation policy.

[28] From time to time, such impacts are also related to non-monitored policy outputs.

[29] Cf. inter alia Klöti 1997: 48ff.; Knoepfel et al. 1997: 103 ff.

fect monitoring, evaluation). This restriction to the non-causal linking[30] of policy outcomes from different phases is a constituent feature of monitoring systems (cf. Kissling-Näf/Knoepfel 1997: 147f). However, it does the system no disservice. On the contrary, thanks to the fact that the monitoring is based on a tried and trusted model from policy analysis and evaluation, the monitoring results provide an ideal basis for (subsequent) evaluation. The formulation of potential problem areas and working hypotheses on the basis of structured monitoring data can be improved and made more salient than it would in the absence of such a structured data basis. For example, no one will undertake an evaluation of the environmental effects of a programme if the monitoring data shows that very few corresponding programme outputs exist. In this case, one would deploy the resources available for the evaluation more profitably by concentrating on a detailed implementation analysis. The classification in the presented system of the indicators to be surveyed in terms of a model from evaluation practice also increases mutual accessibility for the two activities: evaluation should come into play where monitoring leaves questions open or suggest problematic effect hypotheses. This precondition is seldom fulfilled by general environment monitoring systems.[31]

9.3.2 The key link dimensions

In practice, the above-described linking comes up against barriers when the system has to process the outputs of different products with different policy areas with correspondingly varying objectives. Thus, it does not make much sense if a license for the use of hydro-electric power, which is classified as an output, is assigned to the agricultural administrative pro-

---

[30] The direct linking of individual policy products with each other or possible effects is exceptional in the database. For example, it will be possible in individual cases to relate certain actions directly to the influence of specific outputs on the basis of the results of existing studies or our own surveys. Equally, the results of scientific studies available in individual cases could enable the direct (causal) linking of different outcome levels. This is imaginable, for example for the depiction of evident links between the presence of a certain type of vegetation (outcome protected object), soil compaction (outcome location factor), military use (outcome infrastructure) and possible authorisation for operation (output). In individual cases, such links actually depict the policy effects caused by policy products (impacts and outcomes).

[31] This applies in particular also for the "pressure-state-response" models (Meadows 1998), which may include political action dimensions but do not order them on the basis of policy-analysis (but ecological) action dimensions.

gramme. To eliminate such arbitrary linking of policy products to each other or with data on actions or environmental states, the dimensions of space, time, target category, actors and legal rules are included in the data descriptions.

9.3.2.1 Space

The fact that a lot of policy products or the targets and rules they contain have spatially defined dimensions is a basic conceptional idea behind the proposed system consists.[32] Moreover, such spatial targets can be found in each phase of the policy process. Thus, each phase can at least be partly depicted in spatial terms. This reference is often only implicit in laws as it is assumed that the target applies for all areas in the corresponding territory. However, even at this level, it is often possible to find concrete spatial references which define the area of validity. Thus, for example, the Alluvial Sites Decree explicitly lists the individual areas of national importance to be protected. This spatial reference is usually precisely defined, particularly at the level of the outputs. Hence, for example, the validity of the provisions defined in the agricultural contracts is usually defined on a plot basis.

[32] Targets directed at the physical environment always involve a spatial dimension.

Figure 2: Spatial linking of policy products and monitoring

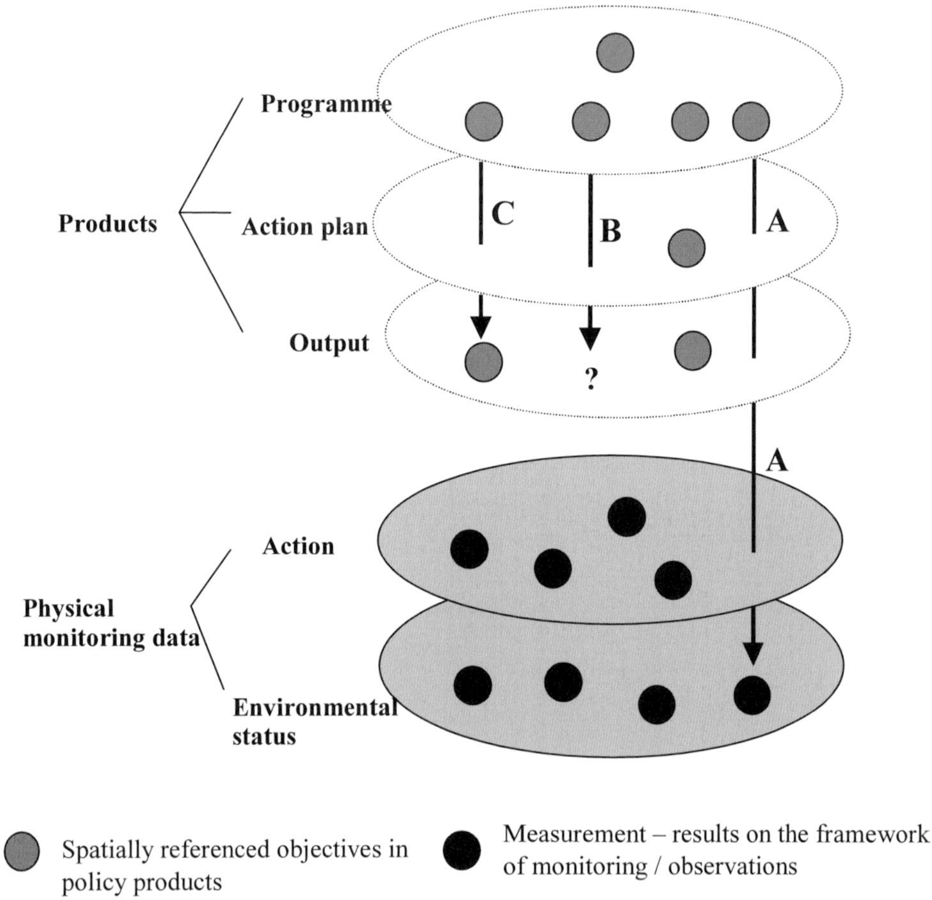

This spatial reference features even more strongly in the monitoring of the physical environment: all data about the state of the environment can, in principle, be associated with a specific location.[33] The same applies for the spatial dimensions of actions, which is also easy to identify in many cases.[34]

It is possible to record the spatial references for the policy products and effects thanks to the integration of the database with GIS software. This precise location of monitored policy products and environmental states (which is sometimes on a level that is even more precise than plot-based location) again enables the linking the observations in relation to their spatial "location". Figure 2 demonstrates how this linking of targets and monitoring defined in different policy products can be used in the overall results of the monitoring project.

This suggests that the system only links policy products and effects which relate to the same area in the GIS. If the Alluvial Sites Decree and the protective measures based on it require, where possible, undisturbed water dynamics (i.e. for water courses in areas of national import), discharge measurements recorded in this area can be compared with the required target status (Arrow A in Figure 2). In the language of policy evaluation, this kind of comparison is known as effect monitoring. At the same time, spatial linking also makes it possible to demonstrate contradictions and inconsistencies in different policy products. Thus, spatial linking can show that defined targets exist for certain areas at the level of the administrative programme but that outputs with corresponding provisions have never actually been produced for these areas (Arrow B in the Figure 2). If defined targets exist at programme and output level for a certain area, thanks to spatial linking, these can be compared with each other (Arrow C in Figure 2). In this way, it could be established, for example, that the volumes of residual water authorised under a license for the use of hydroelectric power in a certain water section contradicts the targets defined in the administrative programme for protection of alluvial sites in the this area. The last two examples are referred to in policy evaluation as implementation monitoring.

---

[33] However, its spatial resolution is often limited due to the surveying method used.

[34] However, actions, which show effects in the monitored test areas, often occur outside the perimeters of the areas mapped in the GIS. The spatial dimension of corresponding actions will not actually be referenced in the context of the proposed policy monitoring but coded in strongly simplified terms as "outside the test area".

9.3.2.2 Time

Time is the most basic and far-reaching dimension for the linking of the different database elements. Hence, all policy products (date passed and enforced, passing of deadlines, invalidation) and all observed actions and environmental states (monitoring time, possible assumed time of advent of monitored state) are described in the database in temporal terms. The significance of the temporal categorisation of policy products and monitored actions (and establishment of possible correlations) is obvious: (causal) elements which occur earlier in the chain of effects or effects model must by definition precede the subsequent (caused) elements. Thus, for example, the above-mentioned correlation between the authorisation of a military firing range, actual military use and monitored soil compaction can only be sustained if the individual elements occur in succession and the corresponding monitoring is actually available. In Figure 3, such a result is only possible in one case (Arrows D1 and D2). It is not possible to check the same facts for a different area as no data is available on the state of the location factor soil prior to military use (Arrows E1 and E2). [35]

[35] This observation clearly demonstrates the importance of the continuous recording of policy production and the development of the environmental status for the meaningfulness of monitoring. With respect to environmental states, therefore, the maximum possible number of staggered "samples" should be made available.

Figure 3: Spatial-temporal linking of observations

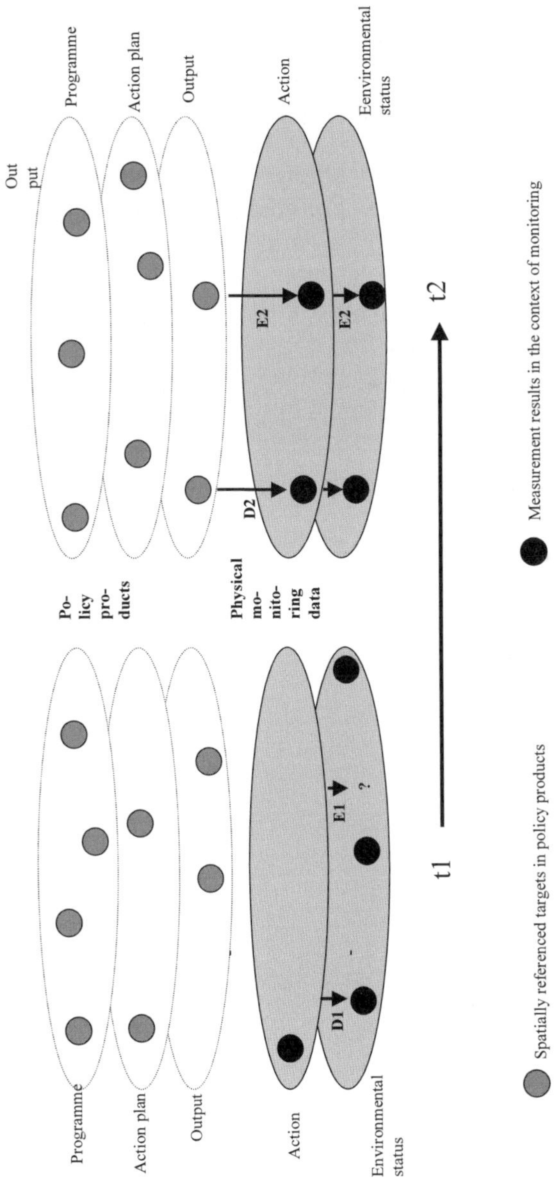

9.3.2.3 Target categories

The linking of policy effects and products exclusively focused on space and time can also give rise to "correlations" which are less useful than those in the examples given above. One such example would involve a plot for which a farming contract exists (output) and for which noise pollution measurement exists prior to and after the conclusion of the contract (Arrow F in Figure 4). The linking of the policy product and monitoring over space and time would suggest a correlation. The (reduction of) noise pollution was probably not included in the targets defined in the farming contract and cannot therefore be used in the evaluation of the effects of the policy product "farming contract". To avoid such mistakes, in addition to space and time we have introduced a third link dimension into the project database, to which all of the recorded policy products and effects can be related in one way or another: this is the substantial or institutional target category.

Figure 4: Spatial-temporal linking of observations and policy products via target categories

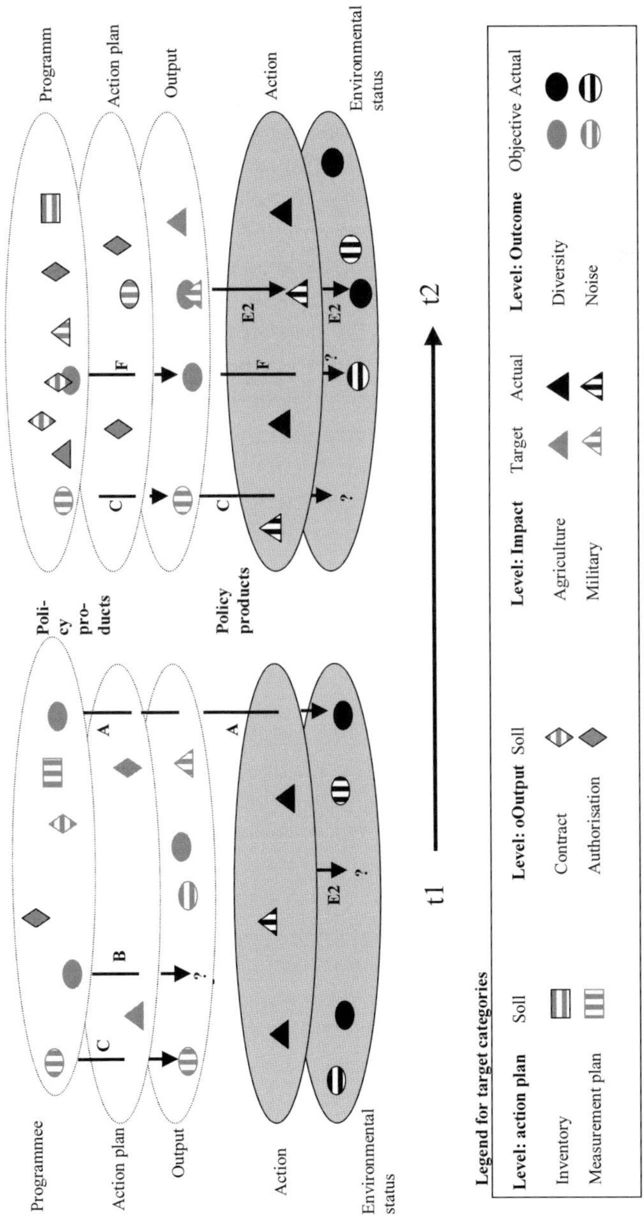

As already mentioned, each policy product (administrative programme, action programme, intermediate output, output) contains more or less precise definitions and formulations of the above-depicted ecological target dimensions. These relate to completely different target levels. Administrative programmes can define either (institutional) rules for subsequent policy products (action plans, intermediate outputs, outputs) or substantial effect targets for the desired impacts and outcomes of one or other of the four outcome levels. In the example given (noise), due to the common spatial and temporal reference, effects on the levels of the location factors (precisely: location-factor noise) were (incorrectly) linked with an output (farming contract), which mainly aims to achieve effects at the outcome level of the protected objects. By assigning all policy products and effects to their precise target categories, the database ensures that the measurement of policy effects will only be linked with those policy products which target the same effect level. This is the case in the previous example given (Arrow 2, Figure 4) whereby the output (authorisation) covers regulation of military use (symbolised by the grey hatched triangle) in an area as well as the tolerated effects on plant species variety (symbolised by the greyed ellipse). As the corresponding measurements are available for the area in question (indicated by the black hatched triangle for the monitored military use and the black ellipse for the mapped diversity of the vegetation), it makes sense to link them. The characteristics of these target categories are derived from the subcategories differentiated in the individual policy products and policy effects. Thus, for example, observations at the outcome level "location factor" are assigned to a more specific category (e.g. individual object Flora.

It is possible for the different policy products to have double links to these target categories. Firstly, policy products formulate substantial targets with respect to various target categories (target reference to target category). Secondly, in themselves, they represent the realisation of (institutional) process targets which are possibly requested in superordinate products (actual reference to target category). If the administrative programme for the protection of alluvial sites stipulates that the cantons must define precisely the borders of protected objects of national importance, a target with respect to the production of intermediate outputs is formulated in the target category "intermediate output inventory". If a canton then zones protected alluvial sites of national importance, it generates a policy product which represents a (partial) realisation of this target category.[36] Substantial targets may also be defined in this inventory which apply spe-

[36] For reasons of clarity, the policy products are not reproduced in the diagram.

cifically for the defined areas, as for example, the ban on issuing new licenses for hydro-electric power production. This means that the intermediate output is firstly a policy product of the category type "intermediate output inventory", and secondly also defines targets for subordinate products (in this case, for the target category "Output concession"). Thus, both policy product references for such multi-layered policy products are shown on their target categories in the database.

9.3.2.4 Actors

The state actors involved in policy formulation and implementation represent another important dimension in the linking of different policy products.[37] Policy products often originate from or are based on institutional actors at subordinate level (e.g. output) who may or may not be specifically defined (i.e. "the department" or section X of authority B). As with the dimension of the target category, policy products have a double link to the official actors named in them. They either sign as the authors responsible for policy products or are named in the policy product as those institutional actors who are responsible for the implementation of the decreed regulations.

This linking of policy products with the actors enables the reconstruction of the structure of the political-administrative arrangement. This is known to indicate who produces or should produce which action programme, intermediate and end output in the course of the implementation process. It is only possible to reconstruct the component of this arrangements which was activated for policy implementation in the monitored area from the relationships between the institutional actors and the policy products created in the various phases of the policy cycled; authorities which are intended to participate in the law but do not do so in reality are not included – a correct course of action from the empiricist's perspective.

In contrast, the system always demonstrates the – mainly social – actors (private persons or legal entities = target groups) in the policy implementation. From the policy perspective, these actors constitute the potential tar-

[37] To facilitate the integrated description more (section X office Z) or less (the responsible department) concretely defined actors are listed in various linked tables in the database. The institutional position in the federal political system (federal/cantonal/local authority institution, para-state organisation, private) and the position of the actor in question within the administration (department, office, section) are also described. Beyond this brief description of the actors, it would be desirable to record the annual finances available per actor. However, this will not be carried out within the described project for the time being.

get group. In general, the intended policy target group is specified in individual concrete policy products along with those responsible for their implementation. The identification of the policy products' target groups and of the actors involved in the monitored action makes it possible to identify whether the policy products are actually aimed at the correct target groups. A comparison of the intended target groups in the (linked) policy products at different levels makes it possible to specify whether, for example, target groups which were identified as important in the administrative programme have been "forgotten" at output level.

With the adoption of this plot-based focus, which has been enabled through the completion in many parts of Switzerland of the digitalization of land registers, the monitoring system also meets requirement of full target group resolution stated at the outset. It actually demonstrates a specific (individual-concrete outputs) and, where appropriate, several non-specific (general-concrete outputs on a certain plot, such as access bans etc.). The authorized user[38] with address can be found at any time. This enables efficient policy management because specifically enables the local authority to avoid time-consuming search processes.

9.3.2.5 Legal rules

All policy products (administrative programme, action programme, intermediate output, output) are related to identifiable legal rules, in which the rights and duties of the policy target groups are defined in the context of the administrative programme. In a state under the rule of law, all (formal) policy products should be (explicitly) based on legal rules.

Hence, in the database we use this easily recordable legislation to convey the formal relationships between the policy products at different levels. Outputs, which relate directly to legal rules, and belong, for example, to the federal administrative programme for the protection of alluvial sites, are recorded *per se* as the concretization of the administrative programme for the protection of alluvial sites. This classification is purely formal and it is not content-based. Hence, for example, a license for hydro-electric power production, which was issued without an explicit reference to the Alluvial Sites Decree or another piece of legislation from the administrative programme for the protection of alluvial sites, is registered as the concretisation of the "hydro-electric power use" administrative programme and not simultaneously also as the concretization of the alluvial sites pro-

[38] A filter will most probably have to be installed here in practice for data protection reasons.

tection programme, although from the perspective of content, this would be completely appropriate.

By recording the systematic number of the federal law[39] or corresponding cantonal legislation, the database also provides simple manual access to the relevant legal texts. The recently (January 2000) completed electronic capture of the systematic bodies of legislation at national level and in one canton (Berne) has guaranteed the direct electronic linking of the database with the full text of the legislation.

9.4. Outlook

Contacts at home and abroad have given rise to the urgent wish for the publication of this paper in environmental and nature conservation circles before the data collection is completed. The system for integrated policy and environment monitoring, whose basic principles are presented here, is now adequately consolidated. After more than ten years work on this topic,[40] we had reached the conclusion that it is only possible to continue to gain further insight into environment policy monitoring through concrete projects in test areas and test policies. We were not getting any further with abstract discussions about indicators, various software products, organisation models and monitoring concepts. This opinion was confirmed to us by all of our foreign contacts who have commented and criticised our project[41] over two years ago when we started our work on the system. The basic tenor of their commentaries was completely positive with respect to our daring to take the leap into a concrete implementation phase. This should be accessible from the political science perspective and sufficiently well founded from a (natural) scientific perspective. Thanks to the multidisciplinary composition of the team (geography, political science, biology, administrative science, computing), we believe we have fulfilled these two requirements. This phase is now complete. Along with this article, we have produced a filled database on four test areas and technical documentation.

39 SR number, as used in this article in references to federal Swiss legislation.

40 Start of work by the Swiss Commission for Environmental Monitoring whose chairman is one of the authors (P. Knoepfel): 1987. The work was published in Grolimund/Peter 1994.

41 Contacts with the Federal German Environment Agency (Bundesumweltamt), the Canadian environment authority, the Austrian and Dutch environment authorities, offers of collaboration from the University of Lüneburg, Germany (Edmund Brandt).

It is clear that many questions remain open and various aspects will be open to criticism. The proposed system is not a traditional social-scientific research project which should be laid to rest *ad acta* on completion of the funding and corresponding publication. It is instead a project which will need ongoing development. Thus, to conclude we would like to outline some practical and theoretical considerations which have arisen from the test runs[42] carried out to date.

9.4.1 Practical applications

Considering the extensive costs involved, a justifiable question arises as to the practical applications of this kind of GIS-supported policy monitoring system. On the occasion of a presentation workshop,[43] staged as part of the research project, the practitioners present agreed on the four following applications.

9.4.1.1 Situation analysis – archiving

The computer program, which is based on View and Microsoft Access, enables the reproduction in temporal sequence on the different outcome levels of all outputs (ordered according to subject area), the relevant actors, the main administrative programmes and the status data for each plot of land that is clicked on. Conversely, it is also possible to retrieve individual administrative programmes to reproduce cartographic images of the plots affected by them in the test area as well as the status data on the effect levels. Finally, the status levels can also be clicked-on for individual plots and the existence or absence of corresponding policy outputs etc. determined. In this way, the responsible administrative actors and interested social groups can quickly obtain an overview of the relevant data for the test area.

When integrated into standard workflows in local-authority administrations, the system facilitates the new and spatially ordered archiving of administrative and status data, which can be accessed rapidly during daily practice. This means that the monitoring system responds to the requirement defined at the outset that it should record all of the public policies

[42] The system was tested as part of the research project using data from a total of four test areas, each only a few hectares in size, in the cantons of Berne and Zurich.

[43] On 27 May 1999 at the Schweizerischen Nationalfonds für die Förderung wissenschaftlicher Forschung in Berne.

and their implementation activities which concern a given area and, hence, enable interested authorities to obtain an overview of the accumulated outputs of federal, cantonal and local-authority bodies. This is far removed from reality in Switzerland today. Local authority officials generally only know about their own outputs and do not know what applies in the agricultural zone (which is usually administered by the state or canton) or in forests and other protected areas. As a result, contradictory decisions and conflicts are not a rare occurrence (for example local-authority water works versus cantonal and state agricultural policy). In this sense, the monitoring system fulfils the role of a policy memory which is threatening to disappear in our short-lived times in the offices of the authorities. Projects already in receipt of multiple subsidies receive further subsidy, things that have been prohibited several times are suddenly sanctioned or a barn structure which was carefully surveyed 20 years ago and for which planning permission was ultimately refused, is reconsidered (on the basis of the same plans) and authorised, despite the fact that the relevant regulations have not changed.

9.4.1.2 Success/failure monitoring capable of communicating

The system enables the visualisation of public policies in (sensitive) areas. Like a highly sensitive probe, it enables the systematic illumination of these areas from the perspective of the (positive and negative) public policies which define them and to project the corresponding results on a large screen. This creates success monitoring which can be experienced, understood, is politically communicative and therefore highly effective. It is a suitable tool for raising the awareness of the responsible officials, interested groups and the affected population. The instrument is relatively easy to use which means that it can also be used in parliaments, special commissions and at public meetings. Experience with Swiss area statistics[44] show that speeded up depictions of spatial changes and their cause are particularly useful in raising awareness. The system can also be used to visualise the – limited – value of costly nature conservation activities (cf. OECD 1998: 111 ff.), which, in Switzerland at least, has declined in recent decades. It reveals both failures and successes without prejudice.

[44] Hectare-scale land-use mapping in time series, which enables for example the demonstration of the rapid expansion of the construction zone at local, cantonal and also national level.

9.4.1.3 Bases for policy evaluation

Given the costs involved, these plot-based monitoring probes can only be set up for a few strategically important "policy fallout areas". As suggested in an earlier proposal compiled by the Swiss Commission for Environment Monitoring (*Schweizerischen Kommission für Umweltbeobachtung (SKUB)*) of the Swiss Academy of the Sciences (*Schweizerischen Akademie der Naturwissenschaften (SANW)*), it would be conceivable to set up eight to twelve such monitoring systems for selected highly sensitive ecosystem types in addition to the "general" Swiss environmental measurement networks[45] (Grolimund/Peter 1994: 205). Comparative data evaluation would then contribute to the generation of hypotheses for causal links, with the help of which periodic evaluations of selected regionally relevant federal policies could be carried out. In accordance with this concept, the monitored areas would represent suitable laboratories for the generation of hypotheses as well as for the definition of generally surveyable key indicators for the entire country.

9.4.1.4 Use in regional development

Recognition of the fact that state steering of regional development is a cross-sectional tasks which can only be carried out through the co-ordinated steering of regionally relevant cantonal and national Swiss sectoral policies is emerging with justification within supra-local authority of regional development, which is subordinate in Switzerland to cantonal directional planning. Experience with the first generation of these directional plans has shown that it is of little use if the cantonal regional development offices are merely satisfied with the creation of attractive plans in all of the key areas in which potential collisions with sectoral policies are demonstrated (co-ordination documents), for which mutual agreement is required without regional development exerting an influence on the design of these sectoral policies at a sufficiently early stage. This – failed – directional planning was limited to amicably settling the discord between competing sectoral policies by means of the "weighing up of interests" in the field.

According to the cantons of Vaud, Berne and Thurgau[46], the new generation of directional plans should concentrate on anticipating the regional effects of these sectoral policies and, where appropriate, correct them in accordance with the concept of "sustainable development". The recent in-

[45] Cf. overview in Grolimund/Peter 1994: 175 ff.
[46] Cf. IDHEAP/Urbaplan 1999 study

tegration of the Federal Office for Regional Development into the Swiss Department for the Environment, Transport, Energy and Communication[47], which has explicitly stated its adoption of the aims of sustainable development[48], is a suitable step in this direction as most of the regionally relevant federal policies are now united in this department. It is a fact, moreover, that at present hardly any of the federal or cantonal authority knows which of their policies exercise specific regionally relevant effects. Thus, it is high time that in the run up to this reorientation of regional development policy monitoring systems be set up in particularly dynamic test areas which correspond to the requirements of the concept presented here.[49]

9.4.1.5 Establishment of sensitive system dimensions

For the test areas themselves, the monitoring system represents a good basis for the establishment of sensitive (ecological) system dimensions and the particularly relevant policy outputs. This facilitates in turn a concentration of policy activities on a few key dimensions and prevents the inefficient allocation of today's scarce administrative resources. Corresponding insights can be directly applied in pending licensing procedures and/or environmental impact studies or in all kinds of regional development decision-making processes. In particular, this makes it possibly to smoothly weave more recent regionally relevant policies into existing "policy carpets" (Knoepfel, Kissling-Näf 1993: 276) for ecologically sensitive areas, in which nature conservation policies in particular are implemented.

9.4.2. Consequences for implementation research

In terms of the theory of implementation research, we draw the following conclusions from the work completed during the development of the presented monitoring system:

---

[47] Decision of 19 January 2000. Integration was implemented in summer 2000 (now Federal Office for Regional Development); the authority used to be part of the Swiss Justice and Police Department.

[48] Cf. model of the Swiss Department (i.e. ministry) for Environment, Transport, Energy and Communication of January 2000.

[49] To this end, it was necessary to develop not only the regional effects as such but also and in particular the sustainability-relevant regional effects (along the three dimensions of economy, ecology and social equality of distribution) for each sectoral policy. A political consensus about these indicators and the minimum levels to be guaranteed then enabled a systematic reorientation of the sectoral policies in question. On the method, cf. IDHEAP/Urbaplan 1999.

9.4.2.1. GIS-supported databases

The analysis of the implementation of regionally effective public policies can no longer function without the help of GIS-based databases. Not only do they offer the advantage of visualising the – speeded up – implementation process, but they also enable the location of the main actors and assignation of corresponding changes in the physical environment to these actors, if the GIS selects the plot as the basic unit. This enables the leap to be made from the policy products to the actors and the changing physical area and its characteristics, which is necessary for all implementation research, in a way that is transparent and comprehensible. In terms of method, the GIS opens a suitable interface on which it depicts the quantitative and qualitative information in terms of the spatial/temporal, actor and legal dimensions.[50] The flexibility of geographical information systems, which must be rigidly structured, makes it possible to make unlimited additions to the data layers and hence also make the transition from relatively simple monitoring to complex policy analyses and evaluations.

Thus, for example, it is possible to record spatial-temporal impact data, including information on economic value added, as well as data on the further qualification of policy networks in the form of a qualitative or quantitative network analysis. The GIS facilitates both synchronous comparisons between different areas at one and the same time tx and, in particular, diachronic analyses of changes in one and the same area between the time tx and tx+1. In our opinion, this latter use will be extremely important as it facilitates the recording of dynamic processes in space, which may have been triggered and are certainly influenced by the implementation of public policies, and which are not often known to the interacting sectoral actors. Ideally, it even enables the identification of the sensitive policies when a clear dominance emerges between the real trend and implementation activities of a (possibly even unexpected) public policy.

9.4.2.2. Bottom-up approach

Traditional policy analysis is correctly criticised in the literature and in the field for its one-dimensionality; in reality, in most cases policy-analysis research places a specific public policy to the foreground and, in their research designs, they record at best two to three potentially competing policies, without leaving sufficient space for "unpredicted third-party poli-

[50] This was the fulfilment of a wish nurtured by the project manager for years to be able to produce a multi-dimensional visualisation of public policies.

cies". This research design[51] necessarily gives rise to an – albeit moderate – top-down approach which can now be avoided by means of multi-policy approach adopted using our monitoring system. This system does not claim superiority for a certain policy, but works on the assumption of a more or less dense "policy carpet" (Knoepfel, Kissling 1993: 276) and does not exclude *a priori* any policy that contributes to changes in the state of the environment or changes in regional development activities. Hence, as we have shown, it is an ideal instrument for the analysis of regionally-relevant policy implementation processes and, hence, also for a correctly focused regional development which considers its main task as concentration on regionally relevant sectoral policies.

9.4.2.3. Standardisation

Every monitoring activity must consciously defend itself against the politically dangerous illusion of permanent evaluation, on the one hand, and resignation in the face of all success monitoring, on the other. Thus, for theoretical reasons, it must work with practicable and theory-led standardisation. The work on our system shows that today policy analysis research is sufficiently consolidated to enable the execution of such standardisation without forcing the complexity of political-administrative and ecological processes into a simplistic pattern/framework. Even if many public policies run counter-sequentially at times, in that for example action plans created under precarious temporal or financial frame conditions represent nothing more than the ex-post legitimisation of previously established or coincidentally spatial-temporal output distributions, in the final analysis, policy implementation processes – at least in the European context – follow a similar pattern over extensive periods which it is easy to record on the basis of the policy cycle used here. Like natural scientists, political and administrative scientists can develop standardised process models in interaction with empirical reality which are accessible to ambitious evaluation research. Even if the monitoring system in question has been developed for an area type with relatively low policy density,[52]it can and should be used

[51] On this "top down" versus "bottom up" debate, cf. Benny Hjern, Christopher Hull, Paul Sabatier, quoted in: Knoepfel, Larrue, Varone1999: 226f.

[52] In retrospect this proved an advantage because the development phase could take place under the laboratory conditions of a less complex reality; the research group would hardly have been able to deal with a higher policy density, as found for example in the highly dynamic peri-urban area, which is why it would have resulted in the abandonment of the task of the building site, as was the case in many other similarly situated projects.

in areas with a higher policy density and desirable with its transition from the basic research phase to the planned practical development phase.

9.4.2.4. Between "causality-itis" and "correlation-itis"

Like the situation of monitoring in the gaping chasm between the illusion of permanent policy evaluation and the – pessimistic – renunciation of all success monitoring, the theoretical construct of non-causal linking finds itself torn between a meticulous "causality-itis" and a "correlation-itis" which is found in the social sciences. While the latter runs the risk of correlating everything with everything else, the former leads in the end result to either resignation with respect to the possibility of explicative social sciences per se (negative variant: clear causalities cannot be proven) or to the excessive straining of the causality concept (positive characteristic: causalities are confused with correlations). Non-causal linking allows the monitored observations to exist in their original "rough-hewn" form and does not therefore run the risk of them being "sanded down" as early as the monitoring stage.[53] Energy is not consumed in the "sanding" but in an unprejudiced and detailed assignation of the observations to a limited number of GIS levels and link dimensions. The link dimensions then form usable and plausible start queries on the basis of which the material can be classified depending on the question raised. The question itself is not a linking hypothesis but a simple ordinal category. The monitoring system enables the formulation of myriad working hypothesis; however, it is based only on a basic hypothesis featuring a specific course for policy and ecological processes.[54]

9.4.2.5. An interdisciplinary not transdisciplinary approach

Policy monitoring and evaluation are a multi-disciplinary or interdisciplinary business. The knowledge that it is in the interest of both the natural and technological sciences to co-operate with political sciences is not new for such projects.[55] The claim that this gives rise to "transdisciplinary practice"[56] which goes beyond the traditional disciplines is frankly a

---

[53] The monitoring project will leave the task of "sanding" to subsequent evaluation.

[54] Cf. Knoepfel, Larrue, Varone 1999: 296 ff. (vol. 2).

[55] Cf. Zieschank et al., Grolimund, Peter 1997, Schweizerische Kommission für Umweltbeobachtung etc.

[56] We are not criticising the inclusion of users in this concept but the lack of professionalism which is often (implicitly) justified as a result.

fairy story. The analysis of policy processes is as incapable of being rendered "scientific" by "transdisciplinary" scientists as a "transdisciplinary" political scientist is capable of changing the laws of ecological effects or natural activities. The strength of the proposed monitoring system as substantiated by the reactions of practitioners from the sciences and administration lies in its accessibility to both areas. In this it differs from highly complex "transdisciplinary" models of interaction between "society, nature and politics" which often work with rather "mysterious" system dimensions and not with real action and effect dimensions, which represent the social realities in the heads of the actors on both sides. It is here that a policy-oriented environment monitoring must start.

9.5. Conclusion

Practical policy monitoring and success monitoring are at the start of a long road. The proposed system is, to our knowledge, one of the few initial steps being taken in this direction. This kind of scientifically reliable success monitoring is undoubtedly more expensive than the legislator assumed when monitoring contracts were first issued. With the careful development of the proposed prototype, significant progress should be made in the area of success monitoring – at least in nature conservation but also in regional development. With their integration into the local-authority and cantonal administration processing and development structures, such systems will decrease in price and there are savings to be made in the area of archiving. Furthermore, they contribute to the raising of awareness and the transparency of the administration. Thus, it is to be hoped that on the basis of its brief but sufficient conceptionalization, which reconciles the needs of practice with theoretical requirements, the proposed solution will also be seriously debated for other areas involving the success monitoring of state action.

References

Bundesamt für Umwelt, Wald und Landschaft (BUWAL), Bundesamt für Landschaft (BLW) 1998a. Konzept zur Vermindering der Nitratbelastung aus der landwirtschaftlichen Bodenbewirtschaftung. Bern (mimeo)
Bundesamt für Umwelt, Wald und Landschaft (BUWAL), Bundesamt für Landschaft (BLW) 1998b. Konzept zur Vermindering der Phosphatbe-

lastung aus der landwirtschaftlichen Bodenbewirtschaftung. Bern (mimeo)

Bussmann, Werner, Klöti, Ulrich, Knoepfel, P. (Hrsg.) 1997. *Einführung in die Politikevaluation*. Basel: Helbing & Lichtenhahn

Grolimund, Peter, Peter, Kathrin. 1994. *Integrierte ökosystembezogene Umweltbeobachtung: Konzept für die Einführung eines Informationssystems*. Zürich: vdf

Hofmeister, Albert, Buschor, Ernst (Hrsg.) 1999. *Verwaltungsreform in der Schweiz – eine Zwischenbilanz, Schriftenreihe der Schweizerischen Gesellschaft für Verwaltungswissenschaften (SGVW)*, Band 39, 1999. Bern: Verlag SGVW

IDHEAP/Urbaplan 1999. *Révision du plan directeur cantonal du canton de Vaud: méthodologie d'évaluation des activités à incidences spatiales de l'administration cantonale vaudoise,* décembre 1999.Lausanne: Urbaplan (mimeo)

Kissling-Näf, Ingrid; Knoepfel, Peter 1997: Evaluation und Monitoring, in: Bussmann, Werner, Ulrich Klöti, Peter Knoepfel (Hrsg.), *Einführung in die Politikevaluation.* Basel/Frankfurt: Helbing & Lichtenhahn): 58–77

Kissling-Näf, Ingrid, Wälti, Sonja 1999. Der Vollzug öffentlicher Politiken, in: Klöti, U., Knoepfel, P., Kriesi, H.-P., Linder, W., Papadopoulos, Y. (Hrsg.) 1999. *Handbuch der Schweizer Politik / Manuel de la politique suisse*. Zurich: NZZ-Verlag: 651–692

Klöti, Ueli 1997. Inhaltliche und methodische Anforderungen wissenschaftlicher Politikevaluationen, in: Bussmann, W., Klöti, U., Knoepfel, P. (Hrsg.) 1997. *Einführung in die Politikevaluation.* Basel: Helbing & Lichtenhahn: 39–57

Knoepfel, P., Kissling-Näf, I. 1993. Transformation öffentlicher Politiken durch Verräumlichung – Betrachtungen zum gewandelten Verhältnis zwischen Raum und Politik, in: Héritier, A. (éd.), *Policy-Analyse, Kritik und Neuorientierung,* PVS-Sonderheft 24. Opladen: Westdeutscher Verlag: 267–288

Knoepfel, P. 1994. Zur Wirksamkeit des heutigen Umweltschutzrechts, in: *Umweltrecht in der Praxis/Le Droit de l'environnement dans la pratique*. Zürich: Vereinigung für Umweltrecht, vol. 8, cahier 4: 201–236

Knoepfel, P.; Bussmann, W. 1997. Die öffentliche Politik als Evaluationsobjekt, in: Werner Bussmann, Ulrich Klöti, Peter Knoepfel (Hrsg.), *Einführung in die Politikevaluation.* Basel/Frankfurt: Helbing & Lichtenhahn): 58–77

Knoepfel, P.; Varone, F.; Bussmann, W.; Mader, L. 1997. Evaluationsgegenstände und Evaluationskriterien, in: Werner Bussmann, Ulrich Klöti, P. Knoepfel (Hrsg.), *Einführung in die Politikevaluation.* Basel/Frankfurt: Helbing & Lichtenhahn: 78–118

Knoepfel, P. 1998. Direktzahlungen aus politologischer Sicht, in: Société suisse d'économie et de sociologie rurale (éds), *Ästhetik und Biodiversität, Jahrestagung 1998 in Einsiedeln: Direktzahlungen an die Landwirtschaft im sich wandelnden Umfeld – Stand und Entwicklung*. Zurich: IAW-ETHZ: 43–63

Knoepfel, P., Larrue, C. Varone, F. 1999. *Analyse de politiques publiques comparée*, Matériel de cours de l'IDHEAP 1/1999, vol. 1 et 2. Chavannes-près-Renens: IDHEAP

Koerner, Ch. 1996. *CO2 beeinflusst die Biodiversität, das Versicherungssystem der Natur.* PAX Forum 1:12–15

Moor, P. 1991. *Droit administratif – volume II: les actes administratifs et leur contrôle*. Berne: Stämpfli

Organisation de coopération et de développement économique (OCDE) 1998. *Examens des performances environnementales, Suisse*. Paris: OCDE

Weidner, Helmut, Zieschank, Roland, Knoepfel, Peter (Hrsg.) 1992. *Umweltinformation: Berichterstattung und Informationssysteme in zwölf Ländern*. Berlin: sigma

10
Environmental Capacity Building in Switzerland (1997)

Peter Knoepfel[1]

10.1 The main environmental problems and improvements since 1970

Significant differences can be observed in the degree of success with which the eight problem areas of classical environmental policy have been dealt over the past 25 years. Least progress has been achieved in the areas of noise, nature and landscape, whereas the most significant improvements can be observed in the areas of waste and the prevention of major incidents. The levels of success attained in the other four environmental policy problem areas (water, air, soil and substances) lay between these two extremes. The situation in the individual areas can be summarised as follows[2]:

- With over 95% of households and industries connected to central sewage treatment plants, Switzerland should be at the top in the European and, indeed, world-wide league table for *water protection*. However, approximately 30% of these treatment plants are currently in need of upgrading; the fourth treatment stage which is required for phosphate precipitation has only been implemented in 25 of approximately 1,000 plants. A significant improvement has been observed in the quality of surface waters, in particular with respect to the problems of organic pollution, over recent years. However, trends show that new pollutants are on the increase. Moreover, the ground water which provides 80% of Switzerland's drinking water is under threat from increasing levels of nitrates, chlorides and chlorinated hydrocarbons. Definitive ground water protection zones have been established at planning level for only

---

[1] in : M. Jänicke et H. Weidner (éds), National Environmental Policies. A Comparative Study of Capacity-Building, Berlin (Springer), 1997, p. 175–198 Copyright: We thank the editor for authorizing the publication of this article in this book.

[2] Cf. summarises: Knoepfel 1994; Knoepfel and Wälti (1996) (UNIVOX survey results).

50% of the ground water sources. The level of pressure exerted by the general public on the authorities in the area of water protection policy can still be described as high. Systematic efforts for the revitalisation of streams and rivers and the redevelopment of dried-up river beds (behind reservoirs in mountain areas) have only recently been initiated.

- Whilst significant improvements have been achieved over the past 25 years in *air quality* with respect to SO_2 and heavy metal contamination, No_x and ozone limit values continue to be massively exceeded in agglomeration areas. Whereas levels appear to be sinking in the case of No_x, ozone levels appear to be stable. "There is still significant room for action in the area of acid and nitrogen pollution in sensitive ecosystems" (BUWAL, 1993: 237). The upgrading of existing plants from the point of view of air pollution and the implementation of the more stringent limit values for emissions in industry, which were defined in 1992, has for the most part been completed. Gaps still remain, however, in the small business sector. As opposed to this, despite the establishment of clean-air measurement plans in all cantons for areas with high levels of air pollution, little success has been registered in the area of traffic. A considerable amount of pressure continues to be exerted by the public on the authorities to do more to improve air quality.
- Significant successes have been recorded in *waste policy* over the past five years: the rise in the levels of recycling (paper, aluminium cans, PET bottles, tin cans), the reduction in packaging volumes, the reduction of mercury content in batteries and the increase in composting all demonstrate the remarkable progress which has been made in this area. Most waste incineration plants have been cleaned up in the interest of preventing air pollution. Sufficient disposal capacity has been created, and excess capacity exists in some regions. The overall situation with regard to waste policy can be defined as very positive.
- In contrast, *noise protection policy* continues to be defined as inadequate. 30% of the population is exposed to excessive noise from traffic. Noise levels exceeding 70 decibels have been recorded on almost 1,000 km of local roads. 4% of the population is exposed to excessive noise levels from railway traffic. The level of noise pollution has, however, remained virtually stable over the past three years. Development measures are lacking above all on cantonal and local authority roads. According to BUWAL, very few traffic restrictions or speed reductions have been imposed. The federal authorities are also extremely lazy (BUWAL, 1993: 254) in their efforts to establish measures for the reduction of noise pollution caused by railway systems and air traffic. It

would appear, however, that the general public has become accustomed to the present situation. Demands that the authorities be seen to make greater efforts in the area of noise protection are few and far between.

- Despite the relatively high degree of pressure from the general public, success in the area of *soil protection* has been limited. Soil monitoring, clean-up of contaminated sites and soil protection measures in agriculture are seen as particularly inadequate. The annual loss of arable land remains high at 24 km^2. Indicative limit values have been exceeded at 17 of 102 monitoring sites (above all for copper, cadmium and lead) and increasing levels of soil compacting and erosion have also been recorded. The clean-up of contaminated sites (approximately 10,000 suspected contaminated sites) is not on the agenda in Switzerland until "the onset of a new environmental discipline" (BUWAL, 1993: 155). The amount of land allocated to housing and roads and hence the "sealing" of land which prevents drainage is still on the increase. Local authority regional planning has not yet succeeded in reversing the trend here.
- The degree of success attained in the area of *chemicals policy* can be defined as average to high. Although it is impossible for the authorities to exercise complete control over small business activity, thanks to the adoption of increasingly comprehensive eco-evaluations and improved eco-toxicological testing procedures, there has been a significant improvement in the levels of self-monitoring carried out by companies. A total of 380 new substances were registered with the corresponding tests with BUWAL from 1986 to mid 1993.
- Significant success has been registered for policy on the *prevention of major accidents*, the implementation of which has been stepped up as a result of the chemicals spillage in Basle in November 1987. The cantons and sectors of industry have developed various aids for the implementation of policy in this area and a risk register is also being compiled. Examples of the success in the area include the reduction in potential danger from chemicals and a new safety philosophy among companies. Deficits still probably exist, however, in the small business sector.
- The very limited success in the area of *nature and landscape protection policy* is regrettable, particularly in view of the high levels of expenditure and effort which have been invested in terms of both costs and manpower. The extension of housing zones and public infrastructure building, the disappearance of streams, the reduction in fruit growing areas and the annual increase in the road and transport networks (800

km per year) speak volumes. The number of animal and plant species which are endangered or already obsolete is frightening. "Despite considerable efforts, the condition of nature and landscape has deteriorated further" (BUWAL, 1993: 277). However, "in some individual areas (...) the negative changes" observed in the 1980s as compared with the 1970s have "diminished somewhat". "The overall level of intervention is however still extremely high. The state with its so-called investments is itself contributing to the destruction of the landscape" (BUWAL, 1993: 276).

A comparison of the efforts involved and outcome of administrative activities in the eight areas which constitute Swiss environmental policy gives the results shown in *Table 1*. In this table, one of the four extremities is clearly occupied, two others are at least partly occupied by corresponding findings.

- *Low effort/low results*. Everyone who is concerned with the implementation of environmental policy in Switzerland knows that, for whatever reason, noise and soil protection are neglected with respect to the allocation of personnel in most cantonal authorities. It is hardly surprising, therefore, that measures in these two areas have shown a very low degree of success despite the fact that in the case of one of the areas (noise), a thoroughly practicable regulation has existed for some time now (seven years[3]).

[3] The Noise Protection Ordinance of 15 December 1986 (RS. 814.41) has been in force since 1 April 1987.

Table 1: Level of effort and results obtained in the eight areas of Swiss environmental policy

<table>
<tr><th>effort \ results</th><th>high</th><th>average</th><th>low</th></tr>
<tr><td>high</td><td></td><td colspan="2">water</td></tr>
<tr><td>average</td><td>waste
prevention of major accidents</td><td>air</td><td>nature and landscape</td></tr>
<tr><td>low</td><td colspan="2">materials</td><td>noice
soil</td></tr>
</table>

According to information from the UNIVOX survey, the pressure on the authorities to take action in the area of noise protection has remained consistently low since 1987, however soil protection is now (1993) top of the list of all environmental areas in this respect. The public is clearly demanding that more be done here.

- *High effort/low results.* This permutation, which is of interest for the research of deficits in policy implementation, is not filled by any of the eight areas of environmental policy. However, the area of water protection is today quite close to this extreme position. This finding would probably be more extreme if the level of financial expenditure were also to be taken into account. The deficits result on the one hand in the inadequacies in performance of the installed treatment plants, and on the other hand – particularly – in the (continued) excessive use of farmhouse manure and pesticides in agriculture. It is of little use to the condition of ground and drinking water if household and industrial sewage is carefully "cleaned and polished" by the local authorities while farmers continue to spread massive volumes of nitrates and phosphorous in the open ground.

- *Low effort/high results.* According to our findings this, the most interesting position from the perspective of policy implementation, is at least partly occupied by environmental chemicals policy. These findings should be an indication of the efficacy of the distinctive principle of cooperation applied in this area and the system of self-monitoring which is based on this principle. The effort on the part of the authorities can also be minimised here through the consistent application of the polluter pays principle just as it was probably maximised in the case of water protection through the community pays principle applied hitherto in the area of water protection" (Knoepfel, 1994: 219 f.).

10.2 The development of environmental policy and the main regulations

10.2.1 Development

The oldest piece of environmental legislation in Switzerland is the Forest Law (*Forstgesetzgebung*) which came into force in 1902. It was passed as a result of the large number of avalanches in the mountain valleys which were caused by excessive tree-felling. The law stipulates the maintenance of the entire Swiss shelterwood area and imposes an obligation that compensation is paid for shelter areas which have been felled. This principle is highly progressive even in today's terms as it is presumably at the core of what can be seen as *sustainable development*. The Federal Act on the Protection of Nature and Habitats originates from the 1960s (1966). This act defines the activities of the nature and landscape protection authorities with respect to the protection of the native animal and plant worlds. It was originally directed at nature reserve policy. Since 1987 it has provided – extended – habitat protection in the planning and construction of state-owned buildings and in the granting of licenses or allocation of state contributions. Landscapes of national importance which are worthy of protection have been recorded in a national inventory on the basis of this legislation. These areas are not to be reduced if possible; projects carried out within the protective boundary have to be passed by a special national commission. Since 1966, local authorities and environmental organisations have the right to take legal action in the case of such projects. The ratification of a popular initiative on the protection of moorlands (6 December

1987)[4] resulted in the introduction of an absolute ban on construction within and modification of such landscapes. The same applies for biotopes recorded in national inventories which must be included by local authorities in their land-use plans.

During the 1970s, a systematic policy for water protection emerged at national level when the Water Protection Act of 1955 was reinforced by effective subsidy mechanisms (1972). In 1975, for the first time in the history of Switzerland, a systematic policy was introduced for the limitation of emissions which prescribes the mandatory raising of emission limit values in the event of critical immission levels being exceeded (Decree on the Discharge of Waste Water). Following the Stockholm environmental conference of 1972, clean-air policy in the area of domestic heating systems was introduced at urban and subsequently at cantonal level. Moreover, the cantonal authorities began to carry out (on the basis of the labour protection legislation of 1961) upgrading and redevelopment projects in industrial and small business concerns where the general public had complained about excessive noise and air pollution emissions. However, within the European context, chemicals, air and noise protection legislation were introduced at a very late stage in Switzerland; the Swiss Environmental Protection Act (*Umweltschutzgesetz*) did not come into force until 1985 (Environmental Protection Act of 1983). Largely divorced from corresponding efforts by the state under previously enacted land use legislation (1979), the most important decrees for the implementation of the act were passed during the second half of the 1980s and began to be implemented by the cantons towards the end of the decade. These decrees are based on standard emission limits which are negotiated with the relevant sectors and which were revised at more stringent levels in the early 1990s. Additional emission limits are defined for specific pollution situations under the relevant decrees.

After 1992, the issue of water protection in agriculture, which had been neglected in the 1980s, was finally addressed with the introduction of restrictions on livestock numbers and regulations on the use of fertilisers. It was during this time that the cantons began to address the extremely difficult and previously largely unsuccessful implementation of clean-air measurement plans for areas with high levels of air pollution (private motor traffic was the main target). At the same time, waste policy took on a new lease of life when strategies similar to those used in chemicals policy, e.g. economic incentive schemes and the inclusion of private actors, were considered. Nature and landscape protection policy also found new impe-

---

[4] 57.8% yes votes (17 cantons and 6 half-cantons); poll participation: 47.7%.

tus in the mid 1990s (special inventories). With the revised Environmental Protection Act of late 1995, the Swiss state now has what can be described in the European context as very modern environmental legislation which provides, in particular, economic instruments for the implementation of environmental policy.

10.2.2 Basic instruments

Up to the mid 1980s, Swiss environmental policy worked almost exclusively with classical command and control instruments. This can be clearly seen in the way that licensing, development and control decisions based on federal and cantonal legislation were not enacted as independent environmental policy procedures but were integrated as (new) features into existing procedures (building permit, planning permission and licensing procedures). With the coming-into-force of the new Environmental Protection Act of 1983 (1985) and the corresponding decrees, new instruments were also introduced. These include the environmental impact assessment (*Umweltverträglichkeitsprüfung – UVP*) which, despite being a component of existing licensing procedures, in its concrete form increasingly assumed a coordinating and integrating function both within and outside environmental policy which goes beyond individual decisions. Up to now (spring 1996), approximately 800 of these assessments have been carried out on large plants throughout Switzerland which are of particular relevance to the environment. The right to take legal action by environmental protection organisations has also resulted in the restructuring of environmental legislation procedural practice in that today (in most cases) such organisations play a significant corrective function in the initial licensing procedures for these plants. The organisations are involved as partners at an early stage in the procedure and are taken seriously as partners by both the national authorities and the project groups. An interesting new instrument which has been developed under the chemicals legislation is that of self-monitoring by manufacturers who have developed an autonomous and increasingly complex system for the monitoring of substances at company level: in keeping with the spirit of the principle of cooperation, manufacturing companies are today far more willing to provide access to test results for new chemicals as part of the registration procedure. Finally, it is possible to observe a trend towards the use of economic instruments at cantonal level (air pollution certificates in the cantons of Basle-Town and Basle-Country). However, the use of these instruments remains insignificant in practice.

During the first half of the 1990s, environmental policy widened its experience with cooperative instruments (particularly in the area of waste policy), negotiated solutions (in cases of conflict over large plants), materials-based eco-reports, company-related eco audits and with campaigns (air pollution, waste). These instruments have proven as successful as the environmental impact assessment in initiating ecological learning processes within the national authorities, environmental protection organisations and industrial and small business concerns. These learning processes could be seen to have led to irreversible changes in environment-related behaviour. Moreover, interpolicy cooperation instruments have also been adopted in individual areas (particularly in the area of agriculture and tourism). This was made possible in the area of agriculture by the fact that, as a result of the GATT Uruguay Round (1992) and internal pressure in the area of water protection policy, official agricultural policy switched to area-related subsidies and combined the direct payments to farmers with increasing incentives[5] for ecological services (Article 31b of the Agriculture Act). In this way, nature and landscape policy finally succeeded in implementing its concerns with respect to agricultural policy.

Environmental instruments in the area of planning also gradually assumed greater influence during the first half of the 1990s. What these instruments have in common is that they focus on general regional pollution problems and aim at the implementation of a phased steering of the reduction in pollution levels through (usually quantitative) strategic aims. Typical examples here include the clean-air measurement plans which will involve the implementation of almost 1,000 coordinated individual measures by the cantons. Most of these measures are aimed at private motor traffic. The general local authority water management plan prescribed under the new water protection is a similar management instrument. This plan aims to put measures for sewage treatment in the larger context of a general water management strategy which involves drinking and rain water as well as still and flowing waters. The new Landscape Concept for Switzerland (*Landschaftkonzept Schweiz,* 1996) and the biotope protection inventories, which are to be included in local authority land use plans in accordance with the revised 1987 legislation for the protection of nature and habitats, are also worth mentioning in this context. Noise protection and contaminated sites policy, which is to be established in accordance with the new Environmental Protection Act of 1995, now also include a strong planning component. The use of these new planning instruments in environmental policy means that environmental policy is moving closer to regional de-

[5] 1993: 7.5%; 1994: 13.0%; 1995: 19.7%; 1996: 24.9% (Source: *Bundesamt für Landwirtschaft*, (Federal Ministry for Agriculture) 24. 01. 1996).

velopment. It is possible, therefore, to refer to the "return of environmental policy to the territory".

Under the provisions of the new Environmental Protection Act, a few economic instruments are also to be introduced over the next few years (levies on heating oil containing sulphur, VOC and possibly other substances), as well as instruments for the promotion of technology and regional development and protection of compulsory catchment areas for waste disposal plants. No new regulation has been introduced in the area of environmental liability law.

10.2.3 Policy style

A significant transformation in policy style has also been under way, particularly in the cantons but also at federal level, which can be seen as reciprocal to the development of these new policy instruments. The former attitudes and behaviour of the old police administration (particularly the water protection authorities) have been replaced by the cooperative style of younger, ecologically-minded officials who have considerably greater responsibility and competencies than was the case with their counterparts in the strictly hierarchical police-run environmental authorities of the previous era. Negotiated solutions began to be introduced as early as the late 1980s when emission limits had to be worked out in collaboration with representatives from industrial and small business sectors. The increasing willingness of these sectors to accept these limit values in the interest of equal treatment led to the disappearance of a previously widespread fear of contact with the authorities. In the early 1990s, the attitudes and behaviour of the 26 cantonal environmental authorities underwent fundamental change yet again as did those of the Federal Ministry for the Environment, Forests and Landscape (*Bundesamt für Umwelt, Wald und Landschaft – BUWAL*), which is organised in divisions with responsibility for individual environmental media. In many places, problem-specific cooperation networks emerged as part of what were known as project organisations, in which the representatives of the polluters, environmental organisations and the individual divisions of the environmental protection authorities cooperated with each other on a largely equal footing with the aim of finding a solution to the relevant problem. This new policy style is most prevalent in the area of waste policy (particularly in the application of the new beverages packaging decree of 1993), in nature protection policy (completion of land-use contracts between agricultural or nature protection authorities and farmers) and in industry-related environmental policy (environmental impact assessment procedure in highly complex, learning-oriented networks).

Long-standing observers of Swiss environmental policy should be united in the opinion that this undoubtedly positive change in policy style, which was brought about by the adoption of a range of new instruments, the advent of a new generation in the environmental authorities and new training programmes, should be seen as the most significant change in this policy during the first half of the 1990s. It could be part of a major trend towards new forms of public management which, however, has also benefited from a series of strong impulses from environmental policy.

10.2.4 Policy evaluation

Systematic empirical analyses of all areas of Swiss environmental policy which also include the implementation activities of the cantons have yet to be carried out in Switzerland. There is also no systematic OECD evaluation on this policy. The first attempts at general evaluation can be found in the two reports on the state of the environment which were published by the Federal Ministry of the Environment, Forests and Landscape (BUWAL, 1990; 1993) and in various cantonal reports on the state of the environment, the second or third editions of which have previously been compiled (in particular by the cantons of Bern, Aargau, Zurich and Lucerne). It is not yet possible to predict the extent to which the initiative for the introduction of an integrated ecosystem-related environmental observation at national level, which was initiated by the Swiss Academy of Natural Sciences (*Schweizerische Akademie der Naturwissenschaften – SANW*), will succeed. It is worth noting, however, that a "Steering and Coordination Body for Environmental Observation" (*Lenkungs- und Koordinationsorgan Umweltbeobachtung – LEKUB*) has been established which aims to establish a comprehensive system of environmental observation in cooperation with the cantons and all other departments within the federal authorities which have relevant data available to them (Swiss Ministry of the Interior, 1995).

In contrast, summary evaluations of the efficacy of environmental protection legislation are available (in particular BUWAL, 1993 and Knoepfel, 1994) as well as systematic evaluations have been carried out on individual topics within the general areas. A case in point is the study for the evaluation of BUWAL which was published in 1991 by the National Council Commission of Enquiry (*Geschäftsprüfungskommission des Nationalrats*). Air pollution control, environmental impact assessment and rural water protection were systematically studied as part of this evaluation (Geschäftsprüfungskommission des Nationalrats, 1991). Studies on the efficacy of clean-air measurement plans (Knoepfel, Imhof and

Zimmermann, 1995), the implementation of clean-air policy in the area of domestic heating systems (Balthasar and Knöpfel, 1995) and on the topic of soil protection (Zimmermann, Achermann and Knoepfel, 1996) exist or will be published shortly. The findings in these studies are taken into account in section 1 above.

In the international context, Swiss environmental policy can still be seen to take a leading position in the areas of air pollution and water protection where critical load limit values have been defined which, as a result of the "eco-centring" of the two acts, are considerably more stringent than those in other countries and in the EU. The adoption of the popular initiative for the protection of moorlands should also place Switzerland in a leading position in Europe with regard to the protection of this type of landscape. As opposed to this, Swiss environmental policy lies somewhere around the average mark for industrialised countries when it comes to noise protection, soil protection and energy-saving policy. In comparison to the EU states, the absence of regulations for free access to environmental information and for eco-audit in Switzerland is a striking deficit. Membership of the European Economic Area would have brought about some progress here, if only the federal authorities promoting the move had suggested the relevant adjustment. As is a well known fact, the electorate rejected this move in December 1992. The fear that as a member of the European Union, Switzerland could be obliged to cut back on its relatively high level of environmental protection is known to have played a role for the Greens in their opposition to Switzerland's membership.

With its new Environmental Protection Act of 1995, Switzerland also made some – tentative – attempts towards the use of economic instruments. However, the revised legislation did not succeed in introducing a CO_2 levy or increased taxation of energy carriers as part of the reinforcement of the energy components of environmental policy. At international level, Switzerland has generally aligned itself with the most progressive countries (Biodiversity Convention, CO_2 reduction, Montreal Protocol etc.). As opposed to this, however, due to internal pressure from the Alpine cantons which were concerned about their economic development, Switzerland did not sign the Alpine Protection Convention in 1995. This is all the more astonishing in view of the fact that in 1994 the electorate surprisingly voted in favour of the acceptance of an Alpine protection initiative[6] which proposed significant reductions in transit traffic through the Alps and the transfer of the transport to the railways. With this decision, the

---

[6] Referendum of 20 February 1994. 51.9% yes (13 cantons and 6 half cantons).

Swiss population contradicted an earlier international contract concluded with the EC (Transit Agreement, ratified in 1992).

10.3 The capacity of proponents of environmental policy

10.3.1 Public institutions

In accordance with the basic principle of executive federalism, which applies equally to all public policies, the federal government is responsible for the formulation of programmes (drawing up and passing of laws, decrees and implementation guidelines) and the cantons are responsible for the implementation of this legislation etc. On the basis of their own legislation, the cantons can delegate individual implementation tasks to (larger) local authorities. This principle also applies in environmental policy, however some implementation tasks remain the task of the federal authorities (regulation of chemicals, testing of appliances and motor vehicles, intercantonal and international implementation tasks).

Personnel, financial and instrumental implementation resources are correspondingly distributed among the three administrative levels. According to my own estimations, at the beginning of 1996, 450 permanent positions were allocated for environmental policy requirements at central federal administration level, approximately 1,500 at cantonal level and 4,500 at local authority level. This total capacity of approximately 6,500 permanent positions (more than one permanent position per 1,000 inhabitants) is considerable, especially in view of the fact that half of this capacity was established over the last decade despite increasing overall budget restrictions. Public expenditure for environmental protection in Switzerland is estimated at approximately 4.3% of the total national budget[7].

When the Environmental Protection Act was first passed in 1983, the Federal Agency for Environmental Protection (BUWAL), which had been developed from the old Swiss Agency for Water Protection, underwent a period of rapid expansion. In 1989, this agency was amalgamated with the former Federal Agency for Forestry and Landscape Protection and in the wake of the change in direction was successively transformed from 1992 in accordance with the recommendations of the National Council Commission of Enquiry. This transformation involved, for example, the establishment of a new department for waste management, a horizontal division for environmental law and a horizontal unit for environmental impact as-

[7] Source: Statistisches Jahrbuch der Schweiz (1996: 394).

sessment, and the general reduction of hierarchical structures. Today, BUWAL has a total of over 314.5 permanent posts and an annual budget of 528.4 million Swiss Francs (1996). During the first half of the 1990s, environmental policy administrative capacities were also expanded in other federal ministries. This applies in particular to the Federal Agency of Health, Energy, Transport, Agriculture, Statistics, Regional Development, Civil Aviation and various sections of the Military Department and the Department of Foreign Affairs. A total of approximately 100 permanent positions are allocated for environmental requirements here.

Even more impressive is the way in which environmental administrative capacities have been developed at cantonal level. Under the environmental protection legislation of the second half of the 1980s, the cantons were obliged to establish eight specialised offices (water, air, waste, materials, major accidents, soil, noise and nature and landscape protection). For historical reasons, the offices for nature and landscape protection are usually located outside the newly established cantonal environmental protection authorities and the other seven services are not found within a single office in all the cantons. As a result of this, the implementation of environmental policy in the cantons shows a varied horizontal degree of centralisation. The cantonal environmental authorities which are structured on a successive and "law-oriented" basis (i.e. in accordance with the phased enactment of various federal decrees) today face increasing criticism. This originates not least from a strong trend in favour of client-oriented policy services. The first canton to establish a second-generation environmental authority was St. Gallen (Baudepartement des Kantons St. Gallen, 1995). In this case, a media-specific basic structure was replaced by a polluter-specific one. This enabled the acceleration of licensing procedures and the introduction of integrated general licenses and also makes it possible to take greater account of aspects of environmental land-use planning.

Considerable administrative resources have also been established in the area of environmental protection at the level of the approximately 3,000 Swiss local authorities in recent years. Thus, in addition to sewage-treatment personnel, local authority administrations in towns of approximately 10,000 inhabitants now also have environmental protection officers who work mainly in the areas of air pollution control and traffic, waste management, nature and noise protection. A conservative estimate puts the total number of permanent positions at local authority level at approximately 4,500. It should also be noted that at local authority level, in particular, environmental protection is often implemented in the form of voluntary official work performed by citizens in their spare time. Hence, the actual capacity at local-authority level is probably somewhat greater.

With the increasing density of regulations under environmental legislation, the courts have gained in significance. Cantonal administrative courts and the Swiss Federal Court have recently proven to be important guardians of the strict application of Swiss environmental protection law in the cantons and local authority areas. This applies particularly to regional development (buildings outside designated construction areas), quantitative water protection, nature and landscape protection (particularly forest protection) and – most recently – the clean-air measurement plans and noise protection. The Swiss justice system exercised a significant influence on the way in which greater coordination of procedures and corresponding reorganisation measures are being considered in the cantons.

10.3.2 Environmental organisations

Table 2 shows the membership of the 20 environmental protection organisations in Switzerland which have the right to take legal action under environmental law enacted at the end of 1995. Although overlap in membership is very common, the membership of this organisation is, however, impressive (almost 10% of the population). It is only slightly lower than that of the trade unions[8] and significantly higher than that of the political parties[9]. The main environmental protection organisations formed an environmental coordination group in the early 1990s which regulates the topic-specific and project-specific responsibilities of the individual organisations. The larger associations have substructures based on cantons and linguistic boundaries. The willingness of the population to contribute to environmental protection in the form of financial contributions or active membership has remained constant since the mid 1980s at approximately 20%–30%; it is slightly higher in German-speaking Switzerland than in French-speaking Switzerland. The former considerable differences in the general political orientation of the different environmental protection organisations, whereby for example the predominantly radical position of the WWF contrasted with the rather conservative policy of the Swiss Association for Nature Protection (*Schweizerischer Bund für Naturschutz*), is generally a thing of the past, at least at national level. Differences can still, however, be found at cantonal and local level.

---

[8] SGB: 429,024 members (1994); CNG: 134,167 members (1995); VSA: 127,103 members (1995) (= Total 690,294).

[9] Total 360,000 members (1990), Statistisches Handbuch, Forschungszentrum für Schweizerische Politik, Bern, 1991.

The environmental protection organisations are represented as recognised and scientifically competent partners at all levels of the Swiss state in the context of licensing, planning and legislative procedures. Moreover, they now increasingly take part in direct discussions with polluters from industry, commerce, agriculture and households and with representatives of state infrastructure policies (in particular: traffic). In the spirit of the principle of cooperation, they enter into contracts with groups of polluters, distribute labels (in cooperation with consumer protection groups), play an important role in practical environmental training, information and consultancy and are increasingly involved in the provision of ecology-oriented services.

The relatively weak position of the Swiss Green Party, which emerged from various groups in the mid 1980s, is striking when compared with the success of the movement in other central European states. This party, which is still vulnerable today to the tensions between the (linguistic) regions and individual cantonal sections, has been represented in the Swiss parliament since 1983; in 1991 it obtained 6.1% of the votes, however, in the last National Council elections of 1995, it lost votes to the Swiss Social Democratic Party. After an initial reluctance, the latter has assumed a strong profile on ecological issues since the early 1990s and is today seen as an important protector of ecological principles in the Swiss four-party government (the party obtained 23% of the vote in 1995). The consultancy procedures for the revision of the Nature and Habitat Protection Act and the Environmental Protection Act during 1995 also showed that there is ample potential for the continuation of the relatively progressive environmental legislation from the 1980s in the middle-class centre-right parties; attacks on the core of this legislation from small business and right-wing middle-class circles were successfully warded off.

Table 2: Register of environmental organisations with the right to take legal action (1995)

Organisation	Membership1993
Rheinaubund (Schweiz. Arbeitsgemeinschaft für Natur und Heimat) (Swiss Association for Nature and Habitats)	800
Schweiz. Vereinigung für Landesplanung (VLP) (Swiss Association for National Planning)	3,609
WWF (Schweiz) (World Wildlife Fund)	170,000
Schweizer Vogelschutz (SVS), Verband für Vogel- und Naturschutz (Swiss Bird Protection, Association for the Protection of Birds and Nature)	6,000
Schweizer Heimatschutz (SHS) (Swiss Habitat Protection)	20,000
Schweiz. Bund für Naturschutz (SBN) (Swiss Association for Nature Protection)	100,000
Schweizer Alpen-Club (SAC) (Swiss Alps Club)	85,000
Schweiz. Gesellschaft für Umweltschutz (SGU) (Swiss Society for Environmental Protection)	8,357
Helvetica Nostra (Our Switzerland)	1,981
Schweiz. Vereinigung für Gesundheitstechnik (SVG) (Swiss Association for Health Technology)	–
Schweiz. Liga gegen den Lärm (Swiss Anti-Noise League)	230
Schweiz. Vereinigung für Gewässerschutz und Lufthygiene (VGL) (Swiss Association for Protection of Waters and Clean Air)	600
Schweiz. Stiftung für Landschaftschutz und Landschaftspflege (SL) (Swiss Foundation for Landscape Protection and Landscape Management)	801
Schweiz. Energiestiftung (SES) (Swiss Energy Foundation)	3,200
Naturfreunde Schweiz (NFS)* (Swiss Friends of Nature)	29,932
Verband Schweiz. Abwasserfachleute (VSA)	12,400

Organisation	Membership1993
(Association of Swiss Sewage Specialists)	
Aqua Viva (Nationale Aktionsgem. zum Schutze der Flüsse und Seen) (National Action Group for Protection of Rivers and Lakes)	–
Schweiz. Fischerei-Verband (SFB) (Swiss Fishing Association)	10,000
Schweiz. Verkehrs-Stiftung (SVS) (Swiss Traffic Foundation)	–
Verkehrs-Club der Schweiz (VCS) (Swiss Traffic Club)	100,000
Total	**552,910**

* Membership numbers for 1993
Source: Direct survey of the different organisations (3/96).

10.3.3 Economic proponents of environmental protection

The contribution of the Swiss ecology industry to the Swiss net national product is currently slightly lower than that of the agricultural sector (approximately 3%, see Pillet, 1988; Conus and Rossi, 1992; Walther, 1990). The ecology industry is gaining in significance in the areas of waste, water protection, air pollution control and noise protection and also represents an important area for potential development. This sector also involves services which are required as part of large-scale environmental impact assessment procedures, the implementation of the clean-air measurement plans and in the landscape and nature inventories. Hence, ecological products and services doubtless represent an area of growth in the Swiss national economy.

Major ecological learning processes have taken place in the other sectors of industry and commerce and – increasingly in the services sector over the past five years. These are reflected in new company-policy models, in company-internal ecological control mechanisms and in the conversion of product ranges (for example in the chemicals industry). It is possible to observe a virtual run on eco-labels and eco-certification of companies (in accordance with the ISO 14'002 standard). This promises not only real improvement in company or product-specific eco-audits but also in the development of important markets for research and consultancy. Environment-oriented training and education programmes are correspondingly being tailored to the future needs of industry and service concerns. Through certification, we have succeeded in doing what eluded official environmental policy for some time, i.e. enabling ecological philosophies to penetrate the core of entrepreneurial strategies. Ecological messages in marketing are now very favourably received by consumers.

This trend can be seen to have reached the agricultural sector by the mid 1990s at latest. Supported by the above-mentioned direct ecological payments, which are intended to motivate farmers to switch to integrated production or even organic production, the agricultural sector and the large distribution chain has strongly promoted the ecological argument in product acquisition. A continued increase can be observed in sales of environment-friendly agricultural products now in early 1996. The two major wholesalers, Migros and Coop, are even reporting supply shortages in the area of organic products. The proportion of agricultural products originating from integrated production today lies at approximately 20%, that of organic products, which cost up to 150% more, is approximately 1.5%[10].

This willingness on the part of the consumer to purchase more expensive environment-friendly products is also reflected in the general willingness shown by the Swiss population to pay for environmental policy measures which has remained relatively high and unchanged in the European context since 1986. A slight decrease was observed during the recession of 1994 followed by a slight recovery 1995 (Knoepfel and Wälti, 1996). *Figure 1* demonstrates the high degree of willingness among the population to alter their behaviour on a non-economical basis.

[10] Individual estimate 1996 from employee at the Federal Ministry for Agriculture (Frau Zajac).

Figure 1: Willingness of the Swiss population to adopt environment-relevant behaviour

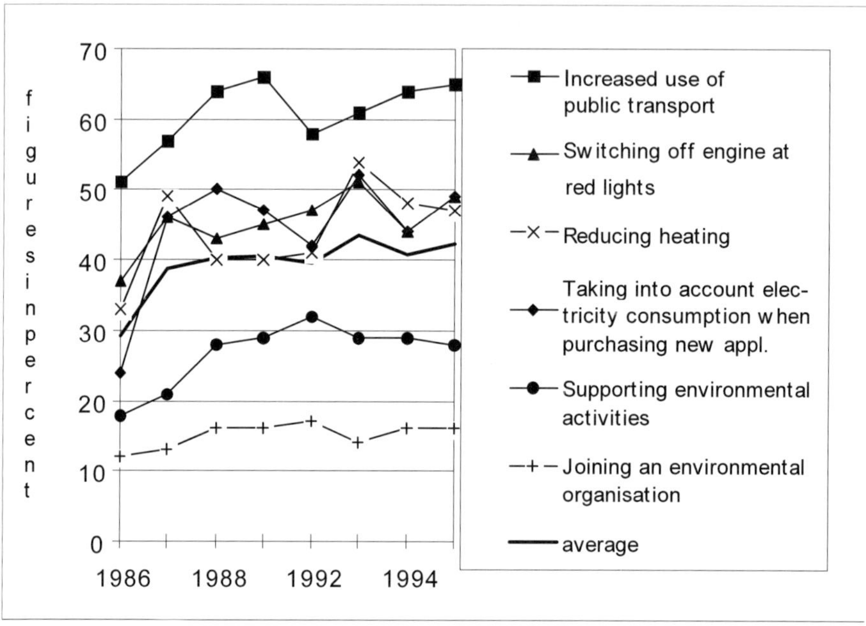

Source: Knoepfel and Wälti, 1996, UNIVOX.

Hence, it is possible to assume that a very high economic potential now exists for ecological policy in 1996 as compared with that which existed in 1985 (coming into force of the first Environmental Protection Act). With respect to its contribution to the creation of wealth and jobs, ecology has today become an important economic factor and its elimination would involve significant economic costs. Hence, ecological policy has succeeded in economically guaranteeing its political potential.

10.4 The changing constellation of proponents

It is possible to distinguish four development phases for environmental policy in Switzerland corresponding to the classification used by *Jänicke* and *Weidner* in this book. As is demonstrated below, the developments in

Switzerland run for the most part parallel to those in federal German environmental policy.

- *Dilution (up to 1975)*

  This embryonic phase, which can be identified in the first hours of life of every environmental policy, is characterised by its exclusive concentration on environmental quality. State intervention is limited to areas with excessive levels of pollution. This intervention can either consist of "improved" spatial-temporal distribution of the emissions (high chimneys, discharge of water into large drainage canals etc.[11]), or in the restriction of measures aimed at reducing emissions to particularly vulnerable areas (low-sulphur heating oil for towns and cities etc.). The protected object is always limited to particularly vulnerable locations with groups of persons or ecosystems at risk; these policies have no preventative component. Such strategies were also implemented in Switzerland up to the mid 1970s at urban and cantonal level; typical examples include the domestic heating controls which were limited to problem areas and the concentration of sewage treatment activities in local authorities with particular water shortages. The end of this period is marked in legislative terms by the Sewage Disposal Decree of 1975. This decree introduced emission limits for the first time which were implemented on a general level and irrespective of the previous level of pollution in the discharge canals of the sewage treatment plants. Around the same time, a similar insight was successfully imposed in the cantonal clean-air policy and in the (case-related) national environmental quality policy. The actor constellation in these dilution policies, which were always implemented by the police administration, is bilateral: you had the police authority on the one hand and the polluting "perpetrator" on the other.

- *Dilution and End-of-the-Pipe Policy (1975–1986)*

  Although it only became an integral part of Swiss environmental policy in 1983, the "two phase concept" of environmental policy, which is still applied today, was introduced at cantonal and national level after the enactment of the above-mentioned sewage discharge decree. According to this school of thought, emissions limitations should be implemented throughout the country, irrespective of previous levels of pollution. In cases where despite application of these limit values, excessive pollution continues with the (increasingly quantified) critical load limit val-

[11] Temporal measures (e.g. reduction of emissions during smog periods) were rare in Switzerland (unlike in France).

ues being exceeded, the emissions restrictions must be reinforced. This policy was initially applied in water protection followed by air pollution control and finally in local and cantonal waste policy, which was increasingly converted to the principles of recycling and energy reharnessing. The origin of this policy can ultimately be traced back to an extended concept of ecosystems, which was able to demonstrate with increasing scientific precision that pollutants accumulate in ecosystem, can travel extensive distances (acid rain) and can combine to form actual material flows which can give rise to effects which go far beyond the original place where discharged or deposited. Hence, general emissions reduction strategies are required which are reflected in an increasingly broad range of emission limit values and in product and processing standards. These standards were developed, applied and gradually intensified in the approximate ten-year phase between 1975 and 1986. Finally at the end of this phase, this development found expression in legislative measures containing the definition of emission limit values for the individual media and economic sectors which were quickly enacted without political difficulties. The developments in this period are largely comparable with those in foreign legislation (in particular the federal German legislation – TA Air and TA Noise). However, in contrast to developments in the Federal Republic of Germany, in Switzerland cantonal environmental policy continued to be predominantly based on a bilateral actor constellation: The critical load limit values were adopted from (scientific) WHO recommendations and environmental protection organisations were not involved in their development; the emission limit values represented the outcome of negotiations between industrial and small business sectors and state authorities with minimum input from the other relevant organisations.

- *"Intense End-of-the-Pipe Treatment" (1986–1992)*

  With the coming into force of the most important decrees and corresponding immission limit values, an important mechanism was initiated in industry which ultimately led to the emergence of the eco-industrial sector described above. Emission limit values became a defining factor in competition policy. The polluters accordingly demanded equal application of the limit values and increasingly identified the economic advantages to be gained in achieving values which are below the legal maximum levels (energy and material savings, new technologies). In 1992, closed cycles and increasingly technologically sophisticated recuperation technology resulted in the, in part, significant tightening of emission limit values in air pollution control. Mandatory End-of-the-Pipe technologies became big business and eventually constituted the key impetus behind industrial and even small business renewal. The position of the environmental organisations was also strengthened during this phase not least as a result of their recognition as partners with a right to take legal action in mandatory environmental impact assessment licensing procedures. The increasing integration of personnel from these organisations with parts of the state administration and with university research teams resulted in their becoming familiar with the importance of emission limit values for effective environmental policy. This resulted in the unlocking of the previous bilateral actor structure, which towards the end of this period took on a definitive trilateral structure in the form of the "iron triangle" (Knoepfel, 1995: 14).

- *Ecological Modernisation (1992–1996)*

  During this last phase (for the present), the above-mentioned economic dynamism led to the political mutation of the topic of the environment: having originated in the form of a demand for economic resources management it became an actual component of economic policy. Supported, no doubt, by the conference in Rio in 1992 and the associated vocabulary of sustainable development, in Switzerland the realisation has spread that environmental policy involves more than the management of emissions. Hence, ecology is penetrating into the core area of economic behavioural maxims (Minsch et al., 1996). This demands of companies, and also of the state and its environment-damaging policies, that they take economic account of the consumption of renewable and non-renewable natural resources. Today, we are merely at the outset of this development. It is unlikely – in Switzerland too – that this message, which is as clear as it is simple, will succeed in spreading over this decade.

10.5 Capacity-building

10.5.1 Actor capacity

The numbers, political weight and competence of official and social, ecologically proactive actors in Switzerland have reached considerable levels. On the side of the political-administrative system, a largely irreversible process was initiated involving the establishment of environmental policy capacities in areas which go far beyond environmental policy in the strict sense. Parliaments and authorities at federal, cantonal and local authority level doubled their capacities in this area in the decade from 1985 to 1995 – a process which is unique to this policy area. However, ecological policy is still relatively weakly represented in the area of state infrastructure projects and in the area of energy policy where it has led to corrective action but not to the kind of fundamental processes of renewal which can be observed in industry and commerce. This explains the previously mentioned low level of success in nature and landscape protection policy.

Industry, commerce and services also developed ecological capacities over the same period, a development which even the most astute observer would not have predicted in 1985. The previously described dynamism of the industrial system is as responsible for this development as a strongly proactive attitude of consumers. Surveys on household waste behaviour and attitudes show a high degree of acceptance of ecologically sound behaviour in this area. Old-style productivity-based agricultural policy was replaced overnight by a new policy which, unlike its predecessor, has a stabilising effect on the ecology situation. The central weak point is and remains the (strongly state-promoted) private and public transport systems. The most important area of conflict today is that between mobility and ecology.

10.5.2 System capacity

The institutional and information structures available to official and social actors have also been considerably reinforced over the past decade. With the overall introduction of environmental impact assessment procedures at project level and different types of mechanisms for establishing consensus at planning level (particularly in energy policy), the possibilities for the real participation of environmental protection organisations, other citizens' action groups and individual persons have increased significantly. These procedures do not only represent decision-making and learning processes. Their positive effect on the ecological moulding of policy output can today be seen as proven.

Interestingly, this observation correlates with a clear reduction in the pressure exerted by the public on environmental policy authorities to "do more" on environmental policy issues. This observation is based on a series of corresponding findings from 1986 to 1995 (cf. *Figure 2*)

Figure 2: Pressure from the public on the authorities ("do more") based on UNIVOX surveys from 1986 to 1995.

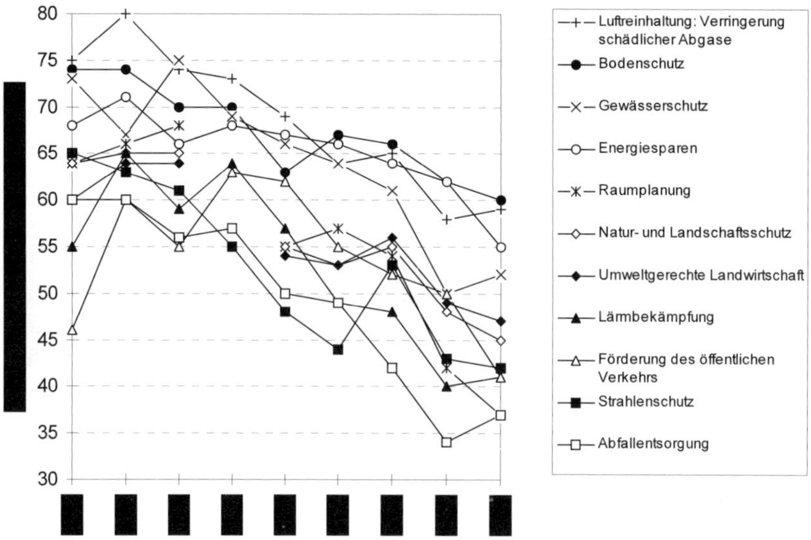

Source: Knoepfel and Wälti, 1996, UNIVOX.

Opportunities for involvement were also improved due to considerable efforts on the part of local authority, cantonal and national authorities to make data on the environment situation more widely available. The – media-based – system of environmental observation in Switzerland is in a leading position, in an international context also (Weidner, Zieschank and Knoepfel, 1992). Attempts at establishing integrated environmental observation have failed for the present. Systematic observation of the performance of environmental policy and non-environmental policy with respect to an ecological evaluation of state activity is only at the very early stages.

Scientific and technological environmental research at universities has expanded considerably since 1985. Several national research programmes and a focal programme of the Swiss National Science Foundation and the Swiss institutes for technology and eight cantonal universities played a key role in this development. Despite some important efforts, insights provided by social-scientific environmental research can still only be described as modest. The graduate and post-graduate studies programmes, which were developed in the second half of the 1980s and which are on offer today in almost all of the institutes for technology and universities, have made an important contribution to the development of environmental capacity. Ecological perspectives are also receiving far more emphasis in the traditional disciplines than was previously the case. Curricula for primary schools, second level and vocational schools have also undergone corresponding changes.

Despite the recession, now in the mid-nineties, the environment dominates the media unlike any other topic. To the delight of the increasing circle of readers and users, numerous new magazines, books, CD-ROMs and a variety of environment-relevant databases have been produced and published over the past five years.

10.6 Capacity for what?

Despite the considerable expansion of capacity in terms of actors and at the system level, it is impossible today to give a conclusive answer to the question as to whether this capacity will be sufficient to bring about the above-mentioned modernisation of environmental policy towards consistent resource management by the end of the century. Environmental economists demonstrate convincingly that important economic resources are becoming cheaper and not more expensive for production (energy and infrastructure). There is no sign of the formation of a solid coalition of actors in favour of consistent price increases for natural resources. The previously mentioned economic dynamism may initially have led to the use of economic instruments, however they are still very much part of state emissions management. At present, there is no room in Swiss policy for solid, widely-based coalitions of representatives from the industrial system, environmental organisations and state authorities which favour a system of artificial scarcity and increased pricing of natural resources in the interest of maintaining their regeneration capacities and in the interest of future generations. Only lone positive signals can be heard from individual political and social groups which are transmitted under the keywords of cost internalisation, true costs and energy tax.

Hence, now at the beginning of the 1996 year of recession, the signs tend to be more negative than positive: the globalisation of markets is forcing the implementation of cost-saving reductions in the area of mobility and the adoption of short-term survival strategies. The situation on the labour markets is used as an argument against more far-reaching environment strategies and for the stabilisation of the ecology industry which is based on emissions management. The ecology industry could also be seeing its position under threat as a result perhaps of a fundamental modernisation of industrial society. Hence the highly promising transition to ecological modernisation in Switzerland which was clearly in evidence in the first half of the decade is now beginning to falter.

References

Balthasar, A., Knöpfel, C. 1994. *Umweltpolitik und technische Entwicklung, eine politikwissenschaftliche Evaluation am Beispiel der Heizungen.* Basel: Helbing & Lichtenhahn (Reihe Ökologie und Gesellschaft Nr. 8)

Baudepartement des Kantons St. Gallen 1995. *Reorganisation des Amtes für Umweltschutz*

BUWAL (Bundesamt für Umwelt, Wald und Landschaft), 1991. *Zum Stand der Umwelt 1990.* Bern: EDMZ

BUWAL (Bundesamt für Umwelt, Wald und Landschaft) 1994. *Zum Stand der Umwelt 1993.* Bern: EDMZ

Conus, J., Rossi, A. 1992. *Le business écologique: définition, structure, évolution et perspectives de l'industrie de la protection de l'environnement de la Suisse.* Bern: Bundesamt für Konjunkturfragen

Eidgenössisches Departement des Innern 1995. *Bericht des Lenkungs- und Koordinationsorgans Umweltbeobachtung (LEKUB) über seine Tätigkeiten der Phase I (Juni 1993–Juni 1995).* Bern

Forschungszentrum für Schweizerische Politik 1991. S*tatistisches Handbuch.* Bern

Grolimund, P., Peter, K. 1994. *Integrierte ökosystembezogene Umweltbeobachtung, Konzept für die Einführung eines Beobachtungssystems.* Zürich: Hochschulverlag AG an der ETH Zürich

Knoepfel, P. 1994. Zur Wirksamkeit des heutigen Umweltschutzrechts, *Umweltrecht in der Praxis/Le droit de l'environnement dans la pratique* 8, 4, 201–236

Koepfel, P. 1995. Approaches to an Effective Framework for Environmental Management, in: Carius, A., Hoettler L., Merckur H. (eds.), *Environmental Management in Kenya, Tanzania, Uganda and Zimbabwe.* Berlin: German Foundation For International Devlopment, Public Administration Promotion Center

Knoepfel, P., Baitsch, C., Eberle, A. 1995: *Überprüfung der Aufbauorganisation des Amtes für Umweltschutz des Kantons St. Gallen,* Schlussbericht (i. A. des Baudepartements des Kantons St. Gallen). St. Gallen: AfU St. Gallen/Chavannes-près-Renens/IDHEAP

Knoepfel, P., Wälti, S. 1996: *UNIVOX-Umfrage Umwelt, Auswertung 1995*. Chavannes-près-Renens: IDHEAP/Adliswil (GfS)

Knoepfel, P., Zimmermann, W. (in coll. with Sailer, G., Matafora, E.) 1992: Evaluation des BUWAL. *Expertenbericht zur Evaluation der Luftreinhaltung, des ländlichen Gewässerschutzes und der UVP des Bundes*, Schlussbericht, 18. November 1991. Bern: EDMZ

Minsch, J., Eberle, A., Meier, B., Schneidewind, U. 1996: *Mut zum Umbau. Nachhaltige Entwicklung durch ökologische Innovationen.* Basel: Birkhäuser

Pillet, G. 1988: *Bilan des coûts économiques de la pollution* (manuscript). Genève: Université de Genève, Centre universitaire d'écologie humaine et des sciences de l'environnement

Schweizerische Kommission für Umweltbeobachtung (SKUB) 1993: *Integrierte ökosystembezogene Umweltbeobachtung, Konzept für die Einführung eines Beobachtungssystems*

Walther, A. 1990: *Die Folgerechnung von Umweltschäden des volkswirtschaftlichen Rechnungswesens am Beispiel der Schweiz*, (dissertation). St. Gallen: Hochschule St. Gallen

Weidner, H., Zieschank, R., Knoepfel, P. (eds.) 1992: *Umweltinformation: Berichterstattung und Informationssysteme in zwölf Ländern.* Berlin: sigma

11 Ecobusiness and the State-Analysis and Scenarios (1996)

Peter Knoepfel[1]

11.1 Introduction

Who would have thought ten or fifteen years ago that one day it would be possible to earn a living from ecology? Who could have predicted the emergence of a whole new sector of the economy producing goods and services whose purpose is to reduce the environmental impact of industry, both large-scale and small-scale, farming, household and state infrastructure? The contributions in this book show that today a number of markets exist which respond to the demand for such goods and services. What we are talking about here is ecobusiness. The existence of the companies which supply these markets is based on the growing demand for technology and expertise which is dictated in turn by the ever-expanding and increasingly exigent body of environmental legislation. Furthermore, these new ecobusiness companies have effectively become the partners of the state in that through their research and development they advance the status of anti-pollution technologies which enable the legislature to raise its standards successively. Thus, it is in the interest of ecobusiness that the state compel industry, farmers, households and the state actors responsible for the construction and maintenance of public infrastructure to exploit its products and consultancy services to the full. This prompts the following questions with respect to the relationship between state services and companies involved in ecobusiness: who controls whom and is using whom?

It would appear that the ecological goods and consultancy markets are partly the product of state legislative intervention: without the legislation which imposes restrictions on all economic actors, there would hardly be any demand for such services. Historically, at least, legislation has anticipated demand. When the first industrial sectors were subject to constraints

---

[1] in: Benninghoff, M., Joerchel, B. et Knoepfel, P. (éds) 1997. *L'écobusiness: enjeux et perspectives pour la politique de l'environnement*. Bâle et Francfort: Helbing & Lichtenhahn (série Ecologie & Société, vol. 11) (= French version française). Copyright: We thank the editor for authorizing the publication of this article in this book.

imposed by legislative programmes aimed at cleaning up air and water pollution and reducing emissions, the most "intelligent" were not content simply to develop technological innovations for their own use and started to market these technologies among their counterparts. Hence, the opportunity to create new profit centres within a stable legislative environment has led to the emergence of a new economic sector: following initial hesitation, the large parent companies established new subsidiaries which with time gained a degree of economic and technological independence. Whereas their development was initially dictated by the immediate needs of the existing economic environment, these new companies subsequently developed their own dynamic to the extent that they formed alliances with the state environmental administrations in the battle against pollution caused by their own parent companies. This veritable Greek tragedy resulted in the emergence of a new triangular actor structure with the polluting companies, the state and its environmental services and ecobusiness at its vertices.

However, this new actor structure had to be introduced into the existing triangular structure which is known as the "iron triangle"[2] and generally comprises a wide range of economic agents whose activities give rise to pollution, the state and its environmental protection services and the environmental protection organisations. Thanks to a series of procedural rights, the environmental protection organisations gained in significance during the 1980s and today constitute an inescapable force in environmental policy, at the level of both the formulation and implementation of a range of public policies. Two questions arise in this context: what is the extent to which the ecobusiness actors are replacing and will replace the other actors who currently enjoy extensive power in the majority of European countries and what role will ecobusiness play in the relationship between polluters and the environmental protection organisations, a relationship which is in the course of being transformed from a highly conflictual to rather more co-operative one?[3]

Based on observations made in our field studies[4], this contribution proposes to analyse in a general way the role of ecobusiness as a new actor

---

[2] Knoepfel, P. 1995a, "Approaches to an effective framework for environmental management", in: Carius, A., et al. (éds.), *Environmental management in Kenya, Tanzania, Uganda and Zimbawe*, Berlin: Deutsche Stiftung für internationale Entwicklung, Zentralstelle für öffentliche Verwaltung, p. 46.

[3] OIKOS / Umweltökonomische Studenteninitiative an der HSG (éd.) 1994, *Kooperation für die Umwelt. Im Dialog zum Handeln*, Coire/Zurich: Rüegger, pp. 103ff.

[4] See: Benninghoff et al. 1997 (note 1).

within the old "iron triangle" of environmental policy actors and to present some conceivable future scenarios. However, it is important to note at the outset of this process that both the analysis of the current situation and the scenarios merely represent an outline based on a very limited number of observations;[5] the following pages make no claim to presenting scientific analyses based on empirical data and they should be taken for what they are – speculation.

11.2 The basic model

Figure 1 presents the position of ecobusiness among the other actors currently involved in the implementation of environmental policy.

11.2.1 Definitions

Along with most of the contributions in Benninghoff et al. 1997, I understand ecobusiness as "the economic sector that produces and sells goods and services whose main intention and priority is to measure, predict, limit or correct damage caused to the environment."[6] This definition does not include economic agents who deliberately produce goods and services which are more environment friendly than those offered by traditional producers. Figure 1 illustrates this phenomenon through a schematic presentation of the polluting economic agents who avail of the services provided on the eco-markets by ecobusiness with the aim of reducing the emissions and waste generated in the course of their production processes (= environmental impacts of production) and/or with a view to marketing products that are more healthy in ecological terms and which will hence be

[5] These empirical observations originate from case studies presented in Benninghoff et al. 1997 as well as examples taken from the following publications: Kissling-Näf, I. 1996, *Lernprozesse und Umweltverträglichkeitsprüfung – staatliche Steuerung über Verfahren und Netzwekbildung in der Abfallpolitik*, dissertation completed at the University of St. Gallen, no 1744, St. Gallen; Kissling-Näf et de Bussy, C. 1995, *Apprentissage dans les politiques publiques: application de l'article 31b de la Loi sur l'agriculture dans les cantons de Neuchâtel, Zoug et Nidwald*, Cahiers de l'IDHEAP, no 152, IDHEAP: Chavannes-prés-Renens.

[6] Conus, J. et A. Rossi 1992, *Le business écologique. Définition, structure, évolution et perspectives de l'industrie de la protection de l'environnement de la Suisse*, Berne: Office fédéral des questions conjoncturelles, p. 3.

more favourably received by the environmental protection organisations and the general public.

Thus, I consider the eco-market as a place of exchange for the supply and demand for ecobusiness products, a place of exchange which is required by polluting economic agents and largely determined by the status of technology and the environmental legislation. These eco-markets must not be confused with the investment products and standard consumption markets which still exist and which, thanks to the services provided by ecobusiness to these agents, will become healthier from an environmental perspective.

Finally, I understand "polluting economic agents" as the producers of goods, services and infrastructure whose products or production processes themselves cause additional impacts on the environment (water, soil and noise pollution, reduction of biodiversity etc.). These environmental impacts must be reduced either by the polluters themselves or with the help of the goods and services produced by ecobusiness. The rates at which these impacts are reduced are imposed by the legislation, contractual commitments undertaken between the economic agents and the state or between themselves, or by means of agreements entered into by the economic agents and the environmental protection organisations.

Figure 1 : Definition of ecobusiness and other environmental policy actors

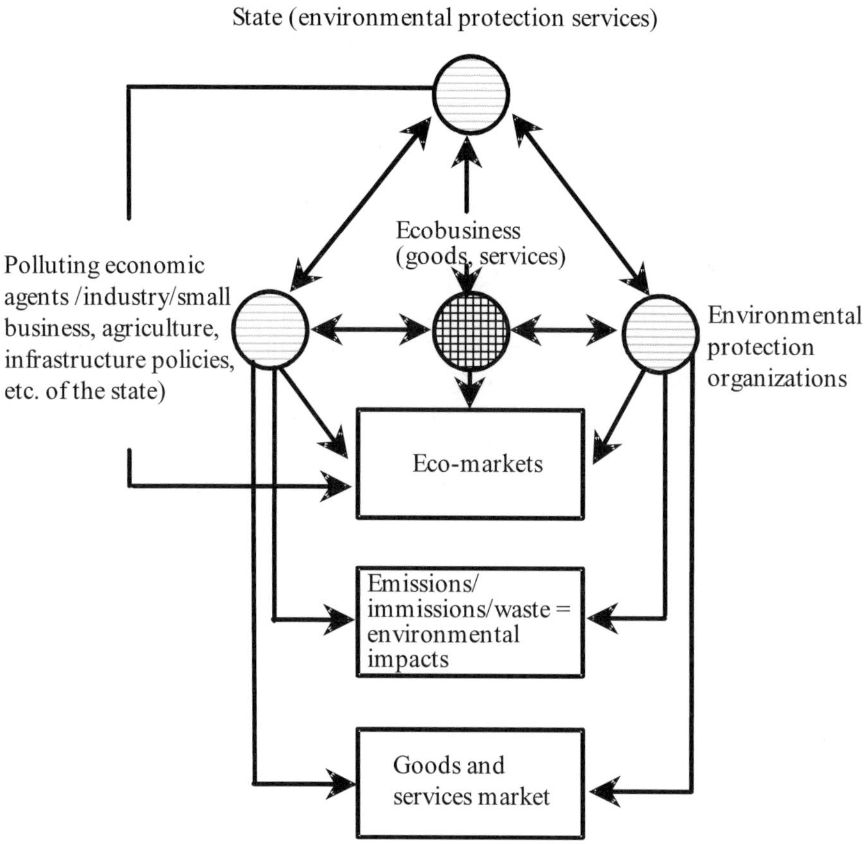

As suggested above, the relationship between ecobusiness and polluting economic agents is strongly influenced by the state legislation which is concretized and implemented by central and decentralised/regional environmental protection services. It is the job of these services to enforce the federal and cantonal legislation among the polluting economic agents under pressure from the environmental protection groups among others. In the process, the state services follow the evolution of the status of the technology, which is strongly influenced in turn by ecobusiness and its research and development activities. The state services oblige the polluting economic agents to observe the increasingly stringent product and production standards.

As several essays in Benninghoff et al. 1997 show, the environmental protection organisations are involved in the ecomarkets by virtue of their recommendation of certain ecological technologies or certain ecological consultancy products (for example, consultancy and popularisation in the area of organic agriculture). They are also involved in the evaluation of and imposition of sanctions for environmental damage arising from production within clean-up and authorisation processes and also try to influence consumer behaviour by means of ecolabels, recommendations and the endorsement of sales campaigns for particularly environment-friendly products. It is important to keep in mind that these organisations also exercise an important influence on the state environment protection services through their participation in formal and informal procedures. Moreover, it is possible to observe an increase in the direct contact between the environmental protection organisations and the polluting economic agents which, in individual cases, may give rise to contracts or "gentlemen's agreements" whereby the polluting economic agents undertake to behave in a more environmentally responsible way than they required to under the state legislation.[7] In exchange, the environmental protection organisations lend their support to the improvement of the image and brand names of the company in question.

11.2.2 The relationships between ecobusiness and the "iron triangle" actors

Figure 2 demonstrates the different relationships that exist between the ecobusiness actors and the polluting economic agents (1), the environmental protection organisations (2) and the state environmental protection services (3).

[7] Weidner, H. 1995, *Twenty five years of modern environmental policy in Germany. Treading a well-worn path to the top of the international field*, FS II 95/301, Papers Forschungsschwerpunkt Technik, Arbeit, Umwelt, Berlin: Wissenschaftszentrum Berlin für Sozialforschung (WZB), pp. 58ff.; Zuppinger, U. et P. Knoepfel 1996, *Usine d'incinération de déchets spéciaux (UIDS), CIBA-GEIGY à Bâle*, Etudes de cas de l'IDHEAP, no 7, IDHEAP: Chavannes-près-Renens.

Figure 2: Position of ecobusiness within the environmental policy "iron triangle"

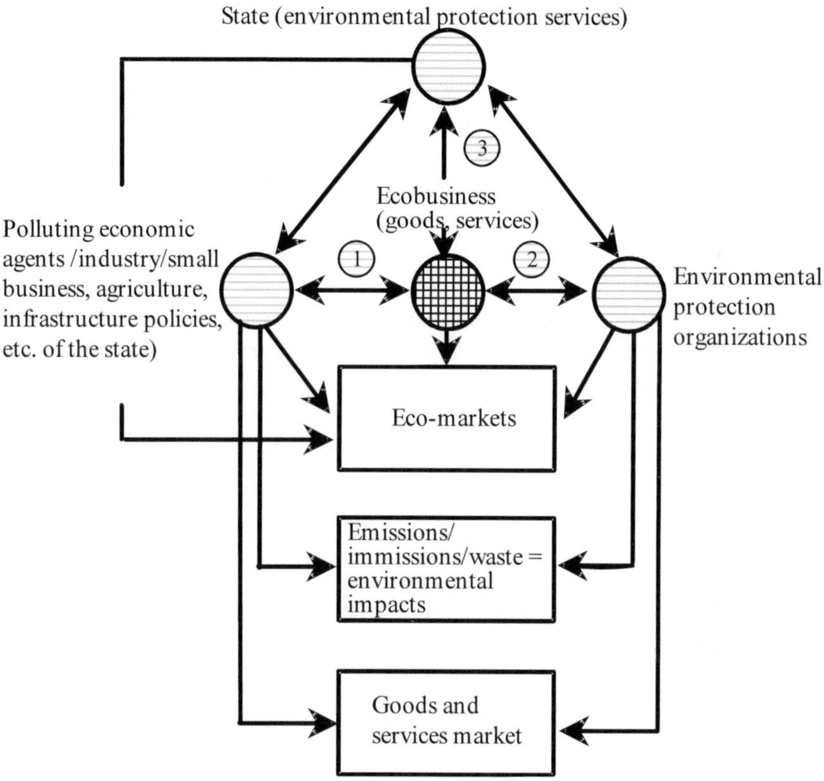

In the following section, I will describe the main characteristics of the relationships that exist between ecobusiness and the actors in its environment:

The relationship between ecobusiness and polluting economic agents (1)
This relationship is essentially commercial in nature as ecobusiness products are generally profitable. The economic agents are special clients in that, sooner or later, they are driven to avail of the services of ecobusiness by the requirements of the environmental legislation. Despite the nature of the demand, however, ecobusiness is obliged to look after its clients and this is not always easy as, in their desire to minimise costs, companies hardly ever choose the services which are the best from the point of view of the environment. Thus, research departments which propose excessively

detailed and hence extremely exigent impact studies run the risk of being eliminated from ecomarkets because they will be considered too "green". Like eco-consultancy, eco-technology does not always receive a warm welcome, particularly from small and medium-sized businesses. It is considered essential; an extra responsibility that comes at a high price. For all of these reasons, in conjunction with its products, ecobusiness should try to convey a powerful ecological message, which will enable its customers to publicly display their environmental awareness, as well as an economic message clearly stating that it is possible to make savings with the help of the new technologies or procedures being recommended.

Given that the state has the tendency to impose new eco-technology procedures on companies which have been developed by ecobusiness, excessively rapid progress in the R&D carried out by the latter is not exactly in the interests of the umbrella organisations of the different economic sectors. Excessive commercial and scientific success on the part of ecobusiness could undermine the survival of old established companies. This ambiguity lies behind a number of sometimes dubious practices, e.g. the purchasing of patents for the purpose of non-application, the reorientation of research by means of the influence of company shareholders etc.

Given the strongly fluctuating nature of the demand for ecobusiness services, ecobusiness's extreme dependency on its clients is all the more prominent in periods of economic crisis. Apart from the fact that during such periods, a number of small and medium-sized ecobusiness enterprises disappear from the market, a weak structure with a latent oligopolistic tendency emerges along with – due to fluctuations – a lack of continuity giving rise to breakdowns in communication (advice) and gaps at the level of after-sales service (repairs etc.).

The relationship between ecobusiness and the environmental protection organisations (2)

None of the authors of Benninghoff et al. 1997 has, alas, succeeded in throwing light on this sociologically fascinating relationship, which is described by a *connaisseur* in the area as an "eco-social complex"[8]. Historically, it is possible to establish that many of the consultancies and small and medium-sized businesses which produce ecotechnologies were founded by individuals who were active in the environmental protection organisations from their inception. Today, numerous personal relationships

[8] Weidner, H. 1993, "Der verhandelnde Staat. Minderung von Vollzugskonflikten durch Mediationsverfahren, in: *Vollzugsprobleme / Problème de la mise en oeuvre des politiques publiques*, Revue suisse de science politique, no 33, pp. 225–244; Weidner, H. 1995, *op. cit.*, pp. 74ff.

still exist between the two camps and they occasionally find themselves accused of complicity or "collaboration". Critical observers accuse certain environmental organisations of intervening in procedures solely for the purpose of procuring work for their colleagues in the environmental consultancies. By all accounts, requests – often considered as excessive – for the detailed examination of a certain aspect of an impact study are sometimes made with the sole aim of providing work for research consultancies. Others claim that ecobusiness circles use the environmental protection organisations to put pressure on state services to make certain ecotechnology products official as this enables them to improve their competitive edge.

To my knowledge, few empirical elements could be identified at present in support of these hypotheses which, however, do not lack a degree of plausibility as they are based on an ambiguity similar to that which governs the relationships between ecobusiness and the polluting economic agents. The focus here is on the relationship between ecobusiness and the environmental organisations: if it is an accepted fact that the effective protection of the environment requires – for the time being at least[9] – technologies and specialised knowledge which must be produced in part outside the polluting companies, it makes complete strategic sense for the environmental organisations to use every means at their disposal to promote the creation and growth of ecobusiness. However, this starting point involves some risks to the extent that this promotion can become an end in itself and, hence, a very simple way of making the knowledge, which has been accumulated over recent years by the leaders of the environmental protection movement through their involvement in their organisations, pay.

Just as in its relationship with its parent companies, to defend itself against such accusations, ecobusiness needs to differentiate itself from the environmental protection organisations and gain a degree of independence vis-à-vis its founding fathers. It does this by means of professionalization based, among other things, on the rules of professional ethics[10] which are intended to regulate its – still ambiguous – relationships with various ecological actors. In so doing, ecobusiness also avails of the support of the state whose legislation contains concrete formulations with respect to the essential independence of consultants and quality standards for ecotechnology services. It is possible to say that ecobusiness will only be able to survive in the medium term if it can attain a guaranteed independent status.

---

[9] Cf. the concept of "modernisation" presented in chapter 11.5.

[10] Cf. the example of the Swiss Association of Ecologists (*Association suisse des écologues*).

The relationship between ecobusiness and state environment protection services (3)
At an initial glance, this relationship seems clearer and less ambiguous than those which exist between ecobusiness and the environmental protection organisations and the polluting economic agents. In effect, there is a certain resemblance between the interests of ecobusiness and those of the state environmental protection services: they both have a stake in the advancement of anti-pollution technologies and the improvement of research and know-how to facilitate a consultancy message that is increasingly specialised, more targeted and easier to apply at the level of the behaviour of ecobusiness clients. Thus, the services of the state have every interest in the existence and promotion of ecobusiness and, moreover, a profitable ecobusiness which is able to sell its services on the ecomarkets, which are the most visible expression of the efficacy of environmental policies. Thus, ecobusiness can be considered as a welcome disseminator of the ecological message enshrined in the legislation and behind the other activities of the state services. Moreover, ecobusiness also acts as a driving force in the accumulation of new knowledge and information about the environment which in turn provide a basis for the subsequent development of the legislation and state political-administrative activity.

Thus, ecobusiness is also a provider of services on behalf of the state which, for various reasons, itself is no longer in a position to create the necessary basis for decision-making by its own means. Similarly, either out of concern of exposing the state to excessive expenses or allowing excessive intervention on the part of the state, under the legislation, the task of developing environmental science is assigned to the polluting economic agents and hence, at least in part, to ecobusiness. Thus, under the regimes governing the self-monitoring of hazardous substances[11] and environmental impact studies,[12] producers or developers are responsible for environmental analysis, however a right of inspection is reserved for the state services. In other cases, the state is effectively becoming the direct client of ecobusiness: for example, it is availing of outsourcing for the purpose of creating federal nature conservation inventories as well as commissioning reports on the status of research, new technologies, waste policies etc. and on the restructuring of certain services etc. Under recent legislation, the

[11] Cf. the Swiss hazardous substances decree (*Ordonnance sur les substances dangereuses pour l'environnement (Osubst)*) of 9 June 1986 (RS 814.013), art. 12, al. 1, l. a.

[12] Cf. the Swiss environmental impact study decree (*Ordonnance relative à l'étude d'impact sur l'environnement (OEIE)*) of 19 October 1988, (RS 814.011), art. 7.

state can even contribute to the development of new ecological technologies through the allocation of subsidies.[13]

In Switzerland there is already a tradition of the state as client at the level of the cantons and local authorities which purchase technological services for the construction and maintenance of water treatment plants and household waste incineration plants. In these two cases, the relationship between the state and ecobusiness strongly resembles that which exists between private enterprise and ecobusiness. The only difference appears to reside in the fact that the state, which is subject to the vigilant and critical eye of the public, cannot allow itself to acquire "second best" technology on the ecomarkets and this in turn prompts ecobusiness to propose increasingly sophisticated and expensive technologies.

In the first case mentioned, the client-state risks becoming either a "master state" or "servant state". The "master state" is in a position to select from several offers and tends to favour established suppliers, which are well known for the quality of their work, over the younger players on the market and, moreover, it is susceptible to political pressure or corruption on the part of ecobusiness.[14] In the second case, the state is dependent on a specific supplier which has the monopoly for a well defined service and does not allow other suppliers to become established on the market. This latter case is all the more worrying as the services required often form a basis for the definition of future services to be provided to the polluting economic agents. Ecobusiness can thus rely on the fact that when it proposes solution "x" to the state, it will be able to resell it a few years later to the cantons or companies obliged to acquire this technology under the legislation. As the facts show, the ecobusiness company can consider its service as an investment purchase and apply a preferential price which enables it to be more competitive than other companies which may be more innovative but less financially solid.

---

[13] Cf. the Federal Council Message concerning a revision of the federal law on the protection of the environment (*Message du Conseil fédéral relatif à une révision de la loi fédérale sur la protection de l'environnement (LPE)*) of June 7 1993 (FF 1993 II 1337); article 49, 3e al. (new) of the environmental protection act (*Loi Fédérale sur la protection de l'Environnement*) (amendment of 21 December 1995) regulating the support provided by the Swiss Confederation for the development of new ecological technolgies was ratified by the Federal Assembly on 21 December 1995 (FF 1996 I 237).

[14] This reproach was raised in the canton of Tessin by the company Thermoselect in the context of the debate surrounding the selection of the best technology for a household-waste incineration plant.

These explanations show that there is no longer any ambiguity in the relationships between ecobusiness and the state. The later can, in effect, expose itself to the criticism of favouring one ecotechnology provider or ecoconsultancy over others and not solely on the basis of the financial advantages but also at the expense of solutions which could be more innovative and which could be enshrined in the legislation. Although it is not possible to substantiate this hypothesis with the help of empirical evidence, a glance at the consultancy companies employed by the cantons for the development of measurement plans for the prevention of atmospheric pollution[15] and compilation of federal nature and landscape conservation inventories appears, however, to underline its plausibility.

11.2.3 A precariously balanced system

Based on the various relationships outlined above, we can conclude that the position of ecobusiness within the basic triangle of environmental policy actors is, to say the least, precarious. The most subtle shift on the part of the polluting economic agents, the environmental protection organisations, and equally on the part of the state services, could undermine the fragile balance which currently exists in this four-actor structure. This is, therefore, an appropriate point at which to consider the possible scenarios for future developments. Seven different scenarios are presented in the next two chapters. In the first four of these, the direction and scope of ecobusiness's activities remain constant while the positions of the three other actors change (chapter 11.3). This is followed by the attribution of an active role to these other actors in the take-over of ecobusiness (chapter 11.4). The questions surrounding the potential future transformation of ecobusiness are raised and discussed in the final chapter.

We would like to emphasise once again that these scenarios make no claim to being comprehensive or being based on empirical observation. However, they should have a role to play in strategic discussions of current difficulties that are taking place in ecobusiness boardrooms.

---

[15] Knoepfel, P. et al. 1995, *Luftreinhaltepolitik im Labor der Städte. Der Massnahmenplan – Wirkungen eines neuen Instruments der Bundespolitik im Verkehr*, Bâle: Helbling & Lichtenhahn (*Reihe Oekologie & Gesellschaft*, vol. 9), p. 112.

11.3 Four scenarios based on the hypothesis concerning changes in the positions of the main environmental policy actors

11.3.1 The "split" scenario

Figure 3: The "split" scenario"

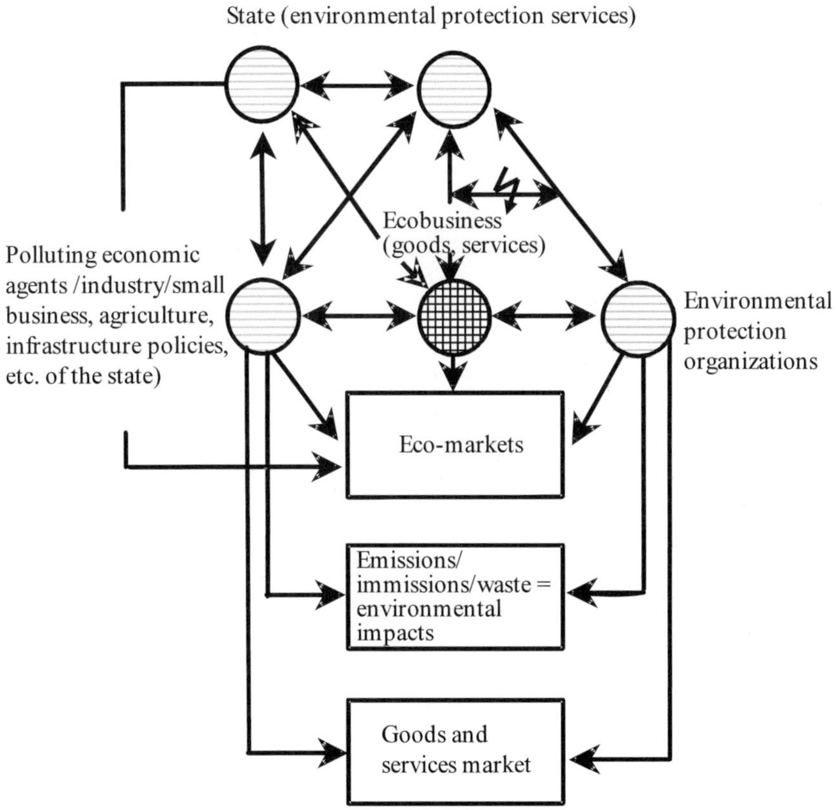

Figure 3 presents a scenario in which a new state actor comes into play. These are state services, which are close to the polluting agents by virtue of the institutional tasks they perform and which are now beginning to develop their own environment policy. Observation of the political-administrative structure of environmental policies in several European countries (Switzerland, France, Germany to mention but a few) shows that, in effect, environmental policy is no longer solely the business of the state service that is explicitly dedicated to the protection of the environment.[16] Thus, in addition to the national ministries of defence and energy, the administrative services in the areas of agriculture and infrastructure (roads, railways, civil aviation etc.) are interested in ecobusiness services on behalf of their "clientele" (state developers, in the case of infrastructure projects, farmers, public railway companies etc.). These "new" institutional actors on the environmental scene are often very powerful due to their terms of reference and the budgets at their disposal.[17] Their behaviour is probably closer to that of the polluting economic agents than that of the state environmental protection services. Armed with this support in terms of ecological services, these new actors will probably be tempted to defend themselves against requirements considered as excessively ecological which are defended by the state environmental protection services and ecological organisations.

Initially, this constellation of interests could trigger a split between these two communities and ecobusiness would find itself either divided or forced to make a choice between the two camps. It is highly likely that this choice will be made in favour of the developers and offices of the powerful public policies. The "split" variant would result in the emergence of two actors on the ecobusiness scene, one of whom would be more "green" while the other would tend more towards producing and providing services for the other non-environmental policies and those developers who could be described as "minimalists" from the ecological perspective. In a situation characterised by open conflict between the two camps, it would probably be difficult for a company to adopt the strategy of serving both camps at once, thus aiming to belong to both communities. In this case, we would probably witness the emergence of more or less unconditional support for the developers' camp on the part of the entire ecobusiness community. Such a development would deprive the environmental protection services of an important resource. The latter should either start by recreating new ecomarkets through the adoption of a solid strategy at the level

---

[16] For France, see Brénac's article on corporatism: Brénac, E. 1988, "Corporatismes et politique intersectorielle: la politique de l'environnement", in: Colas, D. (éd.), *L'Etat et les corporatismes*, Paris: PUF, pp. 127–46; for Germany, cf. Weidner, H. 1995, *op. cit.*

[17] Knoepfel, P. 1995b, "New institutional arrangements for a new generation of environmental policy instruments: intra- and interpolicy-cooperation", in: Dente, B. (éd.), *Environmental Policy in search of new instruments*, European Science Fondation, Dordrecht: Kluwer Academic Publishers, p. 218 = chapter 6 of this book.

of legal requirements, a solution which is made all the more difficult by the fact that they must anticipate the status of the technology without the availability of prototypes, or by producing services for the ecomarkets themselves.[18]

This last option would rapidly give rise to conflict between the environmental protection services and the federal Swiss cartels commission which would denounce the latter for unfair competition as occurred in the case of several federal Swiss offices which established themselves as independent agencies in accordance with the new public management formula (Federal Office for Topography, Swiss Meteorological Institute).[19]

11.3.2. The "retreat of the environmental movements" scenario

Figure 4 shows a second scenario which is similar to the first to the extent that it includes the state services responsible for the implementation of policies, which give rise, at least partly, to additional environmental pollution, as new institutional actors. However, it differs from the first in terms of being based on the hypothesis that the environmental protection organisations will break ranks with ecobusiness having reproached it of being "bought" by the state or the polluting economic agents.

[18] Cf. the "state control" scenario – chapter 11.4.

[19] This problem was discussed during a meeting of the directors of the offices of the Federal Swiss Department of the Interior which took place in Fürigen on 29 April 1996.

Figure 4: The "retreat of the environmental protection movements" scenario

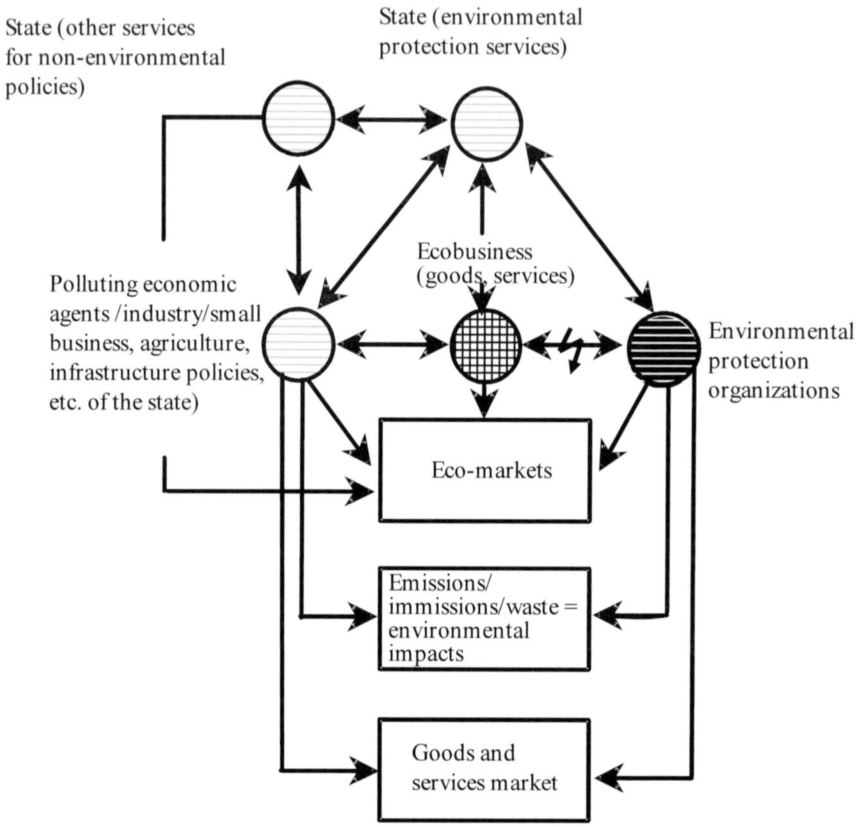

Due to this radicalisation of the position of environmental organisations, the services of ecobusiness would risk losing their value as a badge of legitimacy for the environmental protection services and polluting economic agents during confrontations about new installations or the "ecological" quality of products bearing an ecolabel. This would result in a general ecological "delegitimatization", a loss in the symbolical value of the services of ecobusiness and, hence, a decline in the demand for such services. This scenario which is as "dangerous" for the economic interests of ecobusiness as it is for the political interests of the state environmental protection services could probably only represent a transitory phase because, as in the previous scenario, sooner or later a split would develop in the ecobusiness community with new and more credible products and services being created on ecomarkets.

11.3.3 The "solitary state" scenario

Figure 5: The "solitary state" scenario

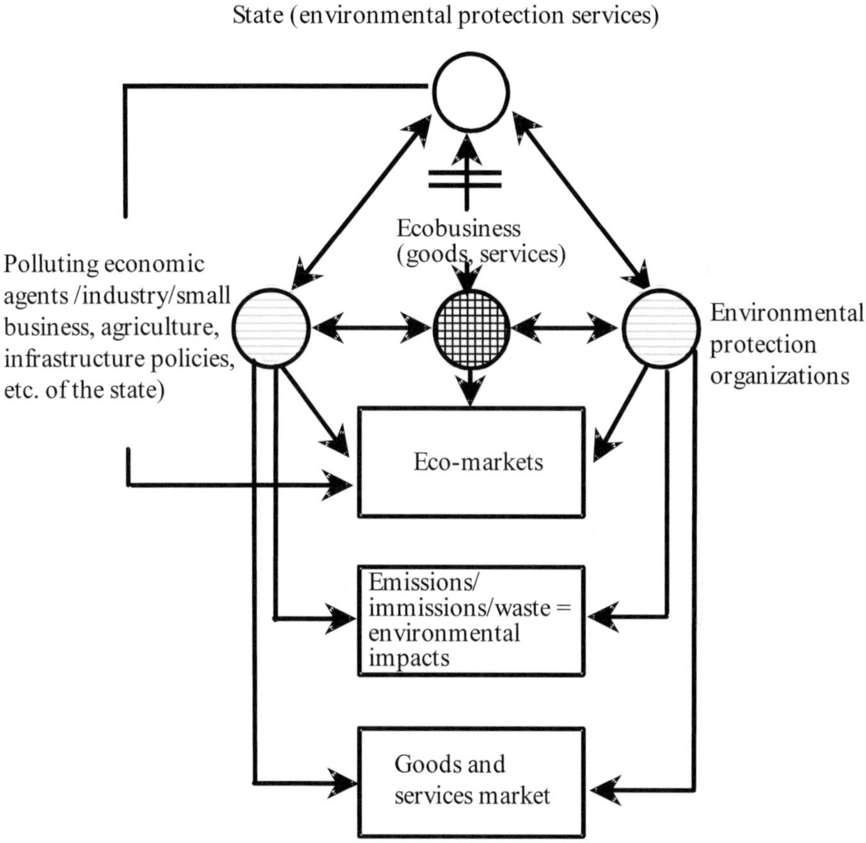

Figure 5 presents the scenario involving the disengagement of the state services, an outcome which is often demanded today by advocates of deregulation. It is possible to identify two variants in this discourse. In the first variant, the state defines minimum frame conditions in its legislation, in the form of quality standards which define the general protection requirements, and encourage the polluting economic agents to implement systematic environmental management which would make it possible to achieve lower levels of pollution than those required under national law. This encouragement may also take the form of financial incentives (subsidies, tax incentives, market mechanisms in the form of emissions certificates etc.), the official assignation of more favourable competitive positions (certification of sites in accordance with the EMAS standard[20], product labelling) or the encouragement of the establishment of contractual regimes between the environmental protection organisations, ecobusiness and the polluting economic agents (for example, state agents) under the form, for example, of a more lenient application of cartel law. This first variant, which is both less radical and more promising contradicts, at least in part, the second which requires a more consistent retreat on the part of the state through the replacement of standards with self-monitoring mechanisms to be implemented by the polluting economic agents themselves. In the process, it banks on their motivation to protect themselves against environmental risk by availing of the services of ecobusiness and, in a more ecological variant, the support of the environmental protection organisations.

In the case of Switzerland and the Member States of the European Union, we can eliminate this second variant, at least in the medium term, on the basis of the environmental law currently in force. In contrast, the scenario involving the solitary state in the form of the first variant would appear to be more or less realistic for several reasons. In effect, the self-monitoring regimes governing hazardous substances (*Osubst*) and prevention of major accidents (*OPAM*[21]), and environmental impact studies (*OEIE*), which attributes the establishment of the impact relationship to the developers, have shown that in many cases the actual implementation goes beyond the requirements of the state legislation.[22] The introduction of the certification of sites in accordance with EMAS, the reinforcement of civil responsibility by means of the partial reversal of the burden of proof in cases involving environmental damage[23] and the current trend towards the certification of industrial procedures and services in accordance with the ISO 14034[24] standard are all moves in the direction of self-monitoring and

[20] Council Regulation (EEC) No. 1836/93 of 29 June 1993 allowing the voluntary participation of companies from the industrial sector in a community eco-management and audit system (better known as EMAS: Environmental Management and Audit System).

[21] Swiss major accidents decree (*Ordonnance sur la protection contre les accidents majeurs (OPAM)*) of 27 February 1991 (RS 814.012), art. 5.

[22] Kissling-Näf, I. 1996, *op. cit.* pp. 271ff. and 282ff.; Knoepfel, P. 1994, "Zur Wirksamkeit des heutigen Umweltschutzrechts", in: *Umweltrecht in der Praxis/Le droit de l'environnement dans la pratique*, Vereinigung für Umweltrecht, Zurich, vol. 8, Cahier 4, pp. 226.

[23] Weidner, H. 1995, *op. cit.*, n7, p. 62.

[24] This new International Standards Organisation (ISO) standard is currently being prepared (Spring 1996).

the permanent adoption of the best environmental technology by the economic agents themselves.

Ecobusiness appears to play a key role in this process because in the long run this regime can only function as a general system if it is permanently backed up by increasingly sophisticated consultancy and technological services. Ecobusiness must also secure the support of the general public, despite the fact that unlike the certification of sites in accordance with EMAS, certification in accordance with ISO 14034 does not formally include the general public as an actor in the process.[25] Nevertheless, the support of the environmental protection organisations is necessary to prevent them from trying to reactivate control by the state services due to a lack of confidence by creating political initiatives heading towards re-regulation. More than the two preceding scenarios, this one, which in reality assigns part of former state activities to ecobusiness and its partners, necessitates the development of a new sector within ecobusiness, i.e. socio-political management. All environmental impacts caused by production activities or the placing of new products on the market involve both a physical phenomenon and a social construct, the result of a – controversial – interpretation based on the specific perceptions of the persons involved, neighbourhood and the environmental protection organisations. These perceptions can vary from one location to another and change over time with the result that the physical pollution may be accepted in a given place and completely rejected in a neighbouring area. Today, the analysis, interpretation and even the management of these perceptions is largely the responsibility of the state environmental services. In the scenario described here, these tasks shall become the responsibility of ecological service providers to an even greater extent. "Social engineering" will become a key element of the services provided by ecobusiness[26] which is forced to develop under the conditions dictated by the "solitary state" scenario which, in my opinion, are highly realistic.

---

[25] Linger, H.-U. 1996, *Umweltmanagement und Vollzugsverwaltungs – ein neuer Weg?*, contribution to the IDHEAP seminars for experts and managers: "Les administrations publiques en transformation: une action publique axée sur les clients", St. Gall, February 1 to 3, 1996.

[26] Pictet, J. 1996, *Dépasser l'évaluation environnementale: procédure d'étude et insertion dans la décision globale*, Lausanne: Presses polytechniques et universitaires romandes, pp. 22ff.

11.3.4 The "symbiosis" scenario

The "symbiosis" scenario, which is presented in Figure 6, differs from the current state of the relationship between the state environmental protection services and ecobusiness in that the state services assign to the latter the role of main negotiator with the other environmental actors. According to this scenario, for both the polluting economic agents and the environmental protection organisations, the main route to the state is through ecobusiness. Thus, the direct relationships between the state services and these two other actors of the "iron triangle" would be reduced to a bare minimum. As in the example of the "state control" scenario (chapter 11.4), what we have here is a kind of "corporatization" of the state within ecobusiness. Contrary to the preceding scenario, the state does not give up its regulatory activities and there is little increase in social self-monitoring. However, the state explicitly allocates the responsibility for the definition and monitoring of environmental standards to ecobusiness, which, with the state's blessing, would attain the status of a monopoly thanks to a more exclusive corporatism and clearer differentiation from the polluting economic agents and the environmental protection organisations than currently exists. In the "symbiosis" scenario, ecobusiness would profit from a guaranteed income (monopoly), economic protection against potential new competitors originating from the economic agents, environmental protection organisations or foreign ecobusiness. This scenario could materialise through rules sanctioned by the state which would define the entry requirements for access to the profession and markets, precise obligations for obtaining a specific ecoservice or ecotechnology from a certain service provider or mandatory supply zones (for the waste collection centres, household waste incineration plants or companies that carry out the pigging of pipelines).

Figure 6: The "symbiosis" scenario

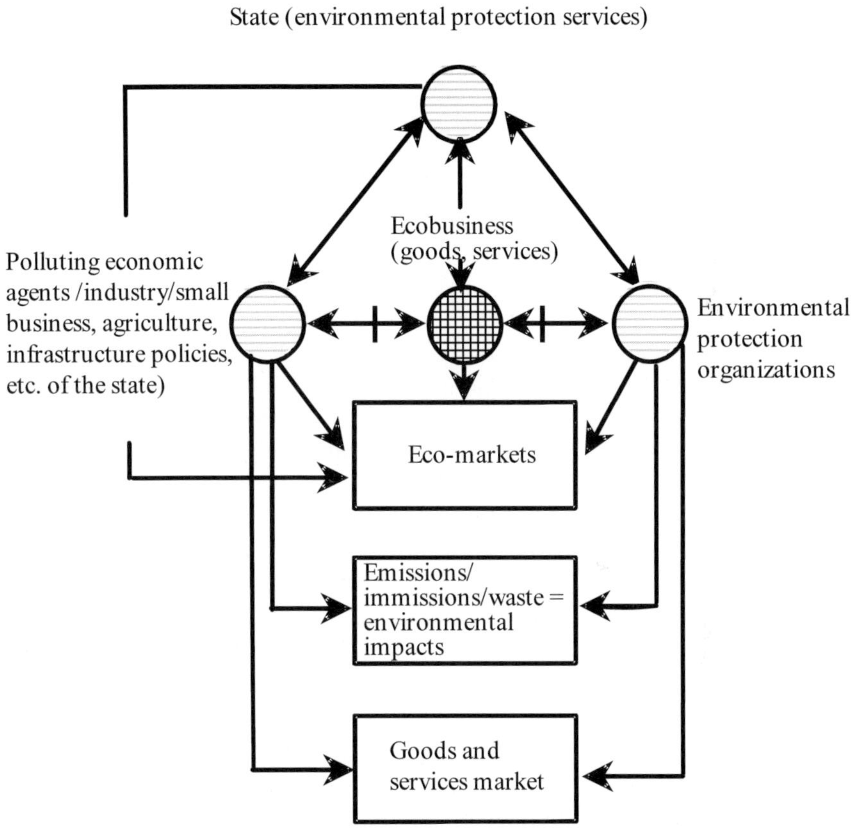

However incompatible it may seem with a market economy, a review of reality in the light of this "symbiosis" scenario show that although it has not yet become the general rule, it can be seen to exist in numerous eco-markets, for example in the area of waste policy (mandatory support zones[27]), water and atmospheric pollution policy (regime governing the control of fuel tanks or cleaning[28]), in the mandatory completion of anti-

[27] Cf. the Swiss decree on the processing of waste (*Ordonnance sur le traitement des déchets (OTD)*) of 10 December 1990 (RS 814.015), art. 18.

[28] The regime for controlling fuel oil tanks is regulated by the decree on the protection of waters against liquids which may alter them of 28 September 1981 (*Ordonnance sur la protection des eaux contre les liquides pouvant les altérer*) , art. 37 (authorisation), art. 53 (monitoring) (RS 814.226.21.) The legal basis

pollution car tests at licensed garages,[29] in the mandatory monitoring of environmental company audits by “state-approved experts” (EMAS) and in the mandatory completion of certain courses to obtain certification as a user of toxic substances[30] etc. The fact of granting the privilege of monopoly to a limited number of persons authorised to provide these ecoservices is still motivated by the desire to protect consumers against ill-qualified service providers and to ensure the availability of high-quality permanent services which are accessible to all (a non-discrimination clause usually imposed on providers of these services). These regimes, which are known as “concessionary”, are very typical of the police policies adopted by the Swiss cantons and local authorities[31] and reflect the high level of cartelization in our economy. It is not surprising, therefore, that this phenomenon can also be found at the level of ecobusiness. Despite what the representatives of new economic policy, who are labelled as anti-cartel, say, in my opinion, it is not beyond the bounds of possibility that the “symbiosis” scenario will develop in Switzerland, given our experiences with the new urban and special waste policies. It represents an immediate response to the problem of the empty state coffers, the new public management postulate of outsourcing and the need to rapidly find solutions to the new challenges arising from environmental policy without being forced to increase the resources at the disposal of the state services. The known risk (loss of flexibility and innovation by a protected and “cartelized” ecobusiness)[32] is seldom raised as a counter-argument in the political debates at cantonal and local authority level, even by those who usually advocate economic liberalism. The preferred argument concerns the security of the supply of these services in the full knowledge that a part of the market will return to either political allies or ecological partisans whom one would like to see integrated and hence rendered inoffensive.

---

for the regime controlling the pigging of pipelines can be found in the cantonal laws governing anti-fire insurance.

[29] Cf. the Swiss decree concerning the maintenance and subsequent monitoring of automobiles with respect to emissions of exhaust gases and fumes of 22 December 1993 (RS 741.437)

[30] See Benjamin's article in this book.

[31] Cf. the Swiss Federal Council Message concerning the federal law on cartels and other restrictions on competition of 23 November 1994 (FF 1995 I 472).

[32] Cf. chapter 11.2 above.

11.4 Three scenarios based on the hypothesis of the weakening of ecobusiness

The following three scenarios propose the hypothesis that ecobusiness will become both a highly lucrative economic activity and, for other reasons, a "dangerous" for one of the three main environmental policy actors. They start from the idea that each of these actors will try to strengthen their control over ecobusiness or to integrate it into their own field of action with the aim of eliminating it as an actor whom they deem too independent. The basis of the three scenarios has already been described in chapter 11.2 which demonstrates the ambivalent relationships which exist today between ecobusiness and each of these three actors. In this chapter, I shall content myself with less detailed sketches than those presented in the previous chapter. It is important to note once more that the disappearance of numerous small and medium-sized businesses from the ecobusiness scene does not correspond to the changes described in the three following scenarios because it is due to the general economic recession which has the polluting economic agents and resulted in a reduction in their economic activities: this, in short, is the reason for the simultaneous decline in their need for ecological services. In contrast, in the scenarios, the disappearance of ecobusiness is not due to a general decline in the goods and services produced by ecobusiness but to the transfer of the relatively independent companies involved to the control of one of the three actors in the "iron triangle". Finally, contrary to the scenario described in the following chapter, the quantitative and qualitative demand for such services remains constant.

11.4.1 Scenario involving the "buyout" of ecobusiness by the polluting economic agents

Figure 7: Scenario involving the "buyout" of ecobusiness by the polluting economic agents

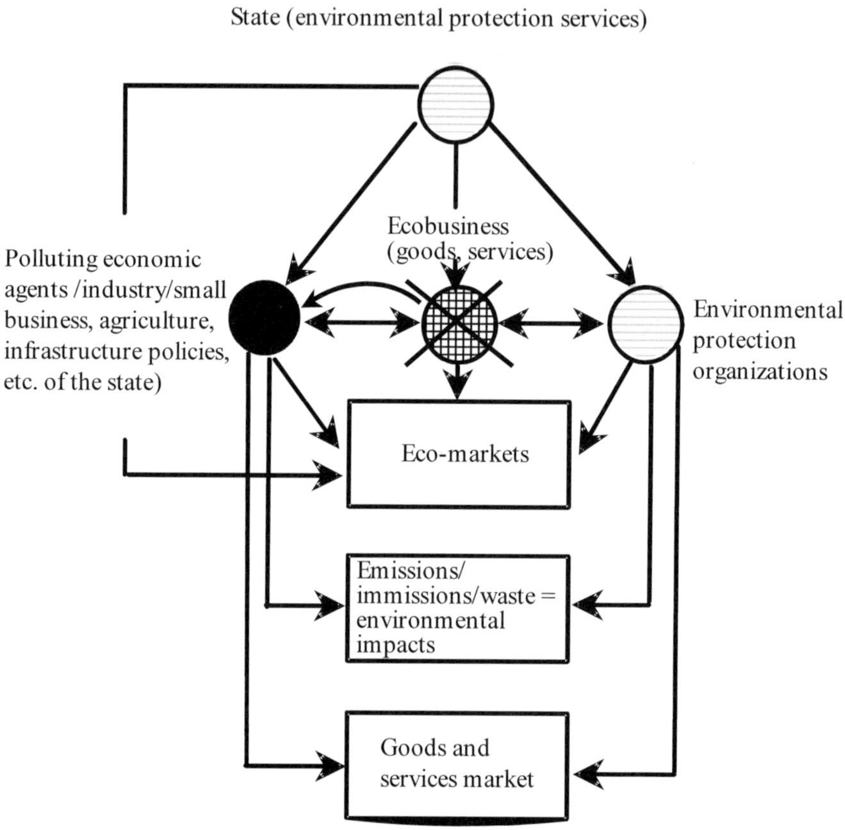

Figure 7 presents the scenario involving the "buyout" of ecobusiness by its clients who, according to the hypothesis presented in chapter 11.2, are often their own parent industries. In this scenario, the polluting companies refuse to comply with the whims of their subsidiaries which are producing the increasingly costly technologies and services dictated by the increasingly stringent state legislation which, in turn, is prompted by the myriad ecobusiness innovations. The "buyout" of the different ecobusiness companies enables the companies which require their services to decelerate, redirect and even halt research and development into specific ecological services.

It is important to note, however, that this shift of ecobusiness towards its old clients does not necessarily have to dampen the increasing interest in the environmental aspects of production because it creates a new proximity, enabling the exertion of a more direct influence by environmental production units on normal production. This new constellation may act as an important impulse for the establishment of mutual apprenticeship procedures[33] which could result in the reduction of costs through the integration of environmental aspects into the research and development of new product generations from the companies involved. Ecoservices are changing increasingly in terms of knowledge, advice and know-how and less and less in terms of technology which is, in fact, rendered superfluous when the advice is followed from the initial planning of products. This process of the "digestion" of ecobusiness could eventually enable the emergence of an "all-ecobusiness" scenario and, consequently, the famous "industrial modernisation" which is discussed in the next chapter.

11.4.2 Scenario involving the "buyout" of ecobusiness by the environmental protection organisations

[33] Finger, M. et al. 1996, "Ansätze zur Förderung organisationaler Lernprozesse im Umweltbereich", in: Ruh. M. et S. Bürgin (éds.), *Förderung umweltbezogener Lernprozesse in Schulen, Unternehmen und Branchen*, Bâle: Birkhäuser, pp. 58ff.

Figure 8: Scenario involving the "buyout" of ecobusiness by the environmental protection organisations

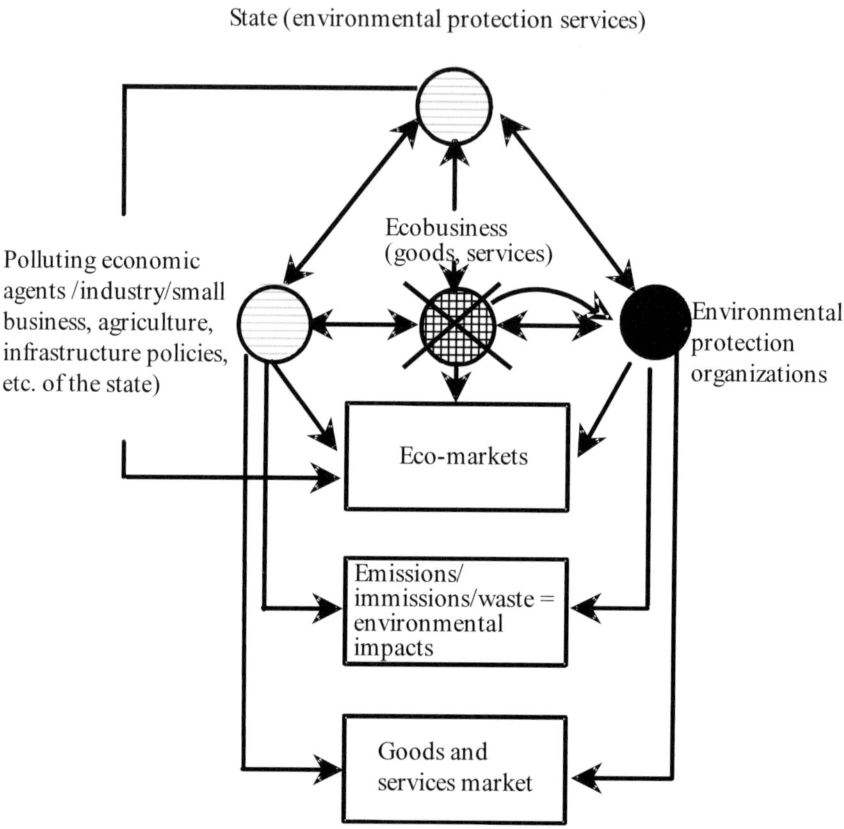

According to the scenario presented in Figure 8, the environmental protection organisations will launch themselves on the ecomarkets through a lucrative or non-lucrative economic involvement in ecobusiness. The WWF, Greenpeace and Croix Verte ("Green Cross") would amend their statues to allow their foundation, sections, national and regional associations to found or acquire companies which supply ecoservices with the aim of accelerating technological progress and reducing the prices charged for such services. Although this scenario is far from becoming widespread today, developments are afoot in certain specific areas[34] which are heading in this direction: some of the contributions in Benninghoff et al. 1997 cite

[34] Eco-consultancy in schools, eco-consultancy for consumers etc.

very telling examples (cf. article by Joerchel). This scenario would result in the establishment of commercial relationships between the environmental protection organisations and the polluting economic agents, the latter becoming the clients of the former.

It is obvious that, as was the case in the previous scenario, such a constellation would challenge the independence of the environmental protection organisations. In the long run, even an ecobusiness organised in the form of a non-profit-making company with a co-operative statute probably would not be able to resist client pressure to soften the strict environmental message in order to make it more compatible with their micro-economic requirements. As tempting as such economic involvement may be, it is important to state that the loss of independence on the part of the environmental protection movement would strongly risk corrupting this movement and, moreover, placing a question mark over the very functioning of the "iron triangle" of environmental policy. A new reading of some examples of such involvement on the part of environmental protection movements would probably reveal that these drawbacks exist already.

11.4.3 Scenario involving the "state control of ecobusiness"

Figure 9 show the third possible variant arising from the buyout of ecobusiness, this time by the state services.

Figure 9: Scenario involving the "state control of ecobusiness"

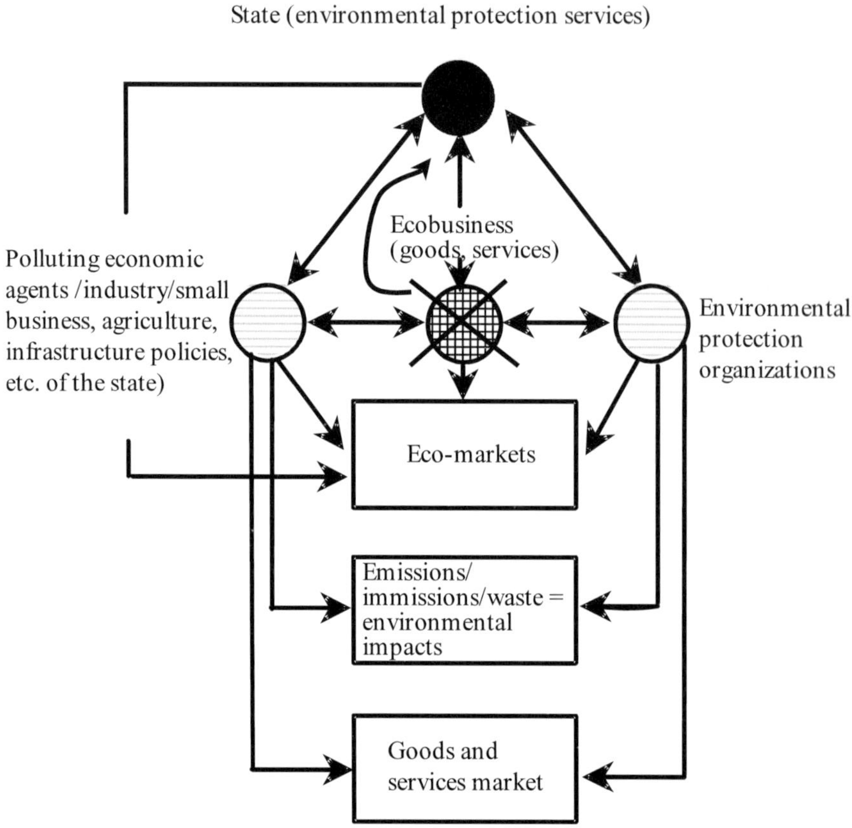

Given the ambiguity which currently reigns in the relationship between ecobusiness and the state services (or in the "symbiosis" scenario), a general outcry demanding at least partial transfer of ecoservices (specifically in the area of consultancy) to the state is conceivable, even in Switzerland. In this case, the state would create additional units within the existing consultancy services[35] whose purpose would be to sell advice to the polluting economic agents for a profit. As part of this perspective, the state could be prompted to take over former private consultancy companies and integrate them into its own consultancy niches. It is interesting to note that at the

[35] Particularly in the area of agricultural consultancy services

level of the state administration, under the new public management,[36] these units which provide services for the ecomarkets should benefit from the same status as an independent state establishment commissioned to provide services, with its own separate accounts and would (probably) be assigned legal status. As already mentioned in the second chapter, such niches already exist today in areas such as agriculture or waste. This is particularly true of the popularisation of agriculture, which has been taken over in the cantons, at least in part, by the agricultural colleges. Identical arrangements can be found in the area of research and development of some advanced ecotechnologies produced within third-level educational institutions and subsequently sold on the ecomarkets.

The reappropriation of the provision of goods by the state normally occurs following the recognition of the collective character of the commodity and the inability of the private sector to produce it. The series of weaknesses displayed by numerous ecobusiness companies in the mid-1990s could in effect prompt politicians with green or left-wing tendencies to propose the state control of at least some branches of ecobusiness. This could hardly be deemed an appropriate solution, however, in a period characterised by a crisis in public finance. Instead, we are now witnessing a movement in the opposite direction, i.e. towards the challenging of the collective character of the ecoservices currently provided by the public sector. There are increasing demands for these services to be made pay their way and for the state to be denied the monopoly for their production. Furthermore, as I have shown in chapter 11.2, the state which provides such services risks intervention on the part of the Federal Swiss Cartels Commission at the request of private service providers who see themselves at a competitive disadvantage on the markets by the presence of state agents providing subsidised services. Thus, the "state control" scenario would appear to have very little chance of materialising in the political-economic landscape of the 1990s in Switzerland.

11.5 Futuristic scenarios

Ecobusiness representatives are now asking themselves whether their sector has a future. The response I present in this final chapter is qualified and

[36] Schedler, K. 1995, Das Modell der wirkungsorientierten Verwaltungsführung, in: Hablützel, P. et al. (éds.), *Umbruch in Politik und Verwaltung, Ansichten und Erfahrungen zum New Public Management in der Schweiz*, Berne: Haupt, pp. 15ff.; Arn, D. 1995, "Organisationsrecht und New Public Management – ein Beitrag aus kommunaler Sicht", in: Hablützel, P. et al., *op. cit.* pp. 283ff.

starts with the hypothesis that in the future, the quality of ecoservices will undergo a significant change which will have inevitable repercussions on the political and economic position of ecobusiness and its actors in the environmental policy stakes. This change will consist in a drastic reduction of ecotechnology services and an increase in ecoconsultancy activities. It seems inevitable to me, as demonstrated by the example of the Swiss chemicals industry,[37] that all advanced production technology and all products will have an integrated ecological component from their conception in the R&D laboratories and workshops. There is no future for technologies which does not take ecological factors into account and to which an "end-of-the-pipe" type complementary technology must be added at the end of the development process in order to make it eco-compatible. If this hypothesis is borne out, the demand for ecotechnology alone will peter out and finally disappear over the next few years. What we are witnessing here is not a buyout of the ecotechnological services of ecobusiness (in the sense of the "buyout by the polluting economic agents" scenario) but a fundamental transformation on the part of the economic agents who must gradually change their image as "polluters". However, this kind of modernisation process[38] requires the extensive expansion of ecoconsultancy in the broadest sense of the term. Such services shall be all the more in demand among industrial and small-business R&D departments, agriculture, households and state infrastructure developers. This new generation of ecoconsultancy would thus increasingly consist of services integrated into the production process (research, development planning), on the one hand, and pluri-disciplinary services (engineering, biology, biochemistry etc. and economic, social and administrative sciences), on the other .

Will this need for integration lead to the complete integration of ecobusiness into the milieus of polluting economic agents and through this to the disappearance of the actor "ecobusiness" at the level of companies and the level of environmental policy actors? Several arguments dictate against this conclusion. Even if it is true that ecology will be even more integrated into the functioning of the economy, given that ecological issues will be more consistently integrated into all university courses concerning national economy and the public sector and environmental policy will thus be transformed from a policy involving the management of emissions to a policy

[37] In particular, the Novartis company.

[38] Jänicke, M. 1996, Erfolgsbedingungen von Umweltpolitik, in: Jänicke, M. (éd.), *Umweltpolitik der Industrieländer. Entwicklung – Bilanz – Erfolgsbedingungen*, Berlin: Sigma, pp. 20ff.; Jänicke, M. 1993, "Über ökologische und politische Modernisierung", in: *Zeitschrift für Umweltpolitik und Umweltrecht*, no 16, pp. 159–75.

involving the management of resources, nonetheless ecological R&D will become more and more specialised requiring its own dedicated courses and production structures.

Furthermore, in order to remain credible, ecology will need a certain degree of independence in addition to markets which work on the basis of the law of competition. Thus, it will be entirely in the interest of ecobusiness that it retain its own markets following the above-described transformation. This interest goes hand in hand with that of the environmental protection organisations and the state environmental protection services and even as part of a futuristic scenario which could be described as "generalised technological ecobusiness". This hypothesis is also a postulate. The functioning of the "iron triangle" of environmental policy actors under the conditions dictated by the market economy undeniably needs ecobusiness. However it needs an ecobusiness which has been transformed in the direction of a concentration on ecoconsultancy and the abandonment of ecotechnology but which may enjoy a certain independence with respect to the three other actors. Under these conditions, it could contribute to the ecological transformation of the economy with the help of impulses from the environmental protection organisations on the one hand and the state services on the other.

References

Arn, D. 1995. Organisationsrecht und New Public Management – ein Beitrag aus kommunaler Sicht, in: Hablützel, Peter, Haldemann Theo, Schedler, Kuno, Schwaar, Karl (éds.), *Umbruch in Politik und Verwaltung, Ansichten und Erfahrungen zum New Public Management in der Schweiz*. Berne: Haupt: 283–296

Brénac, E. 1988. Corporatismes et politique intersectorielle: la politique de l'environnement, in: Colas, D. (éd.), *L'Etat et les corporatismes*. Paris: PUF: 127–46.

Bussy, C. 1995, *Apprentissage dans les politiques publiques: application de l'article 31b de la Loi sur l'agriculture dans les cantons de Neuchâtel, Zoug et Nidwald,* Cahiers de l'IDHEAP no 152. Chavannes-prés-Renens: IDHEAP.

Conus, J., Rossi, A. 1992. *Le business écologique. Définition, structure, évolution et perspectives de l'industrie de la protection de l'environnement de la Suisse*. Berne: Office fédéral des questions conjoncturelles.

Finger, M., Bürgin, S., Haldemann, U.1996. Ansätze zur Förderung organisationaler Lernprozesse im Umweltbereich, in: Ruh, Michel, Bürgin,

Sylvia (éds.), *Förderung umweltbezogener Lernprozesse in Schulen, Unternehmen und Branchen*. Bâle: Birkhäuser: 43–70

Jänicke, M. 1993.Über ökologische und politische Modernisierung, *Zeitschrift für Umweltpolitik und Umweltrecht*, no 16: 159–75

Jänicke, M. 1996. Erfolgsbedingungen von Umweltpolitik, in: Jänicke, M. (éd.), *Umweltpolitik der Industrieländer. Entwicklung – Bilanz – Erfolgsbedingungen*. Berlin: sigma: 9–28

Kissling-Näf, I. 1996. *Lernprozesse und Umweltverträglichkeitsprüfung - staatliche Steuerung über Verfahren und Netzwerkbildung in der Abfallpolitik*, Dissertation der Universität St. Gallen, no 1744. St Gall

Knoepfel, P. 1994. Die Wirksamkeit des heutigen Umweltschutzrechts, in: *Umweltrecht in der Praxis/Le Droit de l'environnement dans la pratique*, Vereinigung für Umweltrecht. Zurich, vol. 8, cahier 4: 201–236

Knoepfel, P., Imhof, R., W. Zimmermann 1995. *Luftreinhaltepolitik im Labor der Städte, Der Massnahmenplan – Wirkungen eines neuen Instruments der Bundespolitik im Verkehr*. Bâle: Helbing & Lichtenhahn (Reihe Oekologie & Gesellschaft, vol. 9)

Knoepfel, P. 1995a. Approaches to an Effective Framework for Environmental Management, in: Carius, A., Höttler, L., Mercker, H. (éds.), *Environmental Management in Kenya, Tanzania, Uganda and Zimbawe*. Berlin: Deutsche Stiftung für internationale Entwicklung, Zentralstelle für öffentliche Verwaltung: 45–72

Knoepfel, P. 1995b. New Institutional Arrangements for a New Generation of Environmental Policy Instruments: Intra- and Interpolicy-Cooperation, in: Bruno Dente (éd.): *Environmental Policy in Search of New Instruments*, European Science Foundation. Dordrecht: Kluwer Academic Publishers: 197–233

Liniger, H.-U. 1996. *Umweltmanagement und Vollzugsverwaltung – ein neuer Weg?*, contribution aux Séminaires pour spécialistes et cadres de l'IDHEAP sur "Les administrations publiques en transformation: une action publique axée sur les clients". St Gall, du 1–3 février 1996

OIKOS / Umweltökonomische Studenteninitiative an der HSG (éd.) 1994. *Kooperation für die Umwelt. Im Dialog zum Handeln*. Coire/Zurich: Rüegger

Pictet, J. 1996. *Dépasser l'évaluation environnementale: procédure d'étude et insertion dans la décision globale*. Lausanne: Presses polytechniques et universitaires romandes

Schedler, K. 1995. Das Modell der wirkungsorientierten Verwaltungsführung, in: Hablützel, Peter, Haldemann Theo, Schedler, Kuno, Schwaar, Karl (éds.), *Umbruch in Politik und Verwaltung, Ansichten und Erfahrungen zum New Public Management in der Schweiz*. Berne: Haupt: 15–30

Weidner, H. 1993. Der verhandelnde Staat. Minderung von Vollzugskonflikten durch Mediationsverfahren, in: Vollzugsprobleme / Problèmes de la mise en oeuvre des politiques publiques, *Schweizerisches Jahrbuch für Politische Wissenschaft* 33: 225–244

Weidner, H. 1995. Twenty Five Years of Modern Environmental Policy in Germany. Treading a Well-Worn Path to the Top of the International Field, *FS II 95/301*. Berlin: Wissenschaftszentrum Berlin für Sozialforschung (WZB) (Papers Forschungsschwerpunkt Technik, Arbeit, Umwelt)

Zuppinger, U., Knoepfel, P.1996. *Usine d'incinération de déchets spéciaux (UIDS), CIBA-GEIGY à Bâle*, Etudes de cas de l'IDHEAP no 7. Chavannes-près-Renens: IDHEAP

Section IV:
New Readings for the 3rd Millennium

This book intentionally deals with research conducted in the past. Nevertheless, as previously noted, the selection of contributions tries to demonstrate how such research has contributed to the present and how it can be considered still valid. This holds particularly true for research showing the limits of environmental policy concepts for further developments in a given time and space. Accordingly, in this last section, we wish to demonstrate how we analytically conceive new approaches to environmental policies today in the framework of comprehensive resource management policies. From the point of view of future research in the area of natural resources, the four contributions in this last section may be considered as starting points. The reason why we nevertheless include them in this publication is to tentatively demonstrate the arrival of new radical questions increasingly raised by a group of scholars who, during the 1990s, remained a minority and more or less marginal. But these questions have led to the increasingly recognized debate on natural resources, property rights, integrated policy approaches and radical questionings about the limits of traditional environmental policies.

In this sense *Rationality Changes in Clean Air Policies (1960–2000) (2000)* demonstrates in a historical perspective how commonly shared truths on the causal models that heavily shaped air quality policy concepts from the past crashed and were replaced by new dominant models over time. Again, these patterns of changing dominant causal models, resulting from new problem definitions, new actor constellations, and causal models highlighting new policy instruments and new interventionist strategies, are not limited to air quality policies but rather are present in the general patterns of changing environmental policies as well as most other sectorial public policies. At the end, the contribution highlights the movement towards resource management policies, which, at the time (1998), was of course not yet an operational model.

Institution Building for Sustainable Urban Mobility Policies (1999) again deals with air quality policies that mainly aimed at reducing traffic pollution. The contribution demonstrates, on the basis of data stemming from an international COST-project, a change in the definition of air pollution to reflect the use of one service (amongst several) the resource air provides: its capacity to receive, store, absorb and transport gaseous pollutants (such as NO_X). This use is in rivalry with the use of another good provided by the same resource, its use for respiration. Urban air quality management should therefore regulate these two uses in a more explicit way, being conscious of the fact that the use of these services is attributed by means of implicit use rights to car drivers to the detriment of the increasingly limited use rights attributed to the residents.

The same basic concept of use rights is developed in *Natural Resource Quotas and Contracts – a New Institutional Regime for our Common Resources (1999)* which again was written in the context of the above mentioned international research community within the framework of a comparative project. It goes one step further because it addresses the idea of annual quotas fixing maximum loads to the environment to be attributed to different environmentally damaging public policies such as transportation, road construction, and air pollution. Especially in the Swiss context, managing scarcity by means of global and individual quotas is not at all a new idea; such quotas have been in use in agricultural, forest, nature protection and land-use policies for decades. The quota idea and the concept of breaking down legally binding global quotas into individual ones attributed to user-actors by means of real use rights will be the central pillars of the new resource management concepts we currently use for re-reading environmental policies presented in the last chapter.

In fact, it was only at the beginning of this century that we systematically started empirical studies financed by the Swiss National Science Foundation testing conceptual elements of the new resource management approach, which is rooted in theoretical concepts of resource economics, institutional economics (property rights) and public policy analyses. *Institutional Regimes for Natural Resources: An Innovative Theoretical Framework for Sustainability (2007)* documents the most recent state of this new research orientation. It combines institutional economics and environmental policy analysis results considering the definition of use rights and their attribution to user-actors acting via regulations of either basic property rights (civil law) or outputs of public policies. This new reading of environmental policies within an enlarged framework again is rooted strongly in research results of the past. Nothing is older than new concepts. This institutional regime (IR) approach has been used within a European Union project on water management (Euwareness) and further developed in the field of various natural resources in Swiss regions (water, air, forests, landscape, land-use, housing stocks) and systematically refined up to today.

12
Rationality Changes in Clean Air Policies (1960–2000) (2000)

Peter Knoepfel[1]

12.1 Background

The suggestion by today's politicians, economics experts or environmental protection practitioners that the problem of increasing air pollution in urban areas could be solved by giving suitable medication to bronchitis sufferers, pregnant women and children would be greeted with outright incredulity. Such an approach would be deemed utterly unsuitable. A similar reaction would greet suggestions to deal with industrial air pollution by means of high industrial chimneys and to tackle water pollution by introducing massive volumes of clean water into the waste water systems. And yet, when I was young, precisely these strategies were considered rational by the vast majority of environmental specialists and politicians. This was the era when collective value concepts which saw the black smoke belching out of industrial chimneys as the very expression of progress, obscuring the fact that they embodied a risk for the population, were only beginning to be described as "irrational". In fact, yesterday's visionaries, a small minority, who, to no avail, hailed the pills and high industrial chimneys as a purely symptomatic policy response, were generally dismissed as irrational zealots arguing on a "purely emotional" basis.

Historians, politicians and philosophers of science have recorded innumerable examples of this phenomenon whereby views previously held to be irrational are suddenly deemed rational and those previously believed rational no longer make sense from a contemporary perspective. All actions which justify the expectation that when carried out by everyone else they will not restrict another person's freedom (through damage) can be described as rational (Kant 1786, 85 B.A: 52). Thus, reason is defined by the perimeter which includes the acting and affected subjects and objects. This perimeter varies from place to place and throughout the time period involved. This variation is closely associated with how the subjects define

---

[1] in: Knoepfel, Peter 2000. *Rationality Changes in Clean Air Policies (1960-2000)*, Cahier de l'IDHEAP no 189. Chavannes-près-Renens: IDHEAP.

themselves and their relationships with others. What is generally perceived as rational changes with the increase in collectively shared knowledge about relationships between acting subjects, the subjects and objects affected by this action and the spatial and temporal perimeter of the corresponding actions? Such rationalities of action[2] constitute the learning and action-guiding reference systems (*référentiels* – Jobert, Müller 1987) which are accepted by the dominant actors as a valid theory and which undergo processes of accumulation and (more or less abrupt) transformation in the course of the life cycle of a public policy. The transformation of such reference systems has also been described extensively in political science (in connection with learning processes, for example, and other attempts to explain policy change – Kissling-Näf 1997).

It is not the aim of this essay to provide a new theory of policy change. What I am aiming to do is to record the factors which can be identified as constitutive to the changing rationalities of environmental policies, which is why it is possible to observe an exchange of rationalities within the transformation of these policies. Moreover, I will try to show that in an international comparison of policy with such exchanges of rationalities, these dimensions are subject to change on the basis of a recurring, internal "regularity".

I define the rationality of an environmental policy as the consensual internal agreement of the specific attributes of its basic elements[3] by the dominant institutional and social actors at a given time in a given policy area. These basic elements can be classified on the basis of the following dimensions:

- the definition of the problem and the policy aim (variable 0),
- the causal hypothesis/hypotheses (identification of target and affected groups – variable 1),
- the intervention hypothesis/hypotheses (instruments – variable 2),
- the main combinations of resources (variable 3),
- the institutional framework (variable 4) and
- the (institutional and substantial) linking of a specific environmental policy with other public policies (variable 5).

[2] I am not using a legal concept of rationality here as held, for example, by Lübbe-Wolff 1996 (modernisation; improved "executability" of environmental law).

[3] Cf. Weidner 1996: 512 ff., where, however, a different definition of basic elements is used (information, participation, equality of legal weaponry).

The following three hypotheses will be examined and substantiated on the basis of a very schematic analysis (restricted to clean air policy) of the varying characteristics of these basic elements over four different periods:

1. The rationality of clean air policy does not exist. The basic elements have been subject to a gradual transformation, in which four major developmental trends can be observed in Western Europe over the past fifty years.
2. The basic elements of the reference system do not change in isolation. The real independent variable is the changing definition of the problem and aims. This transformation regularly gives rise to changes in the five other basic elements which follow a recurring internal rationality.
3. The observable transformation of the rationality of clean air policy is not linked with the constitutional framework conditions of these policies in individual countries. An autonomous line of development can be observed in this transformation of rationality which is independent of the country in which it occurs and which, in West European countries, at least, cannot be prematurely terminated through the omission of a phase.

12.2 Definition of the six basic elements

12.2.1 Definition of the problem and aim

The familiar phenomenon of limited awareness of the existence of a problem due to simple ignorance based on the absence of visible indicators is particularly applicable in the case of clean air policies. It is known that NO_x, O_3 and (more recently) PM_{10} particulate matter are barely perceptible, either sensorily or directly, in concentrations which are damaging to health. Expensive measurement campaigns are needed to detect them. The same is even more true of the greenhouse gases which are mostly non-toxic. However, even where measurement data is available, perception varies significantly according to a range of individual[4], class-specific[5] and cultural[6] factors. Thus, the political evaluation of the "gravity" of the prob-

[4] The relationship of persons affected by the pollution to the source of the emissions, the nature of the emissions source, the emitter's attitude, susceptibilities to illness etc.

[5] Environmental awareness, knowledge of the environment etc.

[6] Anthropocentrism versus ecocentrism.

lem, the quality standards[7] to be attained and the extent of the concretization of these aims differ in an international and interregional comparison. Extensive literature containing international comparisons is available on this topic (Winter 1986; Schwager et al. 1989; Knoepfel, Descloux 1991). The status of knowledge about risks, climatic conditions, chemical transformation and long-term physical transportation play an important role in the definition of policy aims. Significant progress has been made over the past forty years with respect to knowledge in this area on a world-wide basis.

Nevertheless, the definition of the aim of a policy ultimately remains a political decision which will inevitably have both winners and losers. Along with the unequal distribution of the powers of definition among the key actors and their ability to identify "technical practical constraints" and even "natural laws" unchallenged (Bourdieu 1994: 101 ff.), not wanting to know, individual and collective value positions and individual feelings also have an equal role to play here. The former is often expressed in a technicist discourse of measurements (Weidner, Knoepfel 1979). In addition, target definitions usually have multiple levels which consist of positive ("alpine air") or negative ("absence of health hazards") formulations, abstract symbols capable of evoking consensus ("Blue Sky over the Ruhr" - German Social Democratic Party, 1961 - Brüggemann, Rommelspacher 1992) and highly technical chemical formulae (usually immission thresholds). As with other public policies, these formulations of the policy aim are expressed in the language of the problem and not in the language of the behaviour of the key actors which is relevant to the problem. I am referring here to policy outcomes to be aimed at in form of a desired value through the implementation of a series of political-administrative decisions (outputs) and correspondingly altered actor behaviour (impacts). This is particularly strongly reflected in the distinction made in clean air policy between immissions (target factors) and emissions (behavioural factors).

12.2.2 Causal hypothesis

As central factors of the causal model inherent in every public policy, the causal hypothesis formulates "conjectures about the basic structure causing the problem … It responds to the question as to who or what is to "blame" for the situation deemed politically intolerable or who or what is "objectively" responsible (without subjective blame). The response to this question defines the policy target group." (Knoepfel et al. 1997: 79). This con-

[7] Immissions thresholds.

sists of the group of actors "whose behaviour is viewed by the public policy as relevant for to resolution of the problem being approached. Thus, the policy undertakes to alter or stabilise the behaviour of the target group by means of suitable measures (e.g. bans or financial incentives)" (ibid.: 62). As opposed to this, the group of persons "directly or indirectly, positively or negatively affected by the attempt to solve the social problem in question in a particular way as part of a public policy" (ibid.: 63) is defined as the affected group.

In clean air policy, the causal hypothesis simply defines the groups of polluters whose behaviour shall be modified through the introduction of regulations for the reduction of emissions. It is assumed that the emissions caused by these groups are responsible for certain immissions. Potential polluter groups today include industry, business, households (heating systems), agriculture (NO_x) and transport[8]. It is now known that in many cases these emission-producing activities are influenced by measures arising from other public policies (road building, tourism, regional planning etc.).

Analysis of many public policies has shown that they were working on the basis of "incorrect" causal hypotheses and thus "from the outset … were condemned to inefficiency. (...) Powerful social groups can often offload their responsibility on to weaker members of society and this is why the formulation of these basic causal hypotheses is always linked with political evaluations and the nature of the perception of the problem. In many cases, however, science is unable to provide adequately guaranteed information about the mechanisms which cause the problem" (ibid.: 79). Moreover, we know from clean-air policy that the different emitter groups are very concerned about equality among target groups. If industry is to reduce its emissions, the state must also follow suit with its waste incineration plants; if transport is challenged, business must also be called to account etc.[9].

The definition of the affected group, i.e. persons who expect an improvement[10] in the quality of their lives as a result of the reduction in emissions, also varies in accordance with the causal hypothesis. If industrial emissions are reduced, the (other) industrial operations involved and the

[8] Fuel-operated motor vehicles, aeroplanes, ships, locomotives etc.

[9] A good example: the Swiss cantons' clean air measurement plans in which a contribution is requested of all groups. Cf. Imhof 1994.

[10] Or a deterioration (the case of those who profit from air pollution, e.g. sanatoria, environmental doctors etc.). This group is not dealt with under the heading "those affected by policies" in this study despite the fact that it is not insignificant for local coalitions or coalition exchanges. *More recently, we call this "affected group" the "beneficiairies" of public policies.

residents in industrial zones can breathe a sigh of relief. If air quality control activities target transport, the inhabitants of major urban agglomerations and other transport users (pedestrians, cyclists etc.) benefit. If clean air policy falls into line with the (eco-centric) relief of sensitive ecosystems, the flowers and forests will benefit and if climate protection becomes a component of air quality conservation, the potential affected groups have yet to be identified in either geographical or temporal terms (potential victims of climatic change). These groups, which vary qualitatively and quantitatively and are linked with the definition of the causal hypothesis, are extremely important for the efficacy of policy implementation. For they are the first to demand the observation of the policy aims by the state and emitters in political and even legal terms.

12.2.3 Intervention hypothesis

In addition to the causal hypothesis, the causal model on which a public policy is based also contains intervention hypotheses. This indicates the starting points at which, in the opinion of the key actors, state action on the causal mechanism giving rise to the problem should take effect (Knoepfel et al. 1997: 80). Intervention hypotheses in clean air policy can be classified on two levels: firstly, they respond the question as where intervention should take place in the process of damage caused by air pollutants. They range from the protection of objects (e.g. lime input into acidified lakes, the renovation of damaged structures and works of art, the planting of more resistant tree species in forestry) and the modification of the distribution of air pollutants (e.g. industrial chimneys policy) to emissions retention regulations for production processes (e.g. filter technology), regulations for technological optimization (e.g. fluidised-bed combustion) and actual input control (air quality requirements for the composition of fuels). Secondly, intervention hypotheses provide information about the optimum modes of control (regulative, incentive, persuasive, self-regulative).

Both types of intervention hypothesis are linked to the problem perception and causal hypotheses. Thus, an industrial chimney policy is only viewed as rational, if as opposed to their production it is the unfavourable spatial-temporal distribution of the pollutants that is seen as causing air pollution. The intervention will in turn vary according to the size of the target group and the precise extent to which it can be identified[11].

---

[11] Small identifiable target groups: regulative interventions or self-regulation; large target groups that are difficult to identify: incentive or persuasive modes.

The intervention hypothesis often implicitly states the nature of the involvement of the target and affected groups in the political-administrative processes of the application of instruments (adaptation of the rights of participation of target or affected groups depending on the mode of regulation).

12.2.4 Resources

Like the social actors, the institutional (official) actors involved in official policies work with a set of available resources which they combine or substitute in different ways, depending on the aims pursued. The management of public policies consists of the production, storage and sustainable management or renewal of each of these individual resources (personnel, financial and legal management etc.) and in their instrument-specific combination with respect to the production of effective policy outputs at the site of implementation[12]. State actors in clean air policy also have access to the standard set of nine resources which are represented schematically in Figure 1.

[12] Knoepfel et al. 1997: 73 ff.

Figure 1: The nine resources available to state actors in public policy*

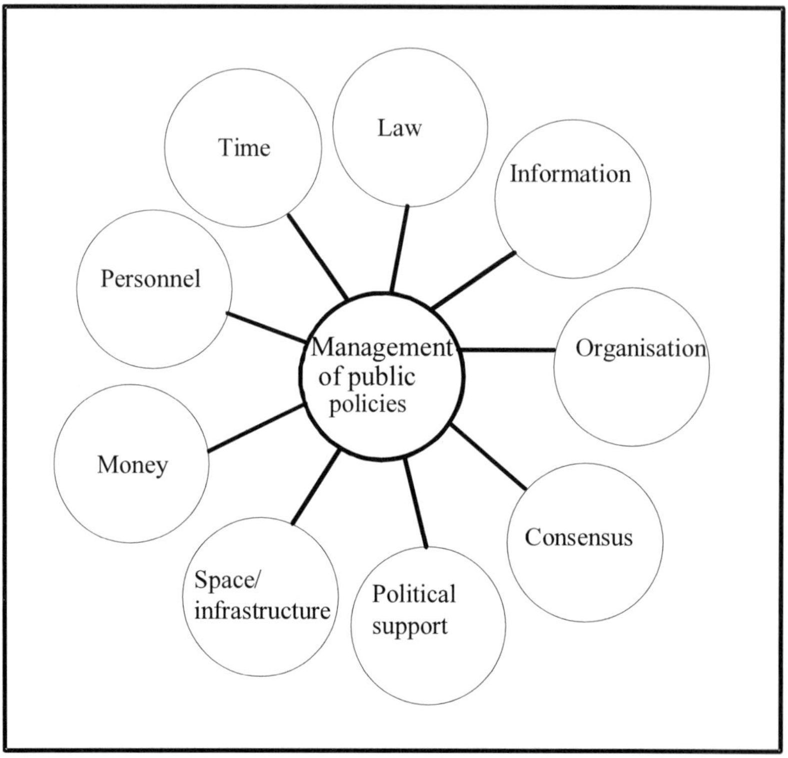

These potential available resources are:

- a set of more or less concrete substantial legal regulations (for example, immissions standards, emissions product or process standards) used as a source of orientation by implementation agencies and target groups;
- information, such as emission inventories, immission observations, damage data and also information about the structure of polluting activities and, finally, data on policy monitoring (output profile, changes in emitter behaviour);

* More recerntly, we added (as a tenth policy resource) "force" and we refomulated the resource "space/infrastructure" including "access to property rights". See Knoepfel et al. 2007: 63 ff.

- organisation, for example a definite, more or less efficient structural and procedural organisation of the responsible administration, established co-ordination processes with other administrations or external structures for the supervision of emitters and the contact with environmental organisations;
- money, which is a type of universal resource with which other resources such as information, organisation and personnel can be produced and which also acts as a basic resource for policies which work on an incentive basis (subsidies, other transfer payments);
- personnel, which in the case of clean-air policy must have (in many cases academic) professional qualifications in the areas of atmospheric physics, pollutants chemistry, biology and economics and requires relevant training;
- consensus in the sense of secondary legitimization of state action through performance (Knoepfel 1996: 160 f.), which is reflected, for example, in the willingness of emitters to implement (voluntary) behavioural modifications or of the environmental organisations to participate in constructive dialogue and which requires active maintenance;
- political support, which in contrast to target and affected groups' consensus is found in the political area of the primary legitimization of public policies and is expressed, for example in the willingness of the legislator to increase the resources of law, money and personnel; this resource also requires intensive management;
- time, which, in the case of environmental policy, has a role to play in both the upholding and failure to uphold deadlines for adaptations, and in its general expression as the allocation of scarce administrative time for certain problems (at the cost of the resolution of other problems);
- space, which in the case of clean air policy consists in the possibility of accessing executive events and enables the spatial concentration of outputs without specific spatial opposition (resource: spatially concentrated consensus or dissent).

Depending on the problem perception and causal hypothesis, clean-air policies require the above-listed resources to varying degrees. Thus, typical combinations of resources can be identified for the four phases described in this study.

12.2.5 The political-administrative arrangement and institutional framework conditions

Political-administrative arrangement is the term used to describe the complete set of institutional actors who are key players in the formulation and implementation of a specific public policy at different national levels. These actors form a structure consisting of procedural co-operation and co-ordination regulations which are centralised or fragmented (horizontal or vertical) to a greater or lesser extent, or more or less open *vis à vis* social actors (Knoepfel et al. 1997: 93 ff.). The institutional actors involved belong to more or less well established, hierarchically structured local-authority, regional or central administrations (administrative institutions), which are characterised by their constitutive task, specific public interests, professional profiles and patterns of perception. Experience has shown that the political-administrative arrangements of clean-air policies vary along the fragmentation-integration axis both vertically (distinctive vertical integration or fragmentation) and horizontally (fragmentation on the basis of emitter groups, industrial, traffic-related etc. clean-air policy). Similar distinctions exist with respect to their openness to target groups, and particularly also affected groups (environmental organisations), and with respect to the institutional roots of their main actors (health policy, trade/plants/factory inspectorate, foreign trade policy and environmental policy).

With respect to the institutional framework conditions, the variance initially occurs at the distribution of competence on the different state levels, to which the key actors of the political-administrative arrangement belong. Thus, specifically local policies can be distinguished from more regional, central-state or even EU policies. This distinction is also directly linked with the nature of the problem perception and the resulting perimeter which, depending on the valid rationality, marks out the recorded actions of subjects and their effects on objects in space and time (Larrue, Knoepfel 1998: 186 ff.). In addition to protection policy aspects, trade-policy aspects also have a role to play in this perception ("competition neutrality"). The general status of knowledge of atmospheric physics or chemistry is of importance here.

12.2.6 External links with other public policies

Like environmental policy in general, clean-air policies also have a varying need to be linked with the other public policies regulating productive or reproductive activities which generate air pollutants (interpolicy co-operation; Knoepfel 1995: 212 f.; chapter 6 of this book). The extent of this interpolicy co-operation with other major public policies varies according to the perception of the problem and the causal hypothesis. Minor isolated clean-air policies are found, for example, wherever air pollution is perceived as a local problem with a limited perimeter. Moreover, it is possible to find highly integrated policies at the level of both legislation and political-administrative arrangements which penetrate deeply into energy, transport and even agricultural and forestry policy. A similar situation applies for the linking of major clean-air policies with institutional public policies. Clean air can indeed become a driving force for institutional reorganisation in the implementation areas. This is the case, for example, if clean air policy succeeds in conjunction with mobility policy in agglomerations in forming new and autonomous political-administrative institutions out of urban and outer conurbation authorities. In this case, environmental policy becomes the forerunner policy for institutional reorganisation (Klöti et al. 1993; Knoepfel et al. 1995: 390 f.). It is possible to observe a similar situation arising under the opposite conditions when the transfer of competence for environmental policy from the central state to the regions is a significant factor behind the formation and consolidation of regional bodies in previously predominantly centralist states (e.g., France, Spain and Italy).

12.3 Different clean air policy rationalities

It is possible to identify the four distinct rationalities described below for clean air policies in West European countries between the 1960s and the turn of the century. This presentation does not aim to provide an empirical account of these four phases and for this reason is rather schematic. Its purpose is to present the basic patterns of the different rationalities and not a detailed and precise description of these phases. Most of the empirical material can be found in Knoepfel, Weidner 1980 and 1986, Héritier et al. 1994 and Jänicke, Weidner 1996.

12.3.1 Clean air policies of the 1960s

Air pollution was initially perceived as a neighbourhood and later local problem involving emissions of smoke, soot and eventually sulphur dioxide which in conjunction with bad weather conditions (inversion) could cause nuisance, damage to health and, in extreme cases, increased morbidity[13]. The problem perception was, therefore, clearly anthropocentric. The cause was identified as outdated commercial and industrial plants and coal or oil-fired household heating systems emitting the above-listed pollutants into the atmosphere at low levels. The problem was mainly observed in poorly ventilated neighbourhoods and "working-class areas" generally located in north-eastern locations near industrial zones. Increased morbidity caused reduced productivity and this translated into a burden on the public health budget. The aim of this policy in Germany was a "Blue Sky over the Ruhr " (German Social Democratic Party 1961 - Brüggemann, Rommelspacher 1992). Quantified immissions limits were only defined in a few cases.

Emissions from household heating systems and commercial plants and from inner-city industrial plants situated too close to residential areas were identified by the causal hypothesis as the cause of the increased concentrations of pollutants in the air. Thus, the target groups of the clean-air policies that emerged were coal and oil-heated households and commercial and industrial operations in urban areas. The poor basic health of employees was also identified as the cause of the problem (nutrition low in vitamins, too little fresh air from holidays etc.). The affected groups included the residents of particularly afflicted neighbourhoods who are at risk from air-pollution (elderly people, pregnant women, children, asthmatics) and also industrial and commercial plants (productivity losses) and certain economic sectors, for which clean air is an important resource (hospitals, tourism, food industry etc.).

The central intervention hypothesis for these clean-air policies focused on improving the spatial-temporal distribution of the pollutants in the air (transmission hypothesis). Thus, what emerged included the infamous industrial chimneys, regional-planning decentralisation (transfer of industry from urban areas) and smog-alarm policies. These smog-alarm policies required a reduction in the output of large heating installations in the event of inversion weather conditions or the conversion of these installations to fuels producing lower levels of pollutants. The core content of these policies remained applicable until the 1990s[14]. They were intended to control the structure of emissions sources in space and time. They imposed agglomeration-specific fuel regulations on the large group of home-heating emitters and individual directives for the increase of chimney heights on

[13] The most important trigger in international terms was probably the smog disaster of 1952 in London.

[14] According to the new French clean-air and energy act (*Loi du 30 décembre 1996 sur l'air et l'utilisation rationnelle de l'énergie* = Law of December 30th 1996 on air and the rational use of energy).

the relatively small number of large-scale emitters. The central mode of intervention was police clauses and bans.

The most important resource of this air pollution control - conceived as "minor police public-health policies" - is law, the regulative density of which was increased over time[15]. Information also assumed increasing significance as a resource, as this kind of immissions-oriented intervention policy requires knowledge of the (portable or noxious) content of the air pollutants in the surrounding air. Such policies proved, therefore, to be increasingly expensive (resource money) and they required the services of specially trained personnel (emergence of the job of "air quality controller"). Special administrative units for air quality control and for the employment of existing organisations in the intermediary area between the state and society (e.g. the "official" chimney sweep) were established in major cities.

The small political-administrative arrangements, which initially had a strong institutional root in the urban health authorities, began to fragment. The established (often regional) industry and trade inspection boards defended their territory for the control of trade and industry practices against the health boards which began to move in on this domain. This initial cell division of air quality control arrangements would be responsible for the corresponding fragmentation over several decades. Conversely, this development of an industrial and commercial clean-air policy resulted in the emergence of a socio-political emphasis through the assignment of its control to the industry and trade inspection boards. For this meant that the traditional protection of workers within companies was, so to speak, extended out to the surrounding area ("protection of the surrounding area") and this surrounding area mainly consisted of working-class neighbourhoods. It was, no doubt, in this way, that clean-air issues made their way into social-democratic party programmes and onto the trade-union agenda. Thus, it is easy to understand how the political-administrative arrangements of clean-air policies in the 1960s were relatively open to trade-unions and socio-political health organisations supported by the trade unions (particularly in France and England).

The institutional assignment of these small clean-air policies was located at local level. It was only in cases where agglomeration associations already existed for regional-planning purposes that local initiatives ultimately led to the competence of conurbation associations (Greater London, *Deutsche Umlandverbände* etc.). This importance of the region is again an

[15] At the end of this period, state standards for authorised immissions, emissions and fuels defined by private standardisation associations gradually replaced the general police regulation which initially served as a legal basis.

expression of the extent to which these clean-air policies were immission-oriented and thus also incorporated strong regional-planning components.

These minor police policies did not maintain systematic interpolicy-co-operation with either other substantial or institutional public policies. This proved even less necessary when as a result of the above-mentioned cell division, industrial and commercial clean-air policy become for the time being part of national industrial policy (France, Italy) or, in more economically liberal countries like Germany and Switzerland, they become national industrial inspection policies. As such, they remained firmly under the control of the factory inspectorates which are close to industry and trade unions in terms of their interests ("cosy relationship" of Alkali and Clean Air Inspectorate in Air Pollution Control - Hill 1983). "Outsiders" could only gain access through employee-friendly health associations. What emerged here was a closed interaction system, whose sole participants were the national inspection authorities and the emitters.

12.3.2 Clean air policies of the 1980s and early 1990s

At this point, the key actors considered the existing and additional total volumes of emitted air pollutants as a collective problem which needed to be brought under control. The decisive change in perception consisted in a shift from the previous immissions orientation to an emissions orientation in clean air policy. This can be explained by the fact that political and scientific arguments were increasingly based on the concept of ecosystems. Thus there was a shift in emphasis to the flow of materials through different environmental media. According to this perspective, on the basis of air chemistry transformation and atmospheric physics transportation processes, the emission of pollutants into the air gives rise to an additional burden on ecosystems. The evidence is found not only in the form of noxious immissions in the air in the immediate surroundings of source of emissions but also in pollutant-rich precipitation (acid rain) at a distance from the source which can lead to the pollution of surface waters and groundwater carriers, and ultimately to damage to soil ecosystems. The extended damage concept applied here includes not only the direct effect on human health but also the functionality of ecosystems close to and far away from the emissions activities. The main trigger for this new perception of the damage was the acidification of Scandinavian and Canadian surface waters and the alarming *Waldsterben* observed in northern Europe in the mid 1980s.

The aim of these clean air policies consisted in defining the global volume of air pollutants emitted and reducing these volumes by means of suitable control measures. SO_2 and dust particles were still the main pollutants, although NO_x and organic pollutants later took centre stage. The aim in urban agglomerations continued to centre on the definition of more precise immissions limits. Given the apparent impossibility of establishing valid correlations between absolutely defined emissions tonnages and the corresponding burden on ecosystems, absolute emission reduction quotas generally continued to apply (in % of the total volume when the quotas are defined) in the description of total target emissions volumes. The aims defined by the Swiss Council of Ministers, whereby SO_2 and NO_x emissions were to have been reduced to the levels of 1950 or 1960[16], were typical of the relative helplessness of such clean air policies. International regimes (pioneer: the Geneva Convention on long-range transboundary air pollution of 1979) also work with relative reductions quotas ("30% Club" etc.).

With the shift in the problem definition, the causal hypothesis also changed and all actual emitters of air pollutants were now seen as causing air pollution. This resulted in the disappearance of the former spatial (urban agglomerations) and temporal (inversion periods) restrictions. Reductions in emissions needed to be achieved "irrespective of the immissions situation and all over the country" ("prevention principle")[17]. The former spatial-temporal definition component of the target group survived, however, in that additional obligations for the reduction of emissions were imposed on emitters from agglomerations if the required immissions limit could not be adhered to, despite the application of country-wide emissions limits (supplementing the emissions-oriented basic strategy with an additional immissions-oriented strategy)[18]. Despite this universal intention of including all emissions activities in the definition of the target groups of clean-air policies, in practice there was an obvious concentration on industry, commerce and households and the transport sector was initially excluded in many countries.

This (as we are aware today, one-sided) causal hypothesis led to a concentration of clean-air measures on industrial and commercial processes and on the technological characteristics of type-tested household heating devices. This in turn triggered an unprecedented growth in technological innovations leading to the insight among target groups that measures for the conservation of air quality can have economic benefits arising from

[16] Cf. Swiss Council of Ministers 1986 (Clean Air Concept).

[17] Article 11, Section 2 of the Federal Swiss Act on Environmental Protection of 7 October 1983 is typical of this rationality (SR 814.01).

[18] For example, Article 9 of the Swiss Clean Air Decree of 16 December 1985 (SR 814.318.142.1) which requests more stringent emissions limits in the event of the immissions limits being exceeded.

reductions in the use of materials and energy (ecological modernisation - Jänicke 1996). As a result, this causal hypothesis gained increasing acceptance among the target groups. It gave rise to new impulses in economic policy, led to the creation of employment and the establishment of an increasingly important ecology sector (ecobusiness - Benninghoff, Joerchel and Knoepfel 1997; chapter 11 of this book). This process was accelerated by the fact that new actors featured in the policy area of clean-air policies who were legitimated from a new expanded definition of the groups affected by the introduction of the relevant policies. The expansion of the target groups was accompanied by a corresponding universalization of the affected groups. The latter no longer consisted solely of local protective (trade union) organisations motivated by an interest in work practices but also included environmental protection organisations working on a national and European basis. The latter succeeded in establishing themselves as the defenders of the ecosystems. By the late 1980s, the initially sharp conflicts between the politically stigmatised target groups and the environmental organisations, formed by the protectors of nature and ecosystems, gradually abated. Partnerships were established ("co-operation principle"), in which reductions in emissions (in excess of the legal requirements) were traded for eco-acceptance. This shift in the nature of the affected groups, which were now mainly recruited among the educated middle classes, meant that clean-air policies relinquished their former sociopolitical bias.

Not least among the factors responsible for the above-mentioned growth in technological intervention was a new intervention hypothesis. This intervention hypothesis assumed that it would be easier to achieve changes in the behaviour of emitters if investment goods and service markets offer the required technologies in temporal harmony with the investment cycles of the different sectors thus enabling target groups to make savings in their economic calculations in the medium term. Global process technology and suitable control of material and energy input was to replace retention technology. This new intervention hypothesis was also reflected in changes in the modes of intervention. While bans and rules remained the preferred instruments, they were increasingly linked with the economic logic of the regulated target groups (negotiation of generally formulated emissions standards within the specific production conditions of different sectors). State regulative output also took long-term operational planning, the capacity of companies for self-regulation and the anticipation of technological innovations in the area of clean air into account. The provision of technical information (persuasive intervention mode) and the direct promotion of new clean air technologies and their application through financial incentives (incentive intervention modes) gained in significance. Conversely, direct regulative intervention was rejected in favour of contract-like agreements between authorities and companies (Knoepfel 1998).

Environmental law, which was extended (new pollutants), intensified (clarifications) and made more stringent with respect to emissions standards, remained the main resource availed of by the official regulative instances. Contrary to the opinion of some politicians, deregulation was not in sight at the end of the period. Moreover, the resources money (direct subsidies or tax relief for new environmental technologies) and time (adaptation of deadlines for redevelopment and introduction of technical innovations to the investment cycles in the different sectors) gained considerably in significance. By the mid-1980s, clean air policies in some countries enjoyed maximum levels of the resource of political support ("*Waldsterben* effect"). The importance of the resource of consensus was initially underestimated. This initially applied in the relationship between the administration and the target groups who, in the early 1980s, were successful in their opposition to supposedly excessive environmental requirements. The same applies for the environmental organisations which staged politically highly visible front-line conflicts with administrations and companies concerning individual projects and planned legislation around the mid 1980s. It was not until the end of the period, that due to the increased environmental awareness and the political and scientific weight of environmental arguments there was clear consolidation of the position of the environmental organisations in the "iron triangle" involving the state, emitters and environmental organisations. This turning point led to a gain in significance for the resource of consensus, also in the relationship between the state and the environmental organisations.

The partial "despatialization" which accompanied the universalization of the emissions limitation strategy and the concentration on industry and business led to the strengthening of the position of national and community actors in the clean air political-administrative arrangements. The centralization was intended to guarantee the harmonization of emissions requirements and hence their competitive neutrality. It is basically impossible for a central administration to implement emissions reductions requirements equally on a country-wide basis. The centralization of the (quantitatively increased) legislation was, therefore, accompanied by a reinforcement of the regional implementation level. The temporary losers in this situation were the local units. The industrial and commercial orientation actually gave rise to the expectation that the position of institutional actors in industry, trade and police authorities would be strengthened. This calculation was, however, thwarted by the increased confidence of the environmental authorities who endeavoured to implement industrial and commercial clean-air policy themselves. Thus, in many cases the corresponding competence shifted from the traditional industry and trade inspection boards to the newly created environmental authorities. The actors responsible for the regulation of industrial and commercial emissions assumed a dominant position in the political-administrative arrangements of these clean air policies which had undergone an institutional "transplant". These arrangements are often very well integrated in vertical terms (co-operation between national and regional level) with fragmentation along this axis existing only down to the local authorities. Horizontal intra-policy fragmentation is, in contrast, rare.

These transfers of competence created a need for new interpolicy co-operation between environmental and industrial or commercial regulations. This was guaranteed at the level of programme formulation through the widespread inclusion of these administrative instances in the definition of emissions and process standards and at implementation level through the development of more differentiated inter-policy networks in the context of the environmental impact assessment procedure (Kissling-Näf 1997). Finally, it is important to recall that the absolute necessity of the regional implementation of the almost precipitously produced international community law and national emissions reductions legislation in some countries (particularly France and Italy) made a key contribution to the formation of relatively autonomous subnational bodies. The implementation of technology-oriented industrial and commercial clean-air policy can therefore be described as one of the focal points for the assumption of autonomy by regional economic policies in these countries.

12.3.3 Clean air policies of the 1990s

A number of collective shock experiences in the 1990s led to the emergence of yet another perception of the problem of air pollution which ultimately transformed into an agglomeration-specific mobility and global climate issue (also strongly influenced by traffic). The shock of the discovery that even in countries which had introduced mandatory use of three-way catalytic converters for their vehicle stock during the 1980s, levels of NO_x pollution, the newly discovered health hazard of the PM_{10} particulate matter and ozone levels had hardly been reduced at all in areas near cities because the reductions achieved had, for the most part, been negated by increases in motor traffic. Clean air policy actors were also severely shaken by the debates surrounding greenhouse gases and climatic change which introduced a new global component. Barely had the pollution caused by industry, commerce and households been brought under control and the traffic file, which has been widely considered as more important, reared its head. Two new dimensions were, therefore, added to clean-air policies: air pollution was perceived firstly as a threat to health and ecosystems caused by traffic in and around urban agglomerations (again spatialized perception) and, secondly, as an initially barely comprehensible global threat mainly arising from increasing fuel-consuming mobility.

This change in perception was accompanied by a recent change in the causal hypothesis. Individual and freight road traffic which was previously for the most part sacrosanct moved to the position of the central target group. Drastic intervention against these two target groups was identified as the price to be paid for "Clean Air in European Cities"[19] and protection against global climatic change. This conviction was articulated in declarations against the increasing collapse of the transport systems of cities and conurbations, the increase in winter and summer smog which is harmful to health and causes material damage in cities and the consistently high ozone levels in valued recreational areas surrounding cities. To this was added the increasing political belief in global temperature increases and variations (El Niño) with their disastrous consequences[20]. With the advent of the greenhouse gas problem, the affected group, which defines itself as a politically legitimated actor, increased significantly. In addition to more active and aggressive pressure groups against commuter and through traffic, non-governmental organisations emerged in Europe and throughout the world as protectors of climatic interests. These NGOs were, in turn, supported by the governments of the potential losers in the game of climatic poker.

The intervention hypotheses also had to be adapted to this articulation of the problem into the local and global levels. Control of the structure of emissions by means of regional planning was increasingly applied in the resolution of the local-supralocal air pollution problem (development of residential settlement areas on the outskirts of cities; infrastructure planning - Snickers. 1998). Moreover, a wide range of interventions were introduced in the area of individual and goods motor traffic in the form of bans and regulations (traffic regulations), incentive systems (traffic taxation, road pricing etc.) and direct infrastructure services (development of public transport in view of the modal split). The heterogeneous target group of vehicle users, which it would be impossible to control using individual measures, became the object of attempts at collective regulation (car fittings, fuel composition), incentive schemes and campaigns[21].

---

[19] Cf. concluding reports of CITAIR, Cost Actions nos. 614-618, Zürich (Synergo), 1998.

[20] NFP 31; Glogger 1998.

[21] Cf. Zimmermann, Wyss and Neuenschwander 1997.

The intensification of traffic-related clean-air policies led to a downturn in the significance of the resource of the law which was overtaken by the resources of information (campaigns, traffic research, traffic education), personnel (control of road traffic regulation), money (incentive schemes, expensive technical traffic regulation systems), consensus (increasingly controversial road-building projects), time (the problem of peak traffic periods) and space (spatial concentration of traffic movement). This list of required resources is extensive and involves correspondingly high costs for the most part carried by the local and regional administrations which by now were plagued with financial difficulties. Also, in view of the increasingly scarce resource of political support (opposition on the part of the automobile associations), tension was inevitable.

With this reorientation in the direction of traffic, both within clean-air policy as a whole and with regard to their external relations, the political-administrative arrangements underwent considerable reorganisation. Internally, the number of local actors increased and there was a gain in significance at national and international level (climate problem). The horizontal fragmentation to the actors involved in industrial and commercial clean-air policy increased because the latter tend to be more active at regional then local level (concomitant vertical fragmentation). Moreover, co-operation between local, supralocal and national actors (national clean air urban policy) increased. Thus, vertical tensions between the local and regional levels, which can be traced back to defensive urban strategies directed against the outer conurbations (which threaten the town or city with commuter traffic), became more common in the political-administrative arrangement[22].

Such clean-air policies can represent very important triggers for fundamental reorganisation at institutional level, however, as they demonstrate the need for a supra-local regulation unit which would be responsible for the urban centres and suburban authorities as an agglomeration spatially defined by commuter movements (Klöti et al. 1993; Knoepfel et al. 1995). Such initiatives for the formation of urban agglomerations are also (in part correspondingly) justified by the fiscal impoverishment of the town and city centres and a corresponding increase in the financial resources of peripheral authorities from which the commuter movements to the town and city centres emerge (Frey 1996: 26 ff.). These clean-air polices can, therefore, become the triggers for institutional innovation in the urban agglomerations.

---

[22] Voter potential which is important for the regional governments is recruited in these outer conurbations.

Such traffic-related clean-air policies can only survive if they maintain intensive interpolicy contact with local and regional road construction, traffic regulation, regional planning and public transport policies. In the past, this classical inter-policy quintet (Knoepfel et al. 1995: 356 ff.) was only complemented by further interpolicy co-operation with local or regional energy policy in a few cases. Such interpolicy co-operation was, however, increasingly observed at national level where clean air policies concerning greenhouse gases were being developed.

In rural areas, there was increased co-operation between traffic-related clean air policy and biodiversity-related nature protection policies. This is hardly surprising in view of the fact that the most important cause in the demise of biodiversity is mobility-promoting infrastructure with its extensive requirement of land (Knoepfel et al. 1996: 76 ff., 301 ff.).

12.3.4 The clean-air policies of the turn of the century[23]

There is much to indicate that the above-described concentration of air pollution problems on the ever-expanding road transport systems in agglomerations (individual and goods traffic) and on the accumulation of greenhouse gases will continue well into the next century. With the support of the general change in the environmental policy paradigm, manifest in the sustainable management of resources[24], the problem addressed by clean-air policy has been transformed into a distribution question. There is increasing recognition of the fact that the authorisation of each motor vehicle and the licensing of industrial and commercial operations which emit air pollutants translates into the distribution of rights for the use of clean air or for the repletion of the atmosphere with greenhouse gases. Despite the compliance of emissions of these gases with the individual restrictions defined in the 1990s, their accumulation in the air in urban agglomerations leads to the massive over-use of the existing absorption capacity of the resource of clean air (sink) with little remaining for competing third-party users. The regime of clean-air use (already excessively burdened by motor vehicle use) sees itself subject to increasingly superseded legitimate claims for third-party use which must be considered for reasons of social peace. Thus, the problem to be addressed by turn-of-the-century clean air policy

[23] The "vision" presented here initially developed in Knoepfel, Grant, Perl 1999 (= chapter 13 of this book) on the basis of contributions in Grant, Knoepfel, Perl 1999; Murswiek 1985 (approaches from a legal perspective) and Bernauer, Kissling-Näf, Knoepfel 1999.

[24] Both natural resources which act as pollutant sinks and productive resources.

consists in the definition of globally available contingents of clean air on the level of expanding urban agglomerations and on the planetary level of the atmosphere and in the allocation of these contingents to competing user groups. Clean-air policy becomes (re)distribution policy.

The central target group of these redistributing and still primarily traffic-oriented clean air policies is the motor vehicle operators and manufacturers. The situation now differs from that in the early 1990s, however, in that it is no longer individual drivers of motor vehicles and their emissions behaviour but the entire fleets of vehicles with their different subgroups that are being held responsible for the over-use of clean air, for the threat to climate (together with industrial firing installations) and for the insufficient use of these resources by competing user groups. Air pollution is seen as (too high) a price to pay for the increasing domination of the use of clean air by motor vehicles. Thus, the affected group, which already underwent extensive expansion during the 1990s, is now becoming the group which claims the clean-air use rights for itself (and at the direct cost of the target groups). It is demanding not only (socially) justifiable reductions in emissions in the area of motor vehicles but the imposition of vehicle mobility restrictions, irrespective of the will of individual drivers. These can take the form, for example of restricted access to motor vehicle ownership, roads and mobility areas in urban centres. Industrial and commercial operations feature as competing consumers and hence parties affected by the pollution in both agglomerative-urban policy (NO_x orientation) and in the planetary clean-air policy (CO_2 orientation). They see their claim to clean air as threatened by the excessive consumption of clean air by traffic. This enables the formation of powerful coalitions which unite productive and reproductive sectors against motor vehicle mobility.

Such clean air policies will have to develop new intervention hypotheses, the starting point for which is the total fleet of motor vehicles authorised for use in the agglomerative air sheds and not the individual owners of motor vehicles. The same applies for the CO_2 question where, in addition to the national motor vehicle stock, all industrial plants etc. that produce greenhouse gases will be subject to control. What was in part practised under the opposite circumstances in the clean air policies of the 1960s - at least at local level - will henceforth become the general intervention philosophy, i.e. the planned management of the resource of clean air in space and time. This should enable the co-ordination of claims for use by motor-vehicle stocks with other claims for the use of non-reproducible clean air. At international level, the planning should define national CO_2 quotas in the context of the global CO_2 absorption capacity in terms of space and time. Such plans can be achieved through regulation (bans and prohibitions), incentives (national incentives) and through the creation of new clean-air markets. In view of the potentially explosive nature of pure market solutions, which can result in the monopolization of current use rights instead of the desired redistribution, a mixed intervention mode consisting of regulative (contingent or quotas: implicit right with limits in the form of bans) and incentive (example: economic incentives for avoiding use of motor vehicles) elements will probably be selected.

In any case, the above-described effect mechanism necessitates the definition of the consumable clean air on offer in an air shed for distribution on the basis of quotas (or clean-air rights formulated in other terms) and of the acceptable level of CO_2 accumulation in the atmosphere of our planet. This definition can be applied on agglomeration level on the basis of existing immission limits; however, more detailed information is required about the current status of immissions, the necessary reduction in immissions (or, in exceptional cases, possible additional immissions) and reliable models for the conversion of immissions to emissions. This could be far more difficult to achieve at global level, given the difficulties with respect to data[25].

Of the resources used by the authorities, access to space based on consensus between the most important target and affected groups in an agglomeration (resource: space) and the resource of organisation (establishment of collective organisations for target and affected groups, establishment of super-local institutions) will gain in significance. The described quota regulations could be extremely radical and thus require a clear legal basis. Even if there is no distribution or redistribution of purely subjective legal claims in the technical sense, the resource of law will become increasingly important in the above-described global control as, in addition to the allocation of individual quotas, determination of group quotas will require legal standardization. Given that what is involved here are redistribution policies, the authorities and social actors will find themselves in highly conflictive zones. There will, therefore, be a scarcity of the resource of consensus at times.

The administrative-political arrangements for such clean-air policies will, no doubt, experience considerable change. The position of institutional actors who, as part of the clean-air agencies, are responsible for the inventory of resources and planning of consumption will gain in significance. This new function would have to be established within the administrative-political arrangement at the level of an air-shed institution (above local-authority level) (Perl 1999). This kind of supra-local institution is highly dependent on democratic legitimization because it is here that the

---

[25] To illustrate: the disputes about CO_2 quotas at the international climate conference in Kyoto (1997).

decisive redistribution processes for local clean air policy are implemented (against the will of individual target groups). It will also be necessary to fight tendencies for vertical fragmentation within the political-administrative arrangement. These tendencies will probably arise the attempt of national authorities to enforce CO_2 reduction quotas imposed on them in international agreements on the regions and urban agglomerations. The latter will react to this by refusing to co-operate (ultimately referring to their own distribution struggles in the area of traffic). Similar fragmentation tendencies may appear at horizontal level reflecting the preservation of the vested rights of the traffic-related clean-air policies and "their" quotas. The introduction of redistributive shifts in these quotas in favour of other users such as pedestrians, urban residents, street residents, industry and commerce is, of course, the aim of these new policies. Fragmentation between administrative actors can only be counteracted by a "strong" and hence direct-democratic legitimised "agglomeration state".

As we have shown in another study (Knoepfel, Grant, Perl 1999 – chapter 13 of this book), such clean-air policies at agglomeration level can only be successful as a component of integrated sustainable mobility policies. Other mobility-relevant policies such as road construction, regional planning, traffic regulation and public transport and both national and regional energy policy are also involved here. With the help of an intelligent interpolicy strategy, clean air policy can become the focal point of this kind of mobility policy despite the fact that it is responsible for the administration of one of the two increasingly scarce natural resources in this area (clean air)[26]. At institutional level, such clean air policy should become an important impetus for institutional reorganisation at both local and international level. The above-described distribution problem can only be solved if new "air-shed institutions" with independent democratic primary legitimization which include urban centres and suburban authorities are established in the agglomerations. At international level, regimes are required which have sufficient legitimization (for example within the framework of UNEP) to enforce the agreed CO_2 reduction quotas on the nations.

[26] The other scarce resource is road surface used by the public, under state or private ownership. The increased use of this resource is now reaching its limits for a wide range of reasons.

12.4 Summary and conclusions

The chemical composition of air pollution has changed in the course of the past fifty years. Nonetheless, throughout this period it has mainly originated from combustion processes which have only been subject to insignificant change with respect to their main polluter groups. Thus, during this period since the 1950s, CO_2 (non-toxic) and nitrogen oxides, sulphur dioxides and dust particles have been emitted into the clean air as a result of combustion processes. Despite this, the problem perception and formulation of clean air policies in West European countries have undergone fundamental change on at least three occasions in this relatively short time. They changed from a focus on the local health hazard caused by house fires, industry and commerce to the potentially ubiquitous threat to ecosystems arising from industrial and commercial emissions, local and global health and climate hazards and finally on the issue of (re)distribution which is concerned with the allocation of competing rights to the local resource of clean air, rights for the use of the CO_2 sinks or for the repletion of the atmosphere with greenhouse gases at global level. This history has taken the clean-air policies of West European countries through a series of rapid and unpredictable changes involving the exchange of both the causal and intervention hypotheses which deploy the necessary public resources for its management and the reorganisation of their political-administrative arrangement, including their institutional framework, on several occasions. The course taken by external relations with other public policies was equally turbulent.

In reality, therefore, we are dealing with four very different policy generations whose only common factor is that their core concern was the fight against the health hazard of air pollution. The interactive arrangement of the five basic elements which react to different problem and target definitions shows a high degree of internal coherence in all cases. These rationalities are updated on a varying but coherent basis through state action in the context of action guiding reference systems for the linking of acting subjects (target groups), objects (affected groups) and a varying spatial and temporal perimeter (local, national, global). It should be noted that the fourth development phase (turn-of-the-century clean air policy) primarily involves prospective speculation[27].

---

[27] Speculation: integration of the economic principle in the management of the resource of clean air with corresponding change of regime. Cf. Bernauer, Kissling-Näf, Knoepfel 1999.

Thus, I believe that I have provided sufficient substantiation for hypotheses one and two formulated at the outset of this essay. There is actually no single "rationality" of clean air policy. On the basis of the six basic elements described, however, it is possible to distinguish four different rationalities, whose changes at the level of these basic elements conform to a specific regularity. This is not only due to the fact that it is a highly technical policy. It is true that the actors' discourse, particularly in an international comparison, is primarily highly technical in nature and thanks to a language capable of generalisation (often English) transfers have taken place between different countries. This is not sufficient, however, to explain the similarities of the rationality structure. Equally significant is the fact that the described policies display a very similar setting of institutional and social actors for each phase. The composition of these "policy operators" changes from one phase to the next depending on the varying target and affected groups and the institutional positioning of the policies with respect to other major policy areas (health, industry etc.). Newly established policy areas produce different substantial policies. The nature of the actor population is decisively influenced by the changing perception of the problem. The latter is clearly less contingent in similarly objective air pollution conditions than in other public policies which often perceive similar problems in very different ways and produce very different actor constellations.

Thus, the change which can be observed in rationality over the past fifty years, at least in West European countries, displays a surprisingly reasonable line of development. This must be mainly explained by the common course taken by the objective problems in these countries, the social and institutional actors involved who populate the policy areas in similar constellations and the gradual learning processes resulting from internationally available information.

Because this statement is restricted to West European countries, it is not possible to identify, for example, the extent to which East European or even Latin American or South-East Asian clean air policies will necessarily display similar patterns in the future. Impressions from Eastern European countries would lead to the assumption that clean-air policies are primarily tackling house-fires, trade and industry and as part of a second phase the traffic sector. Due to the mass resistance of the owners of the very potent symbol of newly-acquired affluence, the motor car, an attack on traffic without previous or at least parallel intervention for trade and industry would face inevitable political failure. This is supported by the increased availability of new industrial clean-air equipment thanks to technology transfer. However, the effect of international transfers in the area of motor vehicles are negative rather than positive in their effect on the environment[28].

It is left to the reader to complete the sketch presented to give a full-scale and detailed portrait. Please forgive me if the heavy pencil lines of the sketch are revealed as inaccurate in places in the course of this detailed work. Whether ultimately the turn-of-the-century clean air policies, whose sketch presented here, seem irrational to some, will actually become a rational reference system in the year 2010, is something I will not be able to judge until I am in my retirement. By then this in turn will be identifiable as a temporary phase and a new (fifth) phase will be under way.

References

Benninghoff, M., Joerchel, B., Knoepfel, P. (Hrsg.) 1997. *L'écobusiness: enjeux et perspectives pour la politique de l'environnement.* Basel/Frankfurt: Helbing & Lichtenhahn (Reihe Oekologie & Gesellschaft, Bd. 11)

Bernauer, Thomas, Kissling-Näf, Ingrid, Knoepfel, Peter 2000. Institutionalized Management of Scarce Resources: Introduction and Overview, *Revue suisse de science politique*, vol. 6(1), printemps 2000. Chur/Zurich: Rüegger: 3—8

Bourdieu, P. 1994. *Raisons pratiques. Sur la théorie de l'action.* Paris: Seuil

Brüggemann, Franz Josef, Rommelspacher, Thomas 1992. *Blauer Himmel über der Ruhr. Geschichte der Umwelt im Ruhrgebiet, 1840-1990.* Essen

Frey, R.L. 1996. Teufelskreis der staatlichen Entwicklung, in: Eidgenössisches Personalamt (Hrsg.), *Alarm um die Städte / Les villes et l'urgence d'agir.* Bern: EDMZ (Schriftenreihe des Eidgenössischen Personalamtes, Bd. 4): 19–39

Glogger, Beat 1998. *Heisszeit, Klimaänderungen und Naturkatastrophen in der Schweiz.* Zürich: vdf, Hochschulverlag AG an der ETH Zürich. Zusammenfassung der wichtigsten Ergebnisse des Nationalen Forschungsprogrammes Nr. 31 zum Thema "Klimaänderungen und Naturkatastrophen"

Grant, W., Knoepfel, P., Perl, A. 1999. *The Politics of Improving Air Quality.* Aldershot: Edward Elgar

[28] Transfer of used vehicles to East Europe with inferior environmental fittings.

Héritier, A., Knill, Ch., Becka, M. 1994. *Die Veränderung von Staatlichkeit in Europa*. Baden-Baden: Nomos

Hill, Michael 1983. The Role of the British Alcali and Clean Air Inspectorate in Air Pollution Control, in: Downing, P.B., Hanf (eds), *International Comparision in Implementing Pollution Laws*. Boston etc.: Policy Studies Journal (vol. 11, no 1): 87–106

Imhof, Rita 1994. *Luftreinhaltung und Verkehr in elf Kantonen – Situation, Akteure und Massnahmenplan*, Cahier de l'IDHEAP no 125. Chavannes-près-Renens: IDHEAP

Jänicke, M. (ed.) 1996. *Umweltpolitik der Industrieländer. Entwicklung – Bilanz – Erfolgsbedingungen*. Berlin: sigma

Jänicke, M., Weidner, H. (eds) 1996. *National Environmental Policies. A Comparative Study of Capacity-Building*. Berlin etc.: Springer

Jobert, Bruno, Müller, Pierre 1987. *L'Etat en action. Politiques publiques et corporatismes*. Paris: Presses universitaires de France

Kant, Imm. 1786, 85. *Grundlegung zur Metaphysik der Sitten*. Riga

Kissling-Näf, I. 1997. *Lernprozesse und Umweltverträglichkeitsprüfung. Staatliche Steuerung über Verfahren und Netzwerkbildung in der Abfallpolitik*. Basel/Frankfurt: Helbing & Lichtenhahn (Reihe Oekologie & Gesellschaft, Bd. 12)

Klöti, U., Haldemann, Th., Schenkel, W.1993. *Die Stadt im Bundesstaat – Alleingang oder Zusammenarbeit?* Chur/Zürich: Rüegger

Knoepfel, P., Weidner, H. 1980. *Handbuch der $S0_2$-Luftreinhaltepolitik. Trends, rechtliche Regelungen und Verwaltungssysteme in den EG-Staaten und in der Schweiz.* Band 1: Problembeschreibung und vergleichende Analyse; Band 2: Vergleichende Länderanalysen. Berlin: Erich Schmidt-Verlag

Knoepfel, P., Weidner, H. 1986. *Luftreinhaltung (stationäre Quellen) im internationalen Vergleich*. Berlin: sigma (6 Bände)

Knoepfel, P., Descloux, M. 1988. *Valeurs limites d'immissions: choix politiques ou déterminations scientifiques?* Cahier de l'IDHEAP no 48. Chavannes-près-Renens: IDHEAP

Knoepfel, P. 1993. *Bedingungen einer wirksamen Umsetzung umweltpolitischer Programme – Erfahrungen aus westeuropäischen Staaten*, Cahier de l'IDHEAP no 108. Chavannes-près-Renens: IDHEAP

Knoepfel, P. 1995. New Institutional Arrangements for a New Generation of Environmental Policy Instruments: Intra- and Interpolicy-Cooperation, in: Dente, B. (ed.), *Environmental Policy in Search of New Instruments*, European Science Foundation. Dordrecht: Kluwer Academic Publishers: 197–233

Knoepfel, P., Imhof, R., Zimmermann, W. 1995. *Luftreinhaltepolitik im Labor der Städte, Der Massnahmenplan – Wirkungen eines neuen In-*

struments der Bundespolitik im Verkehr. Bâle: Helbing & Lichtenhahn (Reihe Oekologie & Gesellschaft, Bd. 9)

Knoepfel, Peter 1996. Plädoyer für ein tatsächlich wirkungsorientiertes Public Management, *Swiss Politcal Science Review* 2 (1): 151–164

Knoepfel, P., Achermann, D. .Zimmermann, W. 1996. *Bilanzstudie Bodenpolitik 1990–1995*, Cahier de l'IDHEAP no 162. Chavannes-près-Renens: IDHEAP

Knoepfel, P., avec la collaboration de Larrue, C., Kissling-Näf, I., Benninghoff, M. 1997. *Analyse comparée de politiques publiques*, Support de cours no 3. Chavannes-près-Renens: IDHEAP

Knoepfel, P. 1998. *Eingriffsverzichte in öffentlichen Schutzpolitiken*, Cahier de l'IDHEAP no 174. Chavannes-près-Renens: IDHEAP (= ders., in: Lenk, Klaus, Prätorius, Rainer (Hrsg.) 1998, *Eingriffsstaat und öffentliche Sicherheit. Beiträge zur Rückbesinnung auf die hoheitliche Verwaltung*. Baden-Baden: Nomos: 125–148

Knoepfel, P., Grant, Wyn, Perl, Antony 1999. Conclusion: Institution Building for Sustainable Urban Mobility Policies, in: Grant, W., Knoepfel, P., Perl, A., *The Politics of Improving Air Quality*. Aldershot: Edward Elgar: 144–167

Knoepfel, Peter, Larrue, Corinne, Hill, Michael 2007. *Public Policy Analyses*. Bristol: The Policy Press

Larrue, Corinne, Knoepfel, Peter 1998. *Protection de l'environnement, principe de subsidiarité et politiques à incidence spatiale*. Rapport à l'intention de la DG XII de l'Union Européenne (Cahier no ENV4-CT96–0238) – version provisoire, mai 1998

Lübbe-Wolff 1996. *Modernisierung des Umweltordnungsrechts*. Bonn: Economica

Murswiek, D. 1985. *Die staatliche Verantwortung für die Risiken der Technik*. Berlin: Duncker & Humblot

Perl, A.1999. Air Quality Policies in the Greater Vancouver Area, in: Knoepfel, P., Grant, Wyn, Perl, Antony 1999. Institution Building for Sustainable Urban Mobility Policies, in: Grant, W., Knoepfel, P., Perl, A., *The Politics of Improving Air Quality*. Aldershot: Edward Elgar

Schwager, St., Knoepfel, P., Weidner, H. 1989. *Droit de l'environnement suisse – CE. Le droit suisse de l'environnement à la lumière des Actes officiels de la Communauté européenne dans le domaine de la protection de l'environnement. Etude comparative des régimes juridiques*. Bâle: Helbing & Lichtenhahn (série Ecologie & Sociéte, vol. 2)

Schweizerischer Bundesrat 1986. *Luftreinhalte-Konzept*. Bern: EDMZ, 10. September

Snickers, Folke (ed.) 1998: *Land-use Planning for Urban and Regional Air Quality*, Proceedings of the International Workshops in Brussels,

December 8th–9th, 1994, and in Stockholm, March 25th–26th, 1996. COST-CITAIR-Programme, Action 616. Stockholm

Weidner, H., Knoepfel, P. 1979. Politisierung technischer Werte Schwierigkeiten des Normbildungsprozesses am Beispiel (Luftreinhaltung) der Umweltpolitik, *Zeitschrift für Parlamentsfragen*, Band 1, Heft 161: 160–170

Weidner, Helmut. 1996. *Basiselemente einer erfolgreichen Umweltpolitik. Eine Analyse und Evaluation der Instrumente der japanischen Umweltpolitik*. Berlin: Edition sigma

Winter, Gerd (Hrsg.) 1986. *Grenzwerte. Interdisziplinäre Untersuchungen zu einer Rechtsfigur des Umwelt-, Arbeits- und Lebensmittelschutzes*. Düsseldorf: Werner-Verlag GmbH (Reihe Umweltrechtliche Studien, Bd. 1)

Zimmermann, Willi, Wyss, Stefan, Neuenschwander, Peter 1997. *Informationskampagnen zur Reduktion der verkehrsinduzierten Luftbelastungen in den Städten Zürich und St. Gallen*. Cahier de l'IDHEAP no 170. Chavannes-près-Renens: IDHEAP

13
Institution Building for Sustainable Urban Mobility Policies (1999)

Peter Knoepfel, Wyn Grant and Anthony Perl [1]

The European Union's COST Action 618 project sought to assess policy initiatives in institution building and information campaigns for urban air quality management. These efforts centred on limiting the growth of metropolitan traffic, which is the main source of urban atmospheric pollution in most major cities. While this volume focuses on the challenges and results of institution building for improved urban air quality, Action 618 has also produced a book that analyses information campaigns that have been conducted to combat air pollution[2].

As discussed in the introductory chapter, the researchers who examined institution building in the Grant et al. book of 1999 adopted a common research design focusing on the question: What factors enable policy net-

[1] Grant, Wyn, Knoepfel, Peter and Perl Anthony 1999. Conclusion: Institution Building for Sustainable Urban Mobilitiy Policies, in: Grant, Wyn, Perl Anthony, Knoepfel, Peter 1999. *The Politics of Improving Urban Air Quality*. Cheltenham UK/Northampton USA: Edward Elgar: 144-167. Copyright: We thank the editor for authorizing the publication of this article with some minor linguistic adaptations proposed by Susan Cox.

[2] The fuzzy definitions of "networks" both as prescriptive and descriptive analytical dimension makes the use of the term often very unprecisely. It is evident that the network concept is a useful tool for describing interactive systems of actors independently of whether these actors are public institutional or societal ones. In this sense, the described policy community will clearly form a policy network composed of social and institutional actors. The above used term of political administrative arrangement refers to that part of the networks which are of composed of institutional actors. The network metaphors then can be used for the description of both interaction modes within this arrangement and between actors of the arrangement and its societal partner actors on the level of social actors. Used in this purely descriptive way, the network notion is an interesting analytical tool. But this use prohibits specific qualifications as defining characteristics of networks in opposition to more hierarchical or more market-oriented coordination systems such as "absence of a centre", or "horizontal instead of vertical coordination", for example.

works to develop and implement feasible and effective control measures for urban air pollution? Rather than considering, these policy networks only as ends in themselves, researchers also evaluated the networks' capacity to enhance urban air pollution control and abatement. In certain policy networks, investigators identified a set of core beliefs and common practices that produced efforts to enhance urban air quality. These initiatives succeeded in introducing improved air quality objectives into policy domains ranging from planning, environmental assessment, policing of parking and road use, traffic management and urban transportation infrastructure development.

In this chapter, we interpret some of the findings presented in the seven case studies of urban air quality initiatives included in the Grant et al. 1999 book. We assess the experience of four countries (Canada, France, Italy and Switzerland) in the light of discussions that took place at a COST 618 workshop on institution building where these results were first presented, along with contributions from other countries (Hungary, Denmark, Great Britain, Spain and Greece). From these analyses, we now highlight the policy dynamics that have demonstrated effectiveness in facilitating urban air quality initiatives. Integrating air quality management and urban transportation objectives has been shown to occur following some combination of: innovation in policy problem definition, the renewal of policy communities, the building of new institutional frameworks, the introduction of new policy instruments, and the facilitation of learning processes. Our objective here is both to summarize some analytical insights regarding how air quality and transportation priorities were reconciled and to synthesize some suggestions for those seeking similar, or even more ambitious, results elsewhere. In summary, this conclusion seeks to fill the need of policy actors who will have to respond to the growing local (and global) damages caused by urban mobility.

13.1 Defining sustainable urban mobility

According to *Our Common Future,* the report of the World Commission on Environment and Development, sustainable development requires economic and social activity 'that meets the needs of the present without compromising the ability of future generations to meet their own needs' (World Commission on Environment and Development, 1987: 34). Such a vision of sustainable development aims to ensure that 'the exploitation of resources, the direction of investments, the orientation of technological development and institutional change, ... are all in harmony and enhance both current and future potential to meet human needs and aspirations' (ibid.: 46). Following this formulation and the growing recognition of the transport sector's pivotal role in facilitating sustainable development, various efforts have been made to develop an understanding of sustainable transportation. Canada's Centre for Sustainable Transportation (1997: 2) offers a broad definition, which emphasizes meeting the basic access needs of individuals and societies in a safe, affordable and efficient manner. Sustainable transportation would do this while limiting emissions and waste within the planet's ability to absorb them and minimizing the consumption of non-renewable resources and land, and the production of noise. The Swiss Agency for the Environment, Forests, and Landscape (1997) has put forward a more focused concept, which expresses mobility in terms of personal CO_2 budgets that need to be balanced in order to prevent climate change. Such an accounting system could be applied to other damaging, effects of mobility.

In order to advance both urban public health and quality of life, sustainable mobility must recognize that the means of movement (vehicles, roads, parking space, public transport and pedestrian facilities) as well as clean air are finite and valuable resources. Realizing an urban region's economic and social potential requires managing both these resources effectively. This means distributing access and the use of transportation infrastructure in a way that balances the reasonable mobility of today's travellers against future mobility needs by internalizing the negative external effects of urban travel (for example, air pollution, noise, accidents). In economic terms, this appreciation of sustainable urban mobility calls for an optimal allocation of two scarce resources: space and clean air. Such allocation would be economically efficient if it created an equilibrium between the marginal cost of each urban trip and the price which each traveller (or shipper) were to pay.

But public policy must also address the equity issues that arise when managing essential resources with many characteristics of a public good (for example, open access, near-universal availability), typical of both urban mobility and clean air. Politically, the trade-offs between clean air and travel opportunities that are implied by sustainable urban mobility trigger a classical distributional conflict among present and future user groups. For a wide range of stakeholders, the significant changes in personal behaviour, corporate activity and the current value of urban assets (property) that must occur in the pursuit of sustainable urban mobility will only be accepted as legitimate when they are seen to be both efficient *and* fair. Some of the European and Canadian policy-makers who are profiled in this book have recognized that road traffic generates the bulk of urban air pollution and arc leading the way towards managing air quality and mobility as interdependent resources. In other cases which we explore, the political legitimacy required to pursue truly sustainable urban mobility remains to be achieved.

Experience to date demonstrates that the equity issue underlying efforts to achieve sustainable urban mobility is crucial to gaining legitimacy. Each policy initiative that seeks to trade off current urban mobility patterns against enhanced air quality must demonstrate scrupulous fairness in respecting the rights of all parties, from frequent travellers to those who are particularly harmed by air pollution (such as the young, the elderly, and those who are ill). Institutions dealing with urban air pollution must be open and robust enough to control this high degree of distributional conflict while still guaranteeing positive and efficient allocation.

Integrating urban travel and air quality management policies is rarely, if ever, accepted as a zero-sum trade-off between drivers and citizens – if for no other reason than most citizens either travel by motor vehicle or aspire to do so. Citizens are wary of changing established urban travel behaviour in order to clear the air they breathe when the mobility alternatives appear unfamiliar and less attractive. Thus policy solutions must deliver a positive sum result of better urban mobility and cleaner air. This means transcending old political conflicts between automotive interests and mass transportation interests (many of which now reside in public bureaucracies). Non-motorized travel options including bicycle, pedestrian and telecommunication alternatives need to be brought into the sustainable urban mobility attainment scheme. Furthermore such efforts need to focus on a geographically complete zone of mobility, not just those trips which lie within urban political boundaries. This means dealing with local, metropolitan, regional and even national mobility patterns, as Lyon's predicament illustrates.

All contributions to the Grant et al. book (1999) underline the necessity of bridging what are today more or less isolated individual policies and their fragmented communities dealing with singular aspects of urban mobility systems such as road construction, traffic regulation, land use planning, or industrial and domestic environmental policies which address emissions that pollute the same airspace as automotive traffic. Each of these policies has a well defined mandate and a more or less exclusive policy community. But the current urban crises in public finances, growing, congestion and increasing air pollution have opened a unique window of opportunity both for intensified cooperation across environmental and transport policymaking, activities and for subsequent integration of a much broader urban mobility policy community including all those who have a stake in mobility and clear air. The Berne and Vancouver cases, as well as the emergence of integrated urban traffic plans in France or Italy, show the way in which this development can evolve. We turn now to pinpointing the changes in problem definition, policy community composition, institutional frameworks and learning processes needed to create this new capacity for sustainable urban mobility.

13.2 A new problem definition

Going through the seven contributions to the Grant et al. book (1999), it becomes evident that, once urban mobility reaches the point at which it becomes a cause of serious air pollution, a wider range of more or less visible social costs will have also arisen. As the Italian cases illustrate most clearly, some of this collateral damage may be perceived as more problematic, or more immediately troubling, than the generalized human health degradation, local and global environmental damages or decay of historical monuments, which arise from urban air pollution. Efforts are needed to fashion a problem definition of sustainable urban mobility that encompasses both the concentrated and imminent impacts of unsustainable travel patterns as well as the diffuse and cumulative effects. Such an amalgamation offers the best opportunity to mobilize support for what are inevitably contentious objectives and instruments. We offer seven components of this encompassing problem definition which can maximize the opportunity for policy innovation.

13.2.1 Congestion

Traffic congestion, a problem explicitly used by the Turin authorities as a more politically palatable proxy for managing air pollution, has been on both the Canadian and European policy agenda since the 1960s. For a long time, the congestion problem was addressed through supply side solutions of building more new roads, or self-contained public transport infrastructure such as underground railways. Such new infrastructure was supposed to reduce congestion either by increasing the physical capacity for traffic or shifting, some part of urban travel from roads to subways. Newman and Kenworthy (1988) document how this approach was ultimately counterproductive because it neglected the effect of new infrastructure on decentralizing urban land use. More infrastructure stimulated longer trips because distant spaces became better linked together, yielding higher traffic volumes, along with greater energy use and air pollution. Lyon's postwar experience with supply-side infrastructure expansion clearly demonstrates the effects on sprawling land use, greater traffic volumes, renewed congestion, and increased air pollution.

By the 1990s, the effects of supply side congestion management policies had reached the financial and physical limits of many European cities. Space and money for the construction of new roads and new public transportation systems became increasingly scarce and urban authorities were compelled to consider new instruments that turned to the management of travel demand. Urban governments introduced various regulatory schemes, as well as more significant parking charges in urban areas as a way to limit the demand for road space. Marlot and Perl's chapter on Lyon (Marlot, Perl 1999: 107–126). shows that demand side management policies have even progressed to the point of experimenting with urban road tolls and consideration of more widespread road pricing or traceable use permits. In so far as demand management strategies reduce the absolute number of cars in use daily, they can also contribute to cutting down on air pollution (Perl and Han, 1996).

13.2.2 Air pollution

This 'core' problem was initially identified and championed from the political fringe. Across Europe, urban pollution has often been used as a political weapon of green parties and ecological movements. Along with noise pollution and road safety, the air pollution problem has become part of a relatively recent, but growing, critique of the automobile and its role in undermining the quality of urban life. This negative perspective on automobility usually spurs calls for restrictive command and control measures to restrict urban traffic – either geographically as in Turin's car free zones, or temporally as in the emergency limits placed on car use during episodes of peak air pollution found in France and Greece. Both the critique of the car and its accompanying restrictions have received little support from two important urban political constituencies. At one extreme, the urban working class still aspires to the automobile, and related material consumption which greens and ecologists denounce as unsustainable. At the other end of the spectrum, urban economic elites embrace car and truck transport as necessities for business. The most receptive social segment to this 'post-material' view of urban life has been the middle class, which is a declining segment of the population of many cities. When mobility and air quality management cross the urban boundary to include the outlying areas where many middle class voters have settled, air pollution is given a higher political priority based on quality of life concerns. For example, political campaigns targeting air pollution were found to have a significant impact in the outer belt of Swiss metropolitan areas (Knoepfel, Imhof, Zimmermann 1995:145 and 365). But in many European cities where traditional political boundaries impeded the politics of air quality management, attention to air pollution often arose only once the need to implement European Union directives became pressing.

13.2.3 Safety

Efforts to enhance urban traffic safety, such as reducing speed limits in cities or traffic calming of roads in particular neighbourhoods, can complement sustainable urban mobility initiatives. The road safety issue has been a longstanding concern for both urban parents and senior citizens. As such, it offers another potential counterweight to the equity concerns that would arise from any efforts to limit or regulate urban mobility. When seen as an equitable sharing of risk, safety concerns also raise the opportunity to gain support for sustainable transportation policies from cyclists and pedestrians seeking a fair share of urban public road space.

13.2.4 Urban fiscal crisis

As shown above, supply side solutions to traffic congestion turned out to be very costly. When cities could afford massive infrastructure budgets, new highways and metros were seen as a mark of affluence. But as many well-to-do taxpayers have moved to outlying jurisdictions (often encouraged by the ease of travel that the new highways and subways created), municipal governments are being compelled to stop splurging on new infrastructure. Some jurisdictions have sought to raise revenue from infrastructure through road and bridge tolls and other user fees, while others have raised taxes. This revenue enhancement strategy offers limited room for fiscal manoeuvre owing to the public's low tolerance of new taxes. Taxpayers' revolts and the rise of various forms of protest parties demonstrate the limits on government in using pricing to support sustainable mobility. Turin's experience illustrates how politicians currently cope – by exempting specific population categories from the new tax obligations. New room for economic instruments might arise through the use of public private partnerships, where the tolls are collected by a private agency. Or governments may succeed in selling urban transportation pricing as a valuable new public service which enhances the quality of life by limiting congestion and pollution.

13.2.5 Discrimination

This issue is particularly delicate and usually implicit in mobility debates. When faced with the prospect of higher prices or restrictions on urban transport, some people will react by blaming 'foreigners' for the problem. To some, these outsiders are the commuters coming from low density areas, while others see recent immigrants (or visible minorities who are assumed to be recent immigrants) as causes of urban degradation. In either case, newcomers are seen as preventing 'legitimate' inhabitants from the full use of 'their' roadways, parking spaces, public places, or clean air. Rather than accepting lower standards in these areas, or changing behaviour, the temptation arises to try and exclude outsiders from urban roads and travel. The discrimination issue becomes even more contentious when drivers challenge other users of urban road space (pedestrians, cyclists, demonstrators, and so on) as being less legitimate; or vice versa.

The discrimination issue reveals different conceptions of property rights on specific roads and parking spaces. It pits those who are 'in' against those from outside and reinforces the typical fragmentation of municipal-based institutional arrangements in traffic, but also in fiscal, physical planning, and cultural policies within the metropolitan area. This dispute between insiders and outsiders takes on a critical dimension in the chapters on Turin and Lyon. When discriminatory policy solutions are adopted for achieving sustainable urban mobility, they create a risk of exacerbating social exclusion and widening the gulf between ghettos of the destitute and enclaves of affluence.

13.2.6 Equity

The equity issue is intimately linked with the discrimination issue. It usually arises in fragmented debates regarding the benefits of scarce urban mobility resources, such as parking spaces, road calming measures, permits for access to traffic restriction zones (Turin), new public transportation infrastructure, or the dedication of some existing road space to public transit through bus or tram lanes. The stakes consist of who will derive the most benefit from new or improved accessibility, whether by public transportation, enhanced parking, or traffic regulation. This issue becomes linked with air pollution when mobility management plans propose large scale realignments of accessibility, which can have serious negative impacts on the value of urban land.

13.2.7 Quality of mobility

As the Italian case study points out, the organization and delivery of urban mobility services can make a difference in the qualitative perceptions of travel alternatives. Public transport that offers seamless travel through, for example, integrated fares, coordinated timetables and accessible stations, can be an attractive alternative to the automobile. But costly, crowded, and limited transport facilities widen the quality cap between the car and its alternatives. Too often, quality measures and standards are disjointed, with one, usually high, set of expectations for automotive travel and another, usually low, set of expectations for all other transport options.

One can argue that, although each of the seven issues described above has its particular social or financial constituency, they belong together under a common framework of sustainable urban mobility. That is because the set of interests behind each interest are actually connected by their need to share a largely finite, and increasingly contested, urban transport infrastructure on the one hand, and their overall responsibility for the cumulative externalities of urban mobility on the other hand. The collective problem at the root of comprehensive urban mobility policies is how to organize urban mobility in a way that optimizes the internalization of social and environmental costs, and equitably distribute the burdens of achieving sustainability (including any remaining external costs) in a way that guarantees minimum mobility standards, minimizes damage (safety, clean air, noise, quality of life), and maximizes quality of life.

13.2.8 New issues – new policy community

Public policy communities are nominally composed of state and societal organizations that share an interest in a common policy issue. Within these policy communities, actors will pursue varying dynamics of cooperation and conflict, to address the issues they share some stake in. Each organization brings its own intellectual and material resources to the community. In the case of sustainable urban mobility, these could range from the strongly held beliefs of radical ecologist groups to the analytical tools of scientists and social researchers to the money and employment opportunities controlled by private industry. One category of participants brings a unique resource to the policy community; state actors, whether bureaucrats or elected officials, have the legitimate use of authority at their disposal. This ability to compel certain actions, or to levy taxes or other mandatory fees, gives the public sector participants a decisive role in policy communities, which can take on quite different structures depending on how, if at all, this authority gets distributed. In most communities, one can identify subcommunities, where public and private organizations cluster together by function, by region, by ideology or by other shared characteristics. These groupings are influenced by the levels of economic and political capacity among societal organizations in relation to their governmental counterparts. The common denominator that both connects these subcommunities, and orients their interaction within the larger community, is their organic connection to the definition of a public problem. Any definition of what is to be done in the name of the public interest, and with the attendant authority and resources of government behind it, will trigger the involvement of some organizations and not others. As the contributions to the Grant et al. book (1999) have demonstrated so well, what gets done about urban mobility and air quality depends on who participates in the public deliberations over objectives, instruments and institutional arrangements that fall under the definition of policy development. And that participation depends on how the problem is presented.

When urban mobility policy is approached from the understanding that traffic congestion is the key problem, we find little or no participation in the policy community by environmentalists, industries which generate emissions from stationary sources, community groups concerned with the safety of children or the elderly, public health agencies, and air pollution control agencies. Conversely, when road safety, urban revitalization, or the equitable access to public roads come to be understood as the key problems arising from urban mobility, then a very different set of actors will take part in the policy community. In this case, the elderly, homeowner associations, banks and financial agencies, even political parties representing particular segments of urban society, will participate in policy debate and development.

For example, the French urban mobility policy community shows a clear division into subgroups focusing on traffic engineering and air quality management. But in France, each reconsideration of the policy problem (say, from the lack of sufficient urban transport infrastructure to an inefficient use of existing facilities) brings new actors into the community. French experience shows that treating air pollution as a problem that is both equivalent to, and interdependent with, traffic congestion will inevitably enlarge the policy community membership with newly mobilized politicians, scientists, nongovernmental organizations, public transport agencies, and even private industry. Each of these actors is interested in having a say in the development of newly mandated urban mobility plans (plans de déplacements urbains; Larrue, Vlassopoulou 1999: 102). In general, linking all seven of the above mentioned problem definitions under the umbrella of a sustainable urban mobility policy will inevitably and considerably enlarge the policy community and render policy development more complex than it is today. Based on what our contributors have discovered in a wide range of urban mobility contexts, we can pinpoint the fulcrum on which a policy community capable of pursuing sustainable urban mobility could be balanced.

Respect for *property rights,* both the rights of private proprietors of the urban space which traffic passes through and the public trustees of commonly held resources such as clean air is the key concept that can engage the range of groups and organizations needed to achieve sustainability in concerted action. The Canadian case study chapter echoes the finding of Ostrom et. al. (1994: 4) that achieving sustainable development demands a preoccupation with institutional economy in general, and property rights in particular. Ostrom's framework differentiates four categories of property rights ranging from the owner's freedom to unilaterally modify the resource in question to a circumscribed autonomy to use a resource in a particular manner and up to a specific quantity based on terms set by some contract or regulatory code.

The legal definition of most transport infrastructure, particularly urban road infrastructure, assumes that this property is a public good whose use cannot be restricted. This public good assumption also implies that no one's use of urban road space will detract from any other use of that space. But this central premise upon which many urban transportation problems have been defined turns out to be defective. In actual fact, both traffic congestion and the exclusion of nonmotorized traffic, such as cyclists and pedestrians, demonstrate that road infrastructure is not truly a public good, but at best a mixed good, and in many cases even a private good being supplied at public expense. By accepting unrestricted 'first come, first served' automotive access for urban roads and enforcing the restriction, and even exclusion, of nonmotorized transport, the state implicitly advances the interests of certain drivers and landowners over other drivers and landowners, as well as cyclists and pedestrians. The same is true for clean air, even though the property right attributed to one user group (travellers using cars and shippers using trucks) to the detriment of others (inhabitants suffering from air pollution) is much less visible and more difficult to communicate. According to this very simplified formulation of institutional economy, sustainable mobility policies will have to define new and more realistic categories of rights for both clean air and road use. This redefinition will entail a major redistribution of these rights which is certain to be politically controversial.

If urban mobility was reconceived as an amalgamation of more-or-less private property rights, with the attendant need to reconcile competing claims, rather than as a public good, then a new policy community composition and configuration would be inevitable. This community would be composed of both the *de facto* owners of property rights which currently govern urban road use and air pollution, as well as those citizens who are negatively affected by current rules of the road through air pollution impacts or restrictions on access to public roads. Among the policy actors who would face significantly changed roles and responsibilities in the new policy community, we find:

- *The state* itself through local, regional, or national governments functions as the primary owner of transportation infrastructure – roads, railways, parking spaces, transportation terminals, and so on. These functions are normally managed by public road authorities, public works departments, or more or less independent road construction and operating agencies.
- *Private corporations* also supply urban transportation infrastructure through building and running facilities such as highways, tunnels, bridges, and even subways, light rail transit (LRT), and tramways (as in Hong Kong and Singapore). Here the full cost of infrastructure, including the cost of capital, is recouped from user fees and tolls.
- *Public transportation agencies* are responsible for operating urban transit systems. In the case of underground and tram systems, these agencies are the exclusive infrastructure operators, and in many cases public transportation agencies are also accorded a monopoly over common carrier passenger transport over city streets as well.
- *Private car owners* are the most common 'consumers' of urban road infrastructure and parking spaces. Car owners link political jurisdictions in the pursuit of urban mobility since the drivers who become part of urban traffic may come from the city, the metropolitan area, or even further away. These car owners are also the most numerous 'producers' of urban mobility externalities since the fuel and vehicle taxes they pay do not cover the full costs of noise, accidents, pollution, capital investment in road infrastructure, and the rest.
- Public or private *parking space owners* lease the right to store vehicles on prime urban land on various terms. The prices charged, and the distribution of parking supply, can have a decisive impact on the efficiency of urban public transit.
- *Urban landowners* more generally have a major stake in the terms of urban mobility. Easily accessible urban subway commands an economic premium as the development around urban metro stations or suburban highway entrances and exits demonstrates. At the same time, transportation infrastructure that passes through, or near to, property

without improving access can reduce its value by bringing externalities (such as noise, fumes, and physical isolation from other property). From homeowners to big corporations and real estate holding companies, property owners pay close attention to transportation planning because of its influence on their assets.

- Local or regional *air quality management agencies* exercise public powers to limit air pollution through regulation, inspection and reporting on the state of the environment. For the most part, these agencies have been called upon to take stewardship of degraded air resources long after the public good concept had been entrenched as a cornerstone of urban mobility. Once the right to pollute excessively had been handed out to urban drivers, these agencies were given the unenviable task of reasserting public authority to limit the unequal distribution of damages (for example, protecting the young and elderly who are less mobile and more harmed by air pollution). Faced with entrenched opposition from drivers who resist giving up their free ride, these agencies would be tempted to compromise by focusing their regulation on less formidable political adversaries including "foreigners", newcomers, future generations and industry.

The transformation of today's disparate policy community environment, where air pollution and urban mobility problems are often addressed by disjointed policy initiatives, appears to depend upon reconsidering the first principles of economic rights and political responsibilities. Such change is unlikely to occur spontaneously, at least not in the first instance. Instead, the problem redefinition required to launch this transformative policy process will need to be championed by a vanguard of policy innovators. Below, we speculate on the groups and organizations with the most reason to advocate an encompassing policy community. These concentrated stakes in sustainable urban mobility may be enough to mobilize their leadership in such a vanguard.

The gains most likely to motivate a vanguard of advocates for sustainable urban mobility rights will arise when the redefinition of today's *de facto* and *de jure* user rights gets translated into a restructured relationship between urban mobility and pollution. This requires either withdrawing some rights from today's travellers or even destroying some portion of right to automobility as Marlot and Perl (1999) explore in the Grant et al. book (1999). They introduce the standard of an ecological carrying capacity for a region's airshed as a principle for reconciling mobility with pollution. Working from this carrying capacity, one can translate a fixed quantity of motorized mobility, which can then be bid upon or rationed. Whichever strategy gets pursued, compensation payments can become a key means to achieving such wholesale change (Small 1992). Until such redistribution occurs, there will be increasingly concentrated losses which can motivate certain groups and organizations to activism. Amongst these *affected* groups we would highlight:

- *Inner city pedestrians, cyclists, and drivers* are each in their own way excluded from mobility options as a result of the current treatment of urban roads as a public good. For each of these travellers, the influx of vehicles from surrounding areas leaves little opportunity to use adjacent road infrastructure. They seek new access to the public domain of urban mobility which could be achieved through either a new institutional economy of allocating mobility rights, or a simple restriction on the use of certain public space – pedestrian zones, protected pedestrian crossings, bicycle lanes, closing roads to through traffic, and so on.
- *Public transit operators* suffer badly from having their buses and trams caught in today's traffic congestion, and will suffer even more from the decentralization and dispersion of population and activity that accompanies growing road infrastructure. Their best chance of offering improved quality to existing passengers and attracting new users comes from ending the "free ride" policy for urban drivers.
- *"Auto-dependent" urban residents* have little choice but to drive or face sharply restricted mobility because they live in areas that are poorly served by public transport (Newman and Kenworthy 1989). Their best hope for a fair share of access to the metropolitan region comes from making public transport a more effective alternative to the car, which depends on realigning mobility priorities.
- *"Excluded" drivers* coming from either less central urban neighbourhoods or the surrounding, low density suburbs are often penalized by traffic restriction schemes which exempt inner city residents. The resulting hierarchy of access can leave residents from outside the city centre doubly excluded in that they are poorly served by public transport, yet blocked from using their cars to access cultural and commercial amenities as well as job opportunities. A more equitable distribution of urban mobility would have strong appeal to individuals who are caught between such restrictions;
- *Inner city residents* face elevated levels of air and noise pollution, which pose their most serious public health threats at high levels of concentration. As a result, they are excluded from enjoying a minimum standard of clean air and tranquillity because these resources are consumed by drivers, and other urban polluters (such as industry). Often there is a correlation between the degree of negative environmental im-

pact and socio-economic status. Inner city dwellers thus have every incentive to bring the social component of sustainability into a redefinition of urban mobility problems;

- *Industry and other fixed source polluters* have the legal right to generate emissions into an urban airshed, within regulated levels. As the experience in Lyon and Vancouver demonstrates, these fixed source polluters identify an economic advantage in shifting the burden of emission reductions on to mobile sources, such as cars and trucks. This motivation is strongest in new industries which must obtain permits for additional emissions, for established polluters who are (often wrongly) accused of "causing" regional air quality problems, and for sources (such as paper mills, cement plants and thermal power generating stations) which seek to expand output to meet growing demand for their outputs. In each case, reducing mobile source emissions could yield concentrated benefits, such as new jobs and higher profits for urban industry.

Having mentioned the groups and organizations that comprise potential vanguard for sustainable urban mobility, we must recognize that there are also organizations which would mobilize to oppose such an expansion of the problem definition and policy community. Among the groups that have traditionally resisted any change to the public good definition of urban road use and its consequences are small and moderately sized urban businesses (often represented by Chambers of Commerce or Boards of Trade). These retail merchants and service providers, who are often quite powerful in local politics, as seen in Lyon, fear a loss of business if the terms of mobility are revisited and current customers are affected. At the other extreme, multinational energy producers and suppliers tend to oppose efforts to reduce transport sector externalities, given its downstream impact on their markets. In between, there will be many other groups that see a broad-based urban mobility policy community as a threat, which is why the vanguard we have identified will have to pursue a strategy that can compensate some of the losses in any redistribution of urban mobility opportunities.

13.3 Building new institutional frameworks ...

What institutional framework is capable of adequately governing, such a sustainable urban mobility policy? When the experience of preceding chapters is factored into the organizational principles and dynamics for achieving a considerable redistribution and realignment of urban mobility outlined above, we see the need for an institutional framework of rules and norms that would fulfil a minimum of four conditions.

First, there must be a *multi-stakeholder forum* composed of pre-existing or new organizations that take responsibility for five key policy components of sustainable urban mobility. These include: public roads (ownership, construction, maintenance and management of transport infrastructure); road police (regulation of access and rules of the road); land use planning (integration of transport infrastructure into the overall scheme of urban development); public transportation planning (development and management of public transportation infrastructure and operation) and, environmental agencies (management of natural resources including clean air, water and soil).

These five agencies would require an appropriate incentive structure to work cooperatively in implementing a political commitment to achieving sustainable mobility. They must contribute their unique administrative and analytical capacities to this common goal. Cooperation should not be confused with organizational amalgamation into one huge mobility management agency. Such an agency would not only be unwieldy, but also threatening to the existing bureaucracy charged with managing various pieces of the urban mobility puzzle. Instead, the existing competency and experience of today's *de facto* urban mobility managers are valuable administrative assets that would be further enhanced through joint action, rather than bureaucratic rivalry. The ability to integrate sustainable urban mobility objectives into related, but otherwise autonomous, bureaucracies responsible for urban development (land use agency), natural resource management (water management influenced by major transport infrastructure, green spaces and so on) or public works agencies (public buildings) appears far more likely through joint action than attempts at formal amalgamation.

A second organizing principle for institutional arrangements ought to be their *trans-jurisdictional structure.* Our contributors have shown that urban mobility management cannot be limited to the administrative jurisdiction of the central city, but must be extended to the whole metropolitan area. The experiences of both Lyon (Communité urbaine Lyonnaise) and Vancouver illustrate how the political debates triggered by congestion and pollution impacts, the equitable distribution of mobility, and the discrimination issue can be addressed most effectively within a jurisdictional framework which is enlarged to include the suburban municipalities. This trans-jurisdictional structure must bridge not only governments, but also private organizations and societal groups throughout the metropolitan area. As with the coordination of multiple stakeholders discussed above, crossing jurisdictional boundaries should not imply amalgamation or integration into regional bodies. The cumulative policy capacity of organized actors throughout a region will be greater if existing organizations work together, rather than be reformulated into four new structures. The Italian and French cases show that the range of jurisdictions can stretch to include regional government (Rhône-Alpes), national government (France and Italy), and even supranational government (European Union).

A sustainable urban mobility initiative will need to adopt our third organizing principle of multipurpose *political-administrative arrangements.* Although policy development should pursue the single and overarching objective of sustainable urban mobility, this goal covers more than one traditional policy issue. As we pointed out above, at least seven controversial issues fall under this objective; each of these is associated with a specialized knowledge base, which can create a varying number and configuration of policy subcommunities. These issues might thus be considered as a sub-goal of the overall objective and then be assigned to a specific policy actor within the coalition or partnership. In general, sustainable urban mobility initiatives will have to develop the level of trust and recognition necessary for such a division of labour and specialization of tasks.

Finally, the great diversity of policy participants, their range across multiple jurisdictions, and their reliance on specialized functions will all require a high degree of *horizontal and vertical integration* complemented by necessary vertical centralization. The utility of highly integrated political and administrative arrangements is demonstrated in most of the contributions to this book. For example, Giuliani (1999): finds that frequent intense contacts between institutional actors seem to be a precondition for coherent policy outputs. We would go even further and suggest that these contacts must not only be vertical, but also horizontal, linking organizations of the same juridical level from public agencies to industry councils in pursuit of the common objective. Whether these contacts are more formal or informal will matter less than their frequency and substance. We agree that frequent intensive contacts will make the arrangement more complex than arrangements of less comprehensive policy domains, such as previously segmented environmental and transportation policies.

Probably the most controversial characteristic that we are proposing is a centralization of the necessary authority needed to empower the many actors and organizations identified above to pursue sustainable urban mobility. We are convinced that even when the most effective and sophisticated policy instruments are used, as discussed below, the redistribution of user rights to clean air and mobility will be highly controversial and politically contentious. In such circumstances, innovation will only succeed if policy actors have the institutionalized authority to discriminate between competing private interests in pursuit of the public interest. An example of such a decision would be compelling a highly mobilized and politically powerful group or organization to give up some share of current urban mobility and/or pay for the damages created by that mobility. This task will require some politically powerful central actors in both metropolitan and urban governments. These actors should exercise their authority through the legal power to tax transportation activities, to regulate land use, and to build (and in some cases dismantle) transport infrastructure. Their powers must be legitimized through democratic elections, both of individuals to the appropriate public office and by the passing of referendums on the policy framework required to implement sustainable urban mobility (for example, full cost pricing of externalities, infrastructure redevelopment programme, and so on). The Italian and Swiss cases demonstrate the power of direct democracy to break down entrenched obstacles to redistributive reform, although Grant's (1996) review of how direct democracy is practiced in California sounds a cautionary note. In the final analysis, we believe that hard policies do require hard institutional arrangements to translate the ideas and options generated by the soft institutions that Giuliani presents into concrete actions.

The sustainable urban mobility policies we are considering do not yet exist. But the contributions to this book reveal some important insights regarding the preconditions of such policies. Breaking through institutionalized barriers to sustainability, which is what the cases profiled here have attempted, is a necessary start to the transition. We would expect that each effort to refashion the disjointed, and often adversarial, policy efforts in air quality management and urban transportation will have to pass through a phase of experimentation akin to those described in preceding chapters. But these initiatives represent only the first stage of a more profound transition to sustainability.

While new ideas can occur spontaneously in the interactions of an encompassing policy community, and while these new ideas can reshape organized interests, there remains a need to institutionalize new patterns of authority to achieve progress toward sustainability. Thus, after problem definitions broaden to correspond more closely to sustainable development, and policy networks are sufficiently in flux to accommodate new patterns of group interaction, there remains a crucial need for institutionalizing the emerging understanding of what needs to be done into a set of organized obligations. By this, we mean conferring the legal and political powers to pursue policy initiatives on one or more organizations, which then gain autonomy within the policy network. We are convinced that these key institutional actors will have to play a dominant role in delivering policy outputs, both through their own initiative and through structuring the activity of other groups in the policy network. Vancouver's experience comes the closest to demonstrating a critical mass of institutionalized authority in action, in the form of the GVRD. But achieving the type of redistribution of opportunity for sustainable urban mobility will require even greater levels of institutionalized authority, which we see as arising through a more widespread and visible democratic legitimation of future lead organizations.

The policy community literature is not clear about how and when organizations become leaders in implementing new paradigms like sustainable urban mobility. Some authors (Lembruch, Schmitter 1982) portray cooperative dynamics, such as corporatism, as naturally emerging when policy communities are given the political space to fill the organizational space between hierarchies and markets. Indeed, the intellectual antecedents of policy community analysis arose from the study of corporatism as a mode of policy development (March, Olsen 1989). In our view, the initially loose, voluntary and informal associational links that arise in an encompassing policy community are necessary, but not sufficient, to achieve sustainable urban mobility. To realize their potential, these new relationships will have to become institutionalized, legitimated, and autonomous so that key actors can take decisive action without the necessary approval of all policy community members. In this sense, the suggestion that soft institutions can truly resolve hard problems is chimerical because the deliberations and negotiations that are implied in the concept actually become counterproductive if left to continue indefinitely. In such cases, certain groups inevitably come to oppose initiatives that were developed collectively, and work to resist or undermine subsequent implementation efforts. In sum, the soft institutions needed to clear obstacles toward sustainable urban mobility are a necessary step on the path of policy development, but they cannot achieve the end in themselves.

13.4 ... capable of using new sets of policy instruments

It is important to understand that redistributional policy objectives like sustainable urban mobility cannot be achieved solely by the use of economic instruments. As important as tools like road pricing energy taxation, and parking charges turn out to be in sending clear signals regarding the cost of urban driving, they must be complemented by other instruments that influence the social and the physical dimensions of urban life. These complementary policy tools include traditional command and control instruments such as land use planning (zoning) or prohibitions of car traffic in specific areas (for example, the restricted traffic zones of Turin) that can be used in more or less draconian redistribution of urban space from drivers to other users. New infrastructure development and finance is another longstanding instrument of creating new transport capacity, which can be targeted to maximizing sustainable mobility. Finally, persuasion and public communication can make inroads into seemingly habitual travel behaviour. The challenge is to discover a mix of all these instruments that optimizes efficiency, effectiveness, and equity.

Both Bobbio and Zeppetella (1999) and Desideri and Lewanski (1999) are probably right when they postulate that regulatory instruments will encounter a much broader and more militant opposition than either infrastructure expansion or economic instruments such as parking charges. This is because restrictive regulations have a more obvious and far-reaching redistributional effect to mobilize potential losers. Economic instruments which enable citizens to buy temporally or geographically specific user lights for public roads or parking space will be more acceptable than their complete reattribution to another category of the urban population (for example, pedestrian zones). But pricing roads and parking at levels that keep emissions within the carrying capacity of a region's airshed might be charge enough to cause secondary redistributional effects in terms of social discrimination. Prohibitive mobility pricing could thus lead to adversary mobilization just as easily as draconian regulation. Studies from Switzerland demonstrate that even the soft instrument of clean air campaigns which raise the issue of redistributing mobility can trigger political opposition to this art of their message (Zimmermann et al. 1997: XII).

One way to assess, and predict, the varying degrees of political mobilization that could accompany different types of policy instruments is to estimate the degrees of freedom and constraints that they would create. For example, the attribution of user rights through regulatory policies that create "car free zones" yields a high level of freedom for new recipients (bicyclists and pedestrians), but this corresponds to a zero-sum imposition of constraints on drivers. Regulatory instruments thus appear extremely discriminatory in that they yield a high decree of constraints on the one hand and great liberties on the other one. In the historic centre of Turin, traffic restriction zones guaranteed both pedestrians and those drivers with special permits exclusive user rights to the detriment of drivers coming from outside.

Regulation's concentration of liberating and constraining impacts, which has often been seen as necessary to create minimum health and safety standards for vulnerable groups, is partially avoidable through the use of economic instruments. Because tolls, traceable permits and other pricing measures only become exclusionary when the target groups are not capable of paying, they offer the opportunity to change mobility patterns, and reduce those with significant negative externalities, while preserving individuals' freedom of choice. Even when pricing schemes restrict total mobility – during rush hours, or air pollution episodes, or around the clock for a given space, for example, they do so without specifying exactly who will be excluded. This makes them more publicly attractive than regulatory approaches.

Economic instruments thus belong in the forefront of sustainable urban mobility, although their pricing parameters might be calibrated based on the same threshold that would inform more coercive regulations, such as the carrying capacity of a regional airshed. As long as air pollution remained within acceptable levels, then more coercive instruments like regulation could be left out of implementation efforts. But during smog episodes or similar environmental emergencies, economic instruments would need to be supplemented by regulatory controls. A permissive complement to pricing instruments would be construction of new infrastructure and redevelopment of existing facilities. Such measures bring popular additions to urban mobility capacity, which if properly planned would yield positive sum gains. These improvements could range from low-cost enhancements to existing public transit facilities (coordinated fare and timetable systems) to incremental infrastructure expansion (new interconnections between existing transport lines or park-and-ride facilities for drivers) to new tram and subway lines. The cost of such efforts could be met by a part of the revenue raised from pricing instruments, Persuasive instruments such as advocacy campaigns for green transportation and information programmes about urban travel alternatives can also be considered permissive, and achieve greatest effectiveness when they target specific user groups for travel behaviour modification. Funding such campaigns from pricing instrument revenues would be unlikely to generate much controversy.

Politicians often believe that the cost of policy implementation will vary inversely with the degree of authority found in the policy instruments. That is, less authoritative policy instruments like persuasion and pricing will cost more to administer than restrictive regulations. The Italian contributions to that Grant et al. book (1999) demonstrate that this is probably a mistaken point of view. Regulation cannot be counted upon to "speak for itself" and the cost of policing restrictions on road or parking access can only be financed in part by raising fees[3]. But collecting user payments also has a price, particularly when new technologies and personnel will be needed to deploy sophisticated marginal cost pricing systems like peak

[3] The city of Rotterdam recently demonstrated that good planning of control activities may use considerable public income and that the number of fees depends largely on the number of policemen. The more policemen a city engages, the more it can increase incomes from fees.

hour road or parking fees. These costs do not disappear if fees are collected by the private sector, since public revenue would be forgone while supervisory and planning activities continue. Turin's experience suggests that private partnership in enforcement or pricing can reduce the legitimacy of such measures. The Swiss case also shows that campaigning for sustainable mobility has both a political and an economic price (Zimmerman, et. al. 1997: 67 and 129). For all these reasons, we conclude that a cost-benefit analysis of the four instrument categories would reveal a far narrower spread in instrument costs than people expect.

Our case studies also reveal that legal constraints on local and metropolitan governments will be a key variable in instrument choice. These constraints normally are found in national or even in EC regulations which local governments cannot ignore without being sued by affected interest groups. The discretion available in crafting local urban mobility policies is currently limited by national road codes stipulating general rules on speed and access for highways, regional roads, and sometimes even local roads. Other traffic and parking restrictions often conflict with higher authority. National and subnational governments are even more jealous of the fiscal prerogatives associated with mobility. National regulations on car taxation, car inspections, and technical requirements for cars can each inhibit local initiatives. The same is true for local governments' autonomy with regard to different forms of intergovernmental cooperation, especially attempts to cross jurisdictional boundaries across a region or metropolitan area. Experience demonstrates that the further local policies move away from a classical regulatory approach towards economic instruments, the more they encounter provincial, regional or national governments' opposition to "discriminating" in the treatment of all citizens in the jurisdiction in question or, in the case of the European Union, of attempting to restrict people's freedom of mobility. As a result, the field of action is usually wider for local governments using persuasive instruments like campaigning or traditional regulatory instruments than it is for the economic instruments which could make such an important contribution to sustainable urban mobility.

Finally, the policy instruments vary also with regard to their possible adaptation to specific urban districts or target groups. Car drivers belong to the famous group of "inaccessible target groups" (Bressers and Ligteringen 1997). For political, social, and economic reasons targeting must be considered as a highly difficult task in this field. The Italian cases demonstrate how both regulatory and economic instruments had to be retargeted after their introduction by means of special permits or exemptions. Studies from Great Britain and Switzerland demonstrate that targeting can make a major contribution to the success of air pollution campaigns (Zimmermann et al. 1997). The highest need for a very sophisticated targeting strategy exists in the field of organizational and infrastructural development instruments. Here targeting is always associated with more or less participatory procedures opening up decision-making, to the concerned local populations (without, however, giving up completely the autonomy of public authorities for their own political choices). For several historical reasons, targeting will be a less complex task in the field of traditional road policy measures such as the creation of pedestrian districts, which are shaped by well established administrative routines and procedures. On the other hand, targeting will be extremely difficult in the application of economic instruments because of their intimate ties with social policy and equity. Political compromises become inevitable in this very nasty business of potentially highly discriminatory decision-making.

The institutional framework of the proposed sustainable urban mobility policies must be simultaneously flexible enough and rigid enough to implement the whole range of the possible instruments simultaneously. The targeting of instruments needed to optimize sustainable urban mobility will require political backing that is strong and uncompromising, but also flexible enough to obtain the minimum necessary acceptance of policy outputs by target and affected groups. Negotiating procedures which facilitate problem-solving will be needed to ward off strategic bargaining (Scharpf, 1993). The institutional framework must contain sufficient financial, personal, and legal resources to implement more or less costly instruments. Furthermore, the need for discriminatory decisions against strong target groups must guarantee that the framework is capable of mobilizing the support of sufficient affected groups by means of additional political legitimacy through direct and indirect democratic procedures (direct election of metropolitan area authorities, referendums). And last, but not least, the metropolitan area policy framework must mobilize support from national government and international bodies in order to efficiently use all potential instruments and not be hampered by regional or central government constraints.

13.5 … and to facilitate policy learning

Given the intellectual orientation of network, sustainability and public management literature, it is not surprising that all contributors bring up the issue of learning in their conclusions. Learning is coupled with "soft instruments" in various ways, but three examples will illustrate the range of possibilities identified by our authors. For Desideri and Lewanski (1999), learning becomes a policy output that arises when policy communities implement persuasive policies and information campaigns. For Giuliani (1999), the interorganizational learning that occurs in encompassing policy communities which are horizontally coordinated by voluntary and spontaneous decision-making creates an important new policy input. And Bobbio and Zeppetella (1999) depict the interaction within policy communities as an ideal place for groups and organizations to discover and refine their economic self-interest in relation to other policy actors.

In each of these perspectives, the authors express their conviction that ideas and knowledge matter for the promotion of policies[4]. While this emphasis might seem to be at odds with our proposition (made earlier in this chapter) that sustainable development policy requires a strong dose of centralized political authority, the two insights are actually quite compatible. While the inevitably redistributive character of sustainability initiatives will require an institutionalized autonomy for implementation, this power needs to rest on a solid and legitimate democratic foundation. Obtaining that legitimacy calls for making sustainable urban mobility an overt political issue, both through referendums on key principles of sustainable mobility (for example, user pay schemes such as marginal cost mobility pricing) and in campaigns for elected representatives who will carry through such programmes. The primary legitimacy thus given to sustainable urban mobility will require all three of the learning modes outlined by our authors. These lessons will help carry the redistributional struggle through the rough and tumble of local politics.

Without active learning among both policy proponents and the urban electorate, we doubt that even a majority of urban voters could impose truly sustainable policy options through episodic democratic "victories" at election time. If the relatively affluent and empowered segment of the population who perceive strong benefits and limited alternatives to currently unsustainable urban transportation arrangements votes overwhelmingly in favour of change, strong minority target groups would subsequently – reject this democratic redistribution of mobility options as an infringement on their rights. Court challenges, and widespread noncompliance could be expected under such circumstances. Instead, the Italian referendum experience demonstrates that political campaigns must contain an effective element of targeted persuasion at key constituencies that will be impacted by the results.

---

[4] For further elaboration on the cognitive or constructive approach advanced in this book, see the growing, literature on policy learning such as Sabatier and Jenkins-Smith 1993; Kissling-Näf, Knoepfel 1998; Knoepfel et al. 1997; Nullmeier 1993.

Changes in beliefs and changes in behaviour are each correlated with learning, but the causal process is hardly straightforward. Empirically, the collective cognitive processes leading to new knowledge, new values, new problem definitions and, finally, to a new "worldview" remain virtually unidentifiable. Scholars agree that social learning is a collective process embodied in complex interactions between all kinds of institutional and societal actors consciously exchanging resources in the interest of solving of a common problem. Such exchanges are not solely intentional, such as in learning from mistakes or even proceeding by "trial and error". Literature suggests a distinction between learning forms can help explain the range of different results. Educational efforts can range from mandatory education efforts (such as military, corporate or government training programmes) through instrumental or conditional learning patterns (for example, "need to know" initiatives that enable specific activities) to more comprehensive efforts (such as systematic investigation and analysis) (Kissling-Näf, Knoepfel 1998). When new polices get implemented, each mode of learning can help explain and legitimize change. Most of these variants can be found in our case studies.

Mandatory education is typical of policy communities that are hierarchically organized, such as when the Italian courts decreed that certain transportation options must be pursued, or when officials in the Rhône-Alpes region of France approved a new express highway. Here, local policy communities are simply taught how to apply, or cease applying, a particular policy instrument without understanding the reasoning behind its use. Instrumental learning is typical of air quality management policy-making. Through both regulatory and economic incentive approaches, polluters learn to change their behaviour as, and only when, they encounter a specific restriction or charge. Turin's development of traffic restrictions follows this instrumental approach; the city's authorities adopted a trial and error mode in adapting their traffic restrictions in response to negative side effects and unintended consequences. Eventually they "learned" to use an incentive system of parking charges to achieve the desired objectives, including air quality improvement. Learning by imitation is also common in the more technical side of air quality management, as local authorities survey the world and try to copy the best available solution, which they identify with success in other cities.

Policy studies can contribute to the understanding of social learning by identifying specific behavioural changes which result from cognitive changes among key policy actors. Changes can be attributed to shifts in the belief structure through the perception, and adoption of new values. But changes can also arise from shifts that have a more instrumental logic, such as the choice among, different policy instruments or short-term policy goals (Kissling-Näf, Knoepfel 1998; Sabatier, Jenkins-Smith 1993: 13–39). Policy studies can connect both changing core beliefs and shifting tactics to variation in the make-up of policy communities and/or political administrative arrangements. Such participatory and structural changes are one indicator of ongoing learning processes. Another such indicator of social learning, processes is a considerable change in policy outputs, especially when it yields more effective policy outcomes.

All contributions to this book dealt with ongoing or imminent changes within policy networks arguing that the process of interconnecting previously insulated networks in urban traffic and air pollution management can be considered to result from learning processes among policy actors. Some contributions also claim to identify chances in the commonly shared values of the actors moving from a supply side towards a demand side approach to management. The evidence presented from Berne and the Greater Vancouver Regional District reveals that several outputs of urban traffic policy became more environmentally sound because of changes in the policy community composition and in the actors' values.

Though still a fuzzy concept at this point, organizational and interorganizational learning will play an important role in sustainable urban mobility policies, which intentionally aim at generating conscious cognitive and behavioural changes in various targets groups. Such learning will have to generate commonly shared frames of reference regarding urban transport and the environment, which will differ significantly from today's urban mobility paradigm. In this transition, people want legitimately to know where urban travel policy is heading before they commit to such change. We probably underestimate this need for security in redistributional processes as radical as the ones proposed by sustainable urban mobility policies. As a car driver, you only agree to learn about giving up your more or less vital mobility rights if others equally affected do the same, and if you know what you will get in exchange. It is therefore absolutely necessary that the new policy comprehensively defines the different categories of those user rights that both target and affected groups will be able to enjoy tomorrow as well as the concrete functioning of the future system. This is the most important precondition for initiating learning processes and moving policy communities towards sustainable metropolitan urban mobility policies. Despite the difficulty of this educational challenge, effective social learning will yield greater safety, more equality, and less physical and psychological struggle than the present daily civil war on our public roads and spaces.

References

Bobbio, L., Zeppetella, A. 1999. Shifting Tools and Shifting Meanings in Urban Traffic Policy: The case of Turin, in: Grant, W., Perl, A., Knoepfel, P. (eds) 1999. *The Politics of Improving Urban Air Quality*. Cheltenham UK/Northampton USA: Edward Elgar: 73–92

Bressers, H., Ligteringen, J. 1997. *What to Do with Non-accessible Target Groups: Policy Strategies for Sustainable Consumption*, presented at the 4th Meeting if the Concerted Action Network, The Ecological State, Barcelona, 7–8 November 1997

Centre for Sustainable Transportation 1997. *Definition and Vision of Sustainable Transportation*. Toronto: Centre for Sustainable Transportation

Desideri, C., Lewanski, R. 1999: Improving Air Quality in Italian Cities: The Outcome of an Emergency Policy Style, in: Grant, W., Perl, A., Knoepfel, P. (eds) 1999. *The Politics of Improving Urban Air Quality*. Cheltenham UK/Northampton USA: Edward Elgar: 52–72

Giuliani, M. 1999. “Soft” Institutions for Hard Problems: Instituting Air Pollution Policies in Three Italian Regions, in: Grant, W., Perl, A., Knoepfel, P. (eds)) 1999. *The Politics of Improving Urban Air Quality*. Cheltenham UK/Northampton USA: Edward Elgar: 31–51

Grant, W. 1996. Direct Democracy in California: Example or Warning?, *Democratization*, 3,: 133–49

Grant, Wyn, Perl Anthony, Knoepfel, Peter 1999. *The Politics of Improving Urban Air Quality*. Cheltenham UK/Northampton USA: Edward Elgar

Knoepfel, P. (Hrsg., im Auftrag der SAGUF) 1995. *Die Lösung von Umweltkonflikten durch Verhandlung / La solution de conflits environnementaux par la négociation*. Bâle: Helbing & Lichtenhahn (série Ecologie & Société, vol. 10)

Knoepfel, P., Imhof, R., Zimmermann, W. 1995. *Luftreinhaltepolitik im Labor der Städte, Der Massnahmenplan – Wirkungen eines neuen Instruments der Bundespolitik im Verkehr*. Bâle: Helbing & Lichtenhahn (série Ecologie & Société, vol. 9)

Knoepfel, P., Kissling-Näf, I., Marek, D., in Zusammenarbeit mit Bussy, C., Gentile, P. 1997. *Lernen in öffentlichen Politiken (Umwelt- und Gesundheitspolitik)*. Bâle: Helbing & Lichtenhahn

Kissling-Näf, I., Knoepfel, P. 1998. Social Learning in Policy Networks, *Policy and Politics*, vol 26, no 3: 343–368

Larrue, C., Vlassopoulou, C. A. 1999. Changing Definitions and Networks in Clean Air Policy in France, in: Grant, W., Perl, A., Knoepfel, P.

(eds)) 1999. *The Politics of Improving Urban Air Quality*. Cheltenham UK/Northampton USA: Edward Elgar: 93–106

Lembruch, G., Schmitter, Ph. (eds) 1982. *Patterns of Corporatist Policy Making*. Beverly Hills: Sage

March, J. G., Olsen, J. P. 1989. *Rediscovering Institutions: The Organizational Basis of Politics.* New York: The Free Press

Marek, D. 1999: CleaN Air and Transport Policy in Switzerland: The Case of Berne, in: Grant, W., Perl, A., Knoepfel, P. (eds)) 1999. *The Politics of Improving Urban Air Quality*. Cheltenham UK/Northampton USA: Edward Elgar: 126–143

Marlot, G., Perl, A. 1999. Lyon's Urban Transportation Policy and the Air Quality Problem: A Policy Network Approach, in: Grant, W., Perl, A., Knoepfel, P. (eds)) 1999. *The Politics of Improving Urban Air Quality*. Cheltenham UK/Northampton USA: Edward Elgar: 107–126

Newman, P. W. G., Kenworthy, J. R. 1988. The Transport-Energy Trade-off: Fuel Efficient Traffic versus Fuel Efficient Cities, *Transportation Research*, 22A: 205–21

Newman, P. W. G., Kenworthy, J. R. 1989. *Cities and Automobile Dependence: A Sourcebook*. Aldershot: Gower

Nullmeier, F. 1993. Wissen und Policy: Forschung. Wissenspolitologie und rhetorisch-dialektisches Handlungsmodell, *Politische Vierteljahresschirft Sonderheft*, 34: 175–96

Ostrom, E., Gardner, R., Walker, J. 1994. *Rules, Games, and Common Pool Resources.* Ann Arbor: University of Michigan Press

Perl, A., Han, J. D., 1996. Evaluating Environmental Component of Automobile Pricing Schemes, in: Hensher, D., King, J., Oum, T. H. (eds), *World Transport Research*. Oxford: Elesvier Science Ltd.

Perl, A., Hargraft, J., Muxlow, K. 1999. When Policy Networks Collide: The Insdtitutional Dynamic of Air Pollution Policy-Making in Two Canadian Cities, in: Grant, W., Perl, A., Knoepfel, P. (eds) 1999. *The Politics of Improving Urban Air Quality*. Cheltenham UK/Northampton USA: Edward Elgar: 13–30

Sabatier, P., Jenkins-Smith, H. 1993. *Policy Change and Learning: An Advocacy Coalition Approach*. Boulder: Westview Press

Scharpf, F. 1993. Coordination in Hierarchies and Networks, in: Scharpf, F. (ed.), *Games in Hierarchies and Networks.* Frankfurt am Main: U.A.P.: 125–65

Small, K. 1992. Using the Revenues of Congestion Pricing, *Transportation*, 19,(4): 359–81

Swiss Agency for the Environment, Forests and Landscape 1997. *Nature in Danger: Facts and Implications of the Greenhouse Effect*. Berne: Swiss Agency for the Environment, Forests and Landscape

World Commission on Environment and Development 1987. *Our Common Future.* Oxford: Oxford University Press

Zimmermann, W., Wyss, S., Neuenschwander, P. 1997. *Luftreinhaltekampagnen in Zürich und St. Gallen.* Cahier de l'IDHEAP 170. Chavannes-près-Renens: IDHEAP

14
Natural Resource Quotas and Contracts – A New Institutional Regime for our Common Resources (2000)

Peter Knoepfel[1]

In this contribution we present an outline of a new institutional regime for natural resources which takes into account the commonly agreed principle of sustainable development and the role public agencies may play in its governance. We start by describing the main features of sustainable development (14.1) and follow this with a discussion of its application in the management of natural resources (14.2). We then present a broad outline of the role public policies actually play in the allocation and consumption of natural resources (14.3). On the basis of these elements we then present an outline of the proposed resource quotas (14.4). This is followed by a more detailed description of the allocation and transfer of quotas within the framework of "Natural Resource Contracts (NRC)" (14.5), a discussion of some aspects of the governance of such a regime (14.6) and, finally, further arguments to support the selective application of such contracts (14.7).

14.1 Introduction – about sustainability

According to the Brundtland report, "Our Common Future", sustainable development means "development that meets the needs of the present without compromising the ability of future generations to meet their own needs"[2]. Such development is based on the presumption that "the exploitation of resources, the direction of investments, the direction taken by technological development and institutional change are all in harmony and enhance both current and future potential to meet human needs and aspira-

---

[1] in: Knoepfel, P. 2000. *Natural Resource Quotas and Contracts - A New Institutional Regime for our Common Resources*, Cahiers de l'IDHEAP no 178, Chavannes-près-Renens: IDHEAP.

[2] World Commission on Environment and Development 1987: 34.

tions"[3]. The most concrete international treaty on sustainability produced today is probably the World Biodiversity Convention, which was enacted on December 29th 1993. Section 6b of this treaty stipulates that the conservation and sustainable use of biological diversity be integrated as far as it is possible and appropriate "into relevant sectoral or cross-sectoral plans, programmes and policies"[4].

According to a current definition[5] the postulate of sustainable development can be summarised in the following five points:

- The exploitation of renewable resources must not exceed the rate at which natural resources are reproduced.
- Additional burdens on ecosystems must not exceed those which can be absorbed by the ecosystems' own natural regeneration processes.
- The integrity of ecosystems is a prerequisite to the sustainable use of natural resources. This only can be guaranteed by the maintenance of a maximum variety of species within the systems.
- The culture and landscape with ecological and social functions must be preserved for human beings and their dignity.
- The exploitation of non-renewable natural resources is only acceptable insofar as the productivity based on the use of natural resources within a country can be raised to a higher level guaranteeing an absolute reduction of the consumption of these resources in spite of eventual economic growth.

With regard to biodiversity, UNEP requires "the use of components of biological diversity in a way and at rate that does not lead to the long-term decline of biological diversity" thereby maintaining its potential to meet the needs and aspirations of present and future generations[6]. Both the green plans stipulated by Agenda 21[7] and the notion of the "use of components of biological diversity" proposed in this quotation and numerous other declarations of international organisations[8] clearly indicate the emergence of a new element within the paradigm for environmental policies. Traditional policies try to achieve the maximum reduction in the emissions produced

[3] Ibid. 46.

[4] Biodiversity Convention of May 1992, passed in Nairobi and actually signed by some 170 states.

[5] Cf. J. B. Opschoor, R. Konstanza 1995: 165 ff.; and T. O. O'Riordan, J. Jäger 1995: 32 ff.; and D. Wachter 1996 on the basis of J. Minsch 1993 and 1994.

[6] UNEP 1995: 15.

[7] Cf. e.g. Dalay-Clayton 1996.

[8] E.g.: OECD 1995.

by industry, agriculture, households, traffic and all kinds of activities linked to the use of infrastructure facilities that is both economically viable and based on the best available technology. According to this new element, however, environmental policies should also contribute to a sustainable management of natural resources.

14.2 The changing paradigm: adding resource management to environmental protection

The traditional approach is aimed at reducing existing or additional burdens to the environment on the basis of immission standards which prescribe minimum environmental quality objectives. These restrictions primarily concern the outputs of human activities (emissions, waste) and not their inputs in terms of the consumption of natural resources. Their main objective is to reduce the damage caused to human health and ecosystems as a result of inadmissible charges.

Although there is no doubt that these traditional environmental policies also aim to protect natural resources, their previous contribution to this dimension was very limited. This is due not only to a wide variety of implementation deficits, which have been studied by academics in the past, but also to the basic philosophy behind these policies. In economic theory, natural resources are understood as goods and services provided to human beings and ecosystems by nature (Sieber 1983: 2). As far as traditional environmental policy is concerned, the goods and services provided by fertile soils, clean airsheds and intact water basins etc. are only protected insofar as they belong to the essential human environment (anthropocentric approach) or precious ecosystems (ecocentric approach) which are under the threat of serious damage. According to the traditional civil-law based philosophy, these threats mainly stem from external physical emission activities which are, or are not, in some way tied to the exploitation of other natural resources.

According to this philosophy, the pollution of the airshed over a city is basically the result of the consumption of one or more other natural resources (normally: petrol, other fuels). Absorption service of this airshed is used in the production of energy by combustion. Air pollution policies are not directly aimed at protecting the non-renewable natural resource petrol. In fact, their instrumentalization for energy-saving policies is only a recent phenomenon. The same is true of water protection policies: the relevant service provided by water sheds is again the absorption and flow-away capacity they supply for gaseous, liquid and solid waste. This cleaning service may be overexploited by the overwhelming demand from households, industry and agriculture. But again, this demand stems from activities which do not use water but many other natural resources as input factors. Thus, protecting water against all kinds of dangerous emissions does not necessarily involve the protection of the resources used in the production of these emissions, although the reduction of emissions will normally be accompanied by a conservation of the relevant raw materials.

Those natural goods and services used in industrial processes could, of course, also be protected by specific (parallel) environmental policies, such as soil and biodiversity protection or landscape conservation implemented at the sites of their extraction. However, even within one country, there is no guarantee that these resources will be duly protected by such (mainly nature and landscape protection) policies because they are normally implemented by other agencies and often at sites other than those where they are actually used for production. Comprehensive protection schemes will only impose serious restrictions on the consumption of such natural goods and services if the reproduction of their basic resource is actually threatened by obvious overexploitation, it is not possible to substitute these goods and services by other natural or technical means and if this fact is politically perceived by the relevant actors as a collective threat. Such situations can, in fact, be due to damage caused to the resource by emissions stemming from outside (externalities). This is not essential, however, as the reduced production of the resource could occur in the absence of any serious "environmental" threat and merely as a result of overexploitation. On the other hand, we know of many cases involving local environmental disasters which are not due to any serious decline in natural resources. Thus, local water pollution, which on the level of the polluted watershed represents an overexploitation of its natural absorption capacity due to excessive use of fertilisers, does not necessarily have anything to do with an overexploitation of the fertilizer production capacity of the fields in the surrounding area. We must, therefore, admit that the phenomenon of natural resource pollution is only identical with that of their unsustainable overexploitation insofar as one of their many services is concerned, namely their capacity to absorb pollutants. For all other goods and services provided by the same resources, there is no necessary link between pollution and the destruction of the resource.

It is true that in the case of small and relatively closed local ecosystems, politically perceived links could exist between the excessive exploitation of specific natural goods and services used by the local community and resulting damage caused to other goods and services produced by the same ecosystem (e.g. hunting versus biodiversity within forests), or even by other natural production processes (e.g. damage to the recovery value of the landscape and the absorption capacity of surface waters due to fluorine extraction). However, there is still no guarantee that the traditional environmental policy approach is capable of providing comprehensive protection of natural resources. And, conversely, the sustainable management of natural resources cannot automatically guarantee a clean environment without any threat to human health and ecosystems.

We must, therefore, assume that the quest for sustainable resource management needs to result in the introduction of (in part) newly created policies for the management of natural resources to complete the overall set of traditional environmental protection policies. As demonstrated, these two sets of policies are not mutually exclusive; they may even be highly interconnected because of the simple fact that most polluting activities are somehow tied to the exploitation of the goods and services provided by nature and that the resulting pollution constitutes one (but only one) possible reason for their deterioration. Even in the field of traditional land use or nature conservation policies, which actually aim at protecting specific resource stocks as a whole against a wide range of exploitation activities, the comprehensive policies on natural resource management cannot be considered as a replacement for traditional environmental policies. The proposed policies will reinforce these existing protection schemes. They aim to tackle the relatively recent causes of unsustainable resource exploitation which consist of an increasing and heterogeneous demand for specific goods and services of resources that runs out of control due to the absence of mechanisms of co-ordination with the (limited) supply.

Economists usually explain the lack of such mechanisms by the sheer (physical) impossibility of creating exclusive use rights for a well defined group of such goods or services to be attributed to identifiable individual owners (lack of exclusiveness). A further explanation lies in the fact that such mechanisms are unnecessary because the goods or services in question are, currently, produced in abundance, thus there is actually no rivalry between different consumers (future generations are not usually considered). These arguments are underlined by analyses of the property rights of the natural resource in question which are rarely private and in most cases state-owned, common pool resources or they are simply devoid of formal property rights in the sense of the Roman "*res nullius*" (Ostrom 1990; Bromley 1991; Devlin and Grafton 1998; Kissling-Näf, Knoepfel, Varone 2000). According to Ostrom (Ostrom 1992; Ostrom 1994), all of these "property rights arrangements" have different rules governing access to the resource, its exploitation, the rights of the property owners to partly or fully transfer use rights, the exploitation duties and also the mechanisms concerning the change of the arrangements, the control and the monitoring of the arrangements and the mechanisms provided for the resolution of conflicts between competing owners.

Most of the resources currently threatened by destruction in Europe are either resources with a no property regime (airshed) or with a state property regime (surface and ground water). Common pool resources are quite rare. As Ostrom (1994) demonstrated, these common pool resources could be managed sustainably under certain conditions. Other resources cause problems because their stock and/or the goods and services they provide have a rather ambiguous status, being exploited under a partly private (e.g. land property for biodiversity) and partly public (state or even world) property regime, e.g. species living on private or state-owned land.

The proposal being presented here is nothing more than a pragmatic attempt at completing the traditional environmental policies through the addition of modified institutional regimes which create exclusive use rights – attributed by means of contracts – to threatened strategic natural goods and services, on the basis of policy plans for sustainable consumption of resources. The proposed quota regime differs from the existing (legal or de facto) private and public property rights which, according to environmental civil law traditions, normally embrace the entire stock and, thus, a multiple set of goods and services provided by one specific natural resource. Moreover, these property rights are quite unspecific with regard to the individual threatened goods and services to be managed under a new public regime. The proposed new policies for sustainable resource management only concern these specific aspects of natural resources and must take into account the (changing) local, regional, national and global sustainable supply of the regulated goods and services of the resource in question. As in the classical cases of national or international fishing regimes, quotas are an instrument for the regulation of individual consumption according to collective global resource policies providing plans which shape the overall supply of the relevant goods and services for a given time and space according the criteria of ecological sustainability mentioned below (Losada 1998). Furthermore, the social and the economic aspect of sustainability must also be covered by both the individual circumscription of the allocated consumption rights and duties as well as their global restriction.

14.3 Public policies and natural resources

Before presenting the above-mentioned quota regime, it would be useful to recall how modern public policies already affect the allocation, the distribution and consumption of goods and services provided by nature. This happens at in least four different ways:

- The implementation of many public policies makes direct use of goods and services provided by nature. The basic requirement for the construction and use of roads, motorways, airports, high voltage electricity lines, pipelines, public buildings, etc. is (private or publicly owned) land. The provision of such land for construction is just one of the goods produced by our soil; the others include crops, fertilisers and various plants. Thus, these other goods and services will partly or completely disappear for the benefit of the new competing use of the land. All of these infrastructure facilities directly extinguish more or less important goods and services provided by nature in a given country which might, one day, disappear. Whereas the availability of such facilities was considered as the main sign of modernity in industrialised European countries during the post-war period, today, at least in some of their most densely populated regions (e.g. Switzerland, The Netherlands), it is possible to observe increasing political pressure to seriously reduce such transformations of the land which pose a threat to biodiversity. In addition to mobilising local, regional and, sometimes even, national opposition to large infrastructure developments which consume vast quantities of land, these movements are also demanding biodiversity compensation which, in many cases in Switzerland, has in fact resulted in an increase in the price of biodiverse land due to the costs of compensatory measures (Knoepfel, Achermann, Zimmermann 1996: 311; Mauch, Buser 1998). According to data from Switzerland, biodiversity is at present mainly under pressure from local road construction for the servicing of newly created peripheral areas of urban agglomerations (FOELF 1997: 311).
- Traditional environmental and land-use policies actually attribute – by means of all kinds of permits – private or corporate use rights to a more or less exclusive consumption of specific goods and services produced by nature (Kissling-Näf, Knoepfel, Varone 2000). In most of the current cases, the affected services involve the absorption capacity of relevant natural resources (airsheds, watersheds, soil) for defined quantities of polluting substances (emissions) which are legally permitted to be deposited in the environment. The contents of these use rights may vary over time and space. It normally depends on both the state of the available technology (defined or not in maximum emission load standards), and the emissions restrictions which are, furthermore, intensified for some substances according to the actual quality of the recipient medium, depending on quality standards which should not be exceeded by

additional pollutants. Nevertheless, these regimes implicitly define and allocate specific rights to use the absorption capacity of given natural resources, mostly among competing operators. Thus, in urban areas, the operators competing for quotas of clean airsheds are, on the one hand, the car drivers, industrial emitters and households and, on the other, the human beings, animals and plants who need clean air. An unequal distribution of the use rights among these competing groups will ultimately give rise to political and social conflicts if a serious scarcity arises in the availability of clean air. This – implicit – distribution aspect of traditional regulatory environmental policies has received little emphasis in the past (Schnaiberg, Watts, Zimmermann 1986; Hartje 1986, Schnaiberg 1986; Bobbio, Zeppetella 1999, Marlot, Perl 1999, chapter 13 of this book). It is, however, quite prevalent in the literature on international environmental resource regimes and conflicts (Young 1997, Underdal 1998, Breitenmeier 1996, Marty 1997).

Such actual use rights play a more central role in the area of land-use policies in European countries. As in the case of state-operated infrastructure facilities, a building permit allocated to landowners or other entitled persons represents nothing less than the exclusive right to destroy existing soils and seal them with waterproof layers of concrete etc. However, under the long-standing tradition of land-use legislation in Britain, for example, property rights owners do not have an automatic right to such construction permits for their land. In contrast to the legal situation after the Second World War, in Britain, as in most other European countries since the 1960's, owners have only been entitled to such permits if their land is within a designated construction zone. In the interest of preserving either biodiversity or the equilibrium of existing water regimes, land-use planning schemes now increasingly make use of sealing quotas which oblige promoters to preserve minimum surfaces even within construction zones in natural (unsealed) conditions (e.g. France, Germany, Switzerland). The circumscription of such construction zones determines the politically defined supply of land available for construction in a given area at a given time. These zones are again usually defined in terms of the value attributed to other goods and services provided by the affected soils, such as rainwater absorption by infiltration, facilities for leisure, agricultural production, biodiversity, etc. In this way zoning is not only an instrument for the public management of urbanisation processes but it also becomes an instrument which can be used to prevent rural land from being urbanised, flooded (natural catastrophe), deforested, etc. In some countries zoning is actually aimed at actively maintaining biodiversity by prompting the concerned land owners (mostly farmers) to implement all kinds of protective measures. Given its explicit starting point in clear-cut and legally-binding restrictions in terms of land available for construction or construction quotas, this land-use policy approach is more comprehensive than the above-mentioned environmental policies. On the contrary, the exceeding of minimum clean-air or clean-watershed "budgets" often has no legal consequences in practice, a fact which in many countries is actually legally sanctioned.

The idea of setting binding maximum exploitation quotas for some goods and services in order to protect the reproduction capacity of the concerned resource is even more prevalent in some (very old) public royal prerogative regimes or international treaties. These policies normally start by placing a legally binding restriction on the overall annual hunting, fishing, wood cutting, gravel extraction, etc. quotas for a given area on the basis of agreed sustainable yield estimations and then attribute industrial or collective use rights within the framework of this "resource budget".

- Furthermore, many public policies have direct or indirect impact on the quality and quantity of goods or services based on natural resources which are consumed by their target groups: when UNEP, or the European Community[9], asks for an integration of sustainability considerations within all sectoral policies, they intend all public policies aimed at promoting, maintaining or reducing human activities which are directly or indirectly connected with the consumption of natural resources[10]. Environmental policy evaluations have revealed that it is these policies which are mainly responsible for the tremendous increase in environmental problems in the last thirty years as a result of their contribution to water, air and soil pollution and the destruction of biodiversity.

To sum up, the state and its public policies constitute a giant machine which consumes more and more renewable and non-renewable natural resources without any comprehensive resource plan or general mechanisms for the fair and legitimate allocation of these resources. The consumption of all kinds of goods and services produced by natural resources is still for the most part considered as a regrettable side effect of the juxtaposed implementation of the rarely co-ordinated public policies of both international organisations and national, regional and local governments.

---

[9] According to the Maastricht Treaty, section 3c.

[10] E.g. agriculture, transportation, tourism, energy consumption, land-use planning, national defense policies, etc.

14.4 A new institutional regime for governing natural resources: the natural resource quotas

The purpose of the natural resource quotas regime proposed in this article is to comprehensively reduce the consumption of those goods and services provided by nature, the production of which is or will be seriously threatened by their increasing exploitation over and above the annually sustainable yield and which could lead to serious threats to the reproduction capacity of the concerned resource. In Switzerland and in many other European countries, most of these goods and services stem from intact surface, ground and sea watersheds, (biodiverse) soils and forests as well as from landscape and intact airsheds. In the future, or in other regions of this planet, such services also might stem from other natural resources, such as the sun, oceans, fauna, the interior of the earth or its protecting ozone layers and they may consist of all kinds of energy production, medical care or protection against skin cancer, etc. The threat to these resources arises from the overexploitation of some of their goods and services in the past, present or in the future, which inevitably leads to increasing rivalry for their use and their potentially exclusive appropriation by individual user groups. This threat is mainly due to dysfunctional property rights arrangements (Ostrom 1992; Ostrom, Gardner and Walker 1994) which do not hold individual consumers economically accountable for their consumption.

This situation leads to the need for both comprehensive restrictive definitions of globally (world-wide, nation-wide, regionally and locally) consumable goods and services produced by the affected natural resources in time and space (general quota), and their concomitant breakdown into well defined manageable use rights to be attributed to economically accountable individual or collective users. The definition of the consumable global quota should be carried out within the framework of existing or newly-created public policies governing the use or protection of specific natural resources. The mechanisms ensuring accountability for goods and services are to be defined by taking into consideration the central (institutional) role played by regimes for natural resources. Such regimes consist of property rights arrangements and related policy designs which define at least some maximum exploitation quotas for given natural resources (Kissling-Näf, Knoepfel, Varone 2000). According to the prevailing economic principle applied to all other goods and services, these regimes must regulate access to use rights by means of market mechanisms allowing the fixing of prices. This price fixing must take into account undesirable losses and (desirable) gains to the owners of the resources, thus facilitating the enhancement of general welfare. This combination of property rights arrangements and public natural resources policies should enable the optimum distribution of the concerned goods and services amongst competing demanders and thus allow a sustainable consumption of these goods both on the level of the individual operator and of regional, national and international ecosystems.

The quotas could include all kinds of goods and services produced by nature. As mentioned, in European countries at least, the latter would probably mainly concern threatened absorption capacities of water, soils and airsheds. Wherever these concerned goods and services stem from resources which, according to the legislation in vigour, belong to private owners (such as soils), the introduction of a quota system will proceed in a similar way to the introduction of the new land-use policies in the post-war-period in most European countries. This process was initiated with increasing restrictions on the scope of private land property rights with regard to some specific goods and/or services which became scarce at that time. The state then proceeded to a redistribute the overall limited and partially tradable use rights within the framework of the general quota either to the current or future land owners or any other potential user groups entitled to purchase them from the landowner against the payment of a market price.

In the case of state-owned resources (e.g. water or mountain landscapes) and of non-existent legal property rights (e.g. airsheds), the responsible state agency would first establish new policy designs defining the global quotas and furthermore establish a partially new "property rights arrangement" defining specific collective and/or private use rights, etc. The government would then put them on the market in order to determine the relevant market prices and put them up for sale among different user groups. According to the economic components of sustainability, an optimum allocation can only be achieved when no operator is excluded from these markets. The social component of the same principle must also guarantee fair access to these markets for economically weak user groups seeking access to the resource.

In the case of Switzerland at least, state-run regimes would probably be the rule in the case of water, state-owned forests, land and airsheds. In practice, such a system requires that whoever wants to buy new land to build houses or infrastructure facilities, for example, must purchase not only the actual land but also a natural resource quota allowing him or her to consume the more or less precious natural biodiversity services etc. This basic mechanism creates market prices for the regulated goods and services of nature which would, for example, increase demand for biodiverse land and thus compensate "biodiversity producers" for their investments to (re-)create or maintain diversity, etc. In this case, the quota could initially be sold by private landowners. Once it is on the biodiversity quota market, it could be traded by other quota owners more or less independently of the legal owner of the land. It is important to note that in many countries local, provincial or national governments (e.g. national defence ministries) and non-governmental environmental protection organisations are also major landowners. They should also be capable of entering contracts with the promoters of all kinds of constructions which, according to the proposed regime, must acquire the necessary biodiversity quota.

Quota systems must be built up by state or regional legislation which defines both the specific natural resource policy design (Kissling-Näf, Knoepfel, Varone 2000) as a whole (general quotas, application instruments, administrative implementation structures) and the specific property-rights arrangements for one or several goods and services produced by the regulated resource. In addition to this compulsory regulatory framework, the appropriation and trade in individual quotas can be implemented on a contractual basis. These contracts would be concluded between the constitutionally-defined owner of the property right to the concerned resource (state or private) and the individual or corporate purchaser of the quota entitling him/her to the legally defined use of specific goods and services produced by this resource. Such purchasers could also be state agencies (e.g. state infrastructure operators) or private firms, households, individual and corporate actors, for example local governments, environmental protection organisations, banking/insurance corporations. As in other markets, the regulations must guarantee a fair market regime (transparency, stringently supervised rules governing access) with mechanisms to avoid the threat of market dominance by specific operators.

It is important to note that such a system requires a related comprehensive public policy for the natural resources which are politically perceived as being under threat from overexploitation. As previously stated, moreover, this regime could never replace the existing traditional environmental protection policies in force because the politically defined general quotas only allow consumption quotas that lie below so-called "sustainable charge levels"[11] or "safe minimum standards"[12]. Finally, the implementation of such new institutional regimes will only work if it is technically possible to define exclusive consumption quotas and if there is actually competition for the consumption of naturally produced goods and services. Rivalry may also exist among currently underprivileged and overprivileged, present and future quota owners, and it is certainly more important in the field

[11] Cf. J. B. Opschoor, R. Konstanza 1995: 182 ff.
[12] D. Wachter 1996: 13.

of non-renewable than renewable resources, especially if it is (technically) impossible to substitute the relevant natural goods or service.

The perception of interpolicy and intergenerational rivalry and rivalry between privileged and underprivileged social groups is a political one. In this way, scarcity becomes a social and political construct rather than being the expression of a real rivalry in economic terms: it depends, *inter alia*, on the political and social power of those actors competing for access to the relevant goods and services. At least in the short term, powerful actors can politically guarantee that rivalry will never occur simply by means of property rights arrangements which stipulate the monopolistic access of one social group to the detriment of other potentially less powerful groups seeking access to the resource. This situation appears to be taken for granted in urban areas where implicit use rights to the relevant airsheds are more or less exclusively attributed to industry, households and individual car drivers to the detriment of pedestrians, cyclists or mere inhabitants of a city (mainly children) (chapter 13 of this book). The same is true for the degradation of the landscape currently occurring in some of the tourist centres in the mountain areas of the European Alps to the detriment of future generations, etc.

Unlike the traditional environmental policies which aim at protecting human health and ecosystems against all kinds of damage or threats, the approach using natural resource quotas must be much more selective. As mentioned above, it does not concern all goods and services provided by nature in a given area, but only those perceived as being threatened by irrevocable overconsumption from a short-term or long-time perspective. In these cases, the relevant policy actors are convinced that the continuing and unrestricted exploitation of the resource in question will inevitably lead to the destruction of the productive stock which is relevant for both the goods and services in question and for other goods provided by the same natural ecosystem. Furthermore, the existing or emerging policy community must be convinced of the need of both introducing a new policy design and a change in the basic property rights arrangement. The quotas regime must be introduced with care and only if the additional conditions mentioned below are met. Its generalisation and undifferentiated use for incomparable goods and services provided by nature could lead to either a lack in economic allocative capacity or problems in its political implementation.

The central political and economic legitimisation of quota regimes lies in an appropriate definition of the available and still sustainable supply of the concerned goods and services produced by nature. Again, this idea is anything but new. It was politically formulated in very clear terms in the early 1970's prior to the birth of the traditional European environmental policies in the mid 1970's by the Club of Rome (Meadows 1972). What seems to be new in the sustainability debate is its integrative approach to natural resources which provide both physical goods (e.g. raw material) and services (e.g. energy production and absorption capacities for pollutants). These goods and services are highly interdependent and were hitherto chaotically exploited by heterogeneous competing social groups in the absence of comprehensive restrictive policy designs and appropriate property rights arrangements to guarantee the application of the economic principle in this field. This new view of natural resources as producing heterogeneous but interconnected goods and services calls for more comprehensive policies with more vigorous overall integrative resource planning and a more conscious co-ordination between policies and the underlying basic property rights regimes (Kissling-Näf, Knoepfel, Varone 2000). In the past, the latter often was taken for granted and property rights debates were limited to the establishment of international regimes which only indirectly affected the day-to-day handling of the use of natural resources within the general economic, ecological, social and political context of European countries.

The request for resource planning leading to precautionary resource policy designs calls for a selective and gradual introduction of quota systems because almost no country knows enough about even its most threatened natural resources. For many resources, there is a lack of knowledge with respect to current levels of consumption and the relevant temporal and regional variations. To obtain this knowledge, it is necessary to carry out accurate inventories on species, actual land uses and watersheds which take into account all ongoing consumption currently covered by related environmental and non-environmental policy permits. The next step consists of defining sustainable yields formulated in terms of overall (annual) consumption quotas which could be regulated within the framework of a national Green Plan according to Agenda 21 and then carefully broken down into different forms of individual quotas to be introduced to the newly created and well regulated markets.

Note that by "individual quota" we simply mean well defined rights to make (quantitatively and qualitatively limited) exclusive use of a specific good or service of nature in a given space and time. These rights should not be confused with the property rights to the resource as a whole, although they are connected in a way to be defined by legislation. Furthermore, the entitled user is not necessarily an individual (physical or legal) person but in many cases will be a collective corporate operator (common pool use right) or public policy agent (e.g. state infrastructure agencies, etc.).

Thus, we can imagine that in Switzerland the National Green Plan would define 10% of the Mittelland (central Switzerland) as areas in which a high degree of biodiversity is to be conserved. This would be followed by the introduction of a new property rights arrangement for biodiversity in the national legislation which would render biodiverse land independent of general (private and state) property rights and create a market for such biodiversity services. In a further step, the National Resource Agency could purchase a certain amount of these right(s) (e.g. 200, 000 hectares), stock part of them in its portfolio and sell an annual quota (e.g. 50,000 ha) to the National Infrastructure Department. The latter could then make use of the land by either destroying its biodiversity or implementing an economically more intelligent policy by recreating the biodiversity by destroying old roads which are out of service, etc.

14.5 The natural resource contract (NRC)

Contracts represent the most appropriate means of transferring individual quotas from one agent to another. Under this system, on the basis of a mutual agreement and within the framework of the law, the quota provider transfers the above-described exclusive use right to the purchaser who undertakes to pay a price fixed by the market. The purchaser basically remains free to accept or to refuse the proposed quota at the given price, opting to look for less expensive offers or simply opting to forgo consumption of the goods or service in question. Within the limits of the law, he or she is entitled to sell the quota to other interested operators or to purchase further quota according to his or her productive (or reproductive) needs. To prevent quota hoarding, it would be possible to implement a system similar to that currently practised in the field of tradable air pollution rights whereby quotas which are not consumed within a specified period are automatically devalued. Depending on specific situations one can also imagine the transfer of only parts of the quota by means of rental contracts, etc. We call these contracts "natural resource contracts" as a reflection of their purpose. The fact that national legislation will bindingly require the acquisition of quotas (e.g. for the consumption of biodiversity, deforestation, hunting, fishing, gravel extraction, etc.) does not basically call this voluntary contractual system into question. The same applies to state-owned resources whereby the quotas for some services and goods based on these resources become tradable independent of the state's basic property rights to the resource itself (e.g. water) and of the fact that no previous property rights existed for some of the concerned resources (e.g. airshed).

There may be good political and constitutional reasons for awarding specific privileges to public agencies in their search for quota when implementing their policies in the public interest. Despite their undeniable political attractiveness, even if perfectly harmonised with the general quotas defined in the national plan, such "reserved public quotas" will not be compatible with the attempt to introduce the consistently economic principle into the field of naturally produced goods and services. In order to establish real market prices, state policy operators must, therefore, be on a completely equal footing with private purchasers. This principle does not exclude a political definition of maximum or minimum annual quotas to be acquired by public agencies who may, or may not, be in competition with private purchasers in terms of binding "natural resource budgets" for specific public policies. Indeed, according to the new philosophy of policy performance management based on public policy agencies which contract rather than commission, in terms of contracts concluded between parliaments, governments and individual governmental agencies (Finger 1997; Mastronardi 1997), it would be possible to imagine the negotiation of specific "governmental natural resource contracts" (GNRC) which empower the agencies to acquire maximum quotas on the relevant markets ("natural resource budgets" as opposed to "global financial, personal, etc. budgets")[13]. These contracts could initially be concluded between national policy operators and, in a more detailed and concrete way, between national agencies and regional implementation authorities which actually construct and operate the regulated facilities.

Such intergovernmental and often multi-level contractual arrangements could allow greater flexibility to be introduced to infrastructure policies than exists under the present system, which often asks for proof of the public interest for a specific plant within the framework of environmental impact study procedures. Indeed, it is possible to imagine the elimination of such "proof-of-requirement" clauses, assuming that infrastructure agencies will be compelled to use their restricted quotas with greater care than they currently invoke the "public-interest" clause. On the other hand, all contractual arrangements in the public sector have the well known disadvantage of predetermining future developments by means of irrevocable commitments. Such rigid contractual commitment of public resource agencies might compromise the options of future generations, particularly in the area of sustainable natural resource management. It would appear, therefore, justified to impose temporal limitations on such contractual ar-

[13] Indeed, the new national resource policy approach will bring about considerable change within the traditional set of public resources (law, finances. personal, trust, time, etc.) by adding a new and sometimes very scarce resource.

rangements (e.g. 5–10 years) and stipulate their complete renegotiation at regular intervals[14].

Finally, it is also important to remember that such governmental contracts can play an important role in cases where the state is either the owner of property rights (state property regime), as in the area of water and land (e.g. national defence ministries), or where the state acquires quotas for specific purposes which become politically controversial. In all such cases governmental contracts could voluntarily reduce the quotas to be put on the markets by the competent public authorities. The motives for such restrictions, which could significantly undercut the currently admissible sustainable yield, could range from a temporary increase in the general quota prices and a specific political awareness for resource economics in the interest of future generations to purely financial considerations.

Given that it is impossible to predict the ultimate market prices for the described quotas, we cannot assess its exact impact on the present state of public finances. We certainly can assume that a shift from no-property resources (such as airsheds) towards state-owned resources and the sale of corresponding quotas to state agencies and private operators (e.g. in the case of roads) will yield extensive public revenue. On the other hand, however, local, regional and national governments will have to spend part of this money in acquiring quotas for the operation of their own infrastructure policies. Applied to the land and biodiversity issue and implemented consistently for both new and existing land-use activities, this system could produce additional state revenue which could be used for restoring the threatened resources[15] and thus enabling their active sustainable exploitation in the future.

14.6. The governance of natural resource quota systems

In view of the fact that the proposed institutional regime needs both new natural resource policies which define, *inter alia*, general quotas for selected goods and services based on natural resources and comprehensive changes in their basic property rights arrangements, its governance cannot simply take place within the existing political administrative arrangements. The only activity which is identical or similar to that currently carried out

[14] This problem of "tied hands" is particularily evident in the field of royalties allocated under various concession regimes to public or private companies in the field of e.g. energy extraction and production facilities.

[15] By a considerably contribution to the creation of new jobs.

by the traditional environmental policy agencies is the comprehensive definition of the general (annual) quotas. The related procedure poses quite similar political, democratic and scientific challenges to the definition of standards by traditional environmental policies (as far as ambient quality standards are concerned) (Winter 1986; Weidner and Knoepfel 1981; Knoepfel, Descloux 1991). Thus, it follows that the job of defining quotas and management of planning processes will be assigned to the environmental quality control divisions within the existing environmental agencies at national and regional level. Given that the quota definition process should consist of a (more general) national level and more detailed and concrete regional plans, will be necessary to involve the relevant bodies at regional level.

This integration of general quota definition into the core activities of the traditional environmental policy agencies should also guarantee close co-operation between natural resource management and the ongoing activities of environmental protection policies for each governmental level, because the one cannot be considered as a replacement for the other. This need for coherence mainly concerns those resources which are both protected against all kinds of externalities by traditional regulations for admissible charges and against overexploitation by the quota systems. Such co-ordination is also needed on the level of the concomitant implementation of permit systems and the contractual attribution of individual resource exploitation quotas at regional or local level.

Besides this possible integration of core functions of the natural resource policies, it does not make sense to put the governing of the functions connected with the quotas trade (acquisition, purchasing, general market creation and control) in the hands of the potential contracting partners. In order to avoid a conflict of interests, the state management of quota markets and the trading in state-owned quotas should be allocated to a newly created more or less independent state agency: this could be a section of the finance ministry or a real natural resource bank operating under a similar regime as the national monetary bank. Only a relatively independent status could prevent either the fiscal agencies from appropriating state income from natural resource quota sales to fund general state expenses or the environmental agencies from manipulating quota markets to the disadvantage of other potential public or private quota purchasers.

More specifically, at the level of the environmental agencies, an adequate public governance of a quota regime needs a strong administrative unit which is capable of compiling precise inventories and accounting systems for the relevant goods and services at local, regional, national and international level. Furthermore, these administrative bodies would negotiate the governmental natural resource contracts with interested governmental agencies (mainly those involved in infrastructure). They will also have to control their gradual implementation by means of further governmental contracts concluded with local and regional agencies and ultimately their translation into definite individual quotas. Last but not least, these resource divisions will have to continuously monitor the impacts of the (authorised) consumption of the natural resources on their evolution and compare these consumption rates with the goals defined in the global quotas and related planning decisions. They could also be authorised to award "green labels" to particularly successful public policy agencies who achieve significant reductions in their quota consumption.

In political terms, the governance of such systems presupposes a minimum equilibrium of political and economic power amongst the present (legal or de facto) owners of use rights and present or future social groups who are actually (currently) excluded from but legitimately interested in purchasing goods and services of nature. Hitherto weaker groups must, therefore, be empowered to participate in the newly created markets. Undoubtedly, the most critical aspect of the proposed quota system would be the absence of social sustainability due to exclusion. It is important to recall that the key argument in favour of this system is the fight against the overexploitation of natural resources due to non-existing or inappropriate general quotas (natural resource policy design) and inefficient market mechanisms which merely consist in the more or less exclusive, monopolistic position of current users. Every new system must care about creating greater social equality and, thus, enhance redistribution in favour of the users who are currently excluded of today and of tomorrow. As in other areas, fairly and strictly implemented market regulations will bring about both improved (economic) welfare and greater social sustainability.

14.7 Concluding discussion

Even if it is applied on a highly selective basis, as recommended above, the governance of the proposed quota system will be rather complex. Given the high political and managerial costs of this new institutional regime, the legitimate question to be raised at the end of this article is whether the practical application of this new market-oriented instrument for natural resource management is desirable. Given the present state of concrete (global) sustainable development schemes, it is our belief that we must continue thinking in this direction.

It is true that in theory we could simply continue to operate with traditional environmental policies by extending the present command-and-control policies beyond more or less stringent emission reductions into the area of strong and comprehensive natural resource conservationist policies (irrespective of whether the undesired overexploitation is due to external emissions or internal economic activities involving the exploitation of natural resources). This (more traditional) approach would, in fact, involve the redesign of existing or emerging natural resource (protection or use) policies without introducing any changes to the existing basic property rights arrangements.

However, we now know about the limits of these approaches. They are and will be helpful in reducing or even preventing more or less visible dangers and threats to vital naturally produced goods and services stemming from identifiable activities, for which the property rights owner of the threatened resource is not responsible. In this sense, command-and-control systems remain rooted in the typical realm of defensive (police-) policies aimed at protecting property right owners against (illegitimate) threats caused by third persons. However, they are not stringent enough to protect the interest of third persons (legitimately interested in the sustainable use of the resource) against abusive exploitation by their (legal or de facto) owners. What the traditional system lacks is economic accountability for such excessive exploitation of resources because of the increasing short-term oriented exploitation strategies of the owners which would appear to have been intensified by today's world-wide globalisation movement (destruction in defiance of their long-term economic interests). We all know that the economic consequences of administrative sanctions such as reminders or penalties are normally so weak that they cannot seriously be considered as real economic incentives for the prevention of the legally prohibited destruction of resources.

In summary, the main disadvantage of the current system is the lack of awareness of the economic value of goods and services provided by nature as reflected in real monetary market prices. It is interesting to note that in the case of damage caused by emissions to natural goods and services belonging to private or corporate actors protected by property rights, this economic valuation is increasingly perceived as necessary (and enhanced by a renaissance in civil law under the form of compensation mechanisms in environmental policies). This is still not the case, however, when it comes to the overexploitation of similar goods and services by their own legal property rights owners or by those who dispose – de facto – of comparable use rights on the basis of their political power or explicit state regulations (royalties or all kinds of environmental regulation permits). The need for sustainability-oriented intergenerational solidarity in the ecological, social and economic spheres undoubtedly calls for the inclusion of natural resources in such an economic valuation as the disappearance of these resources is due not only to emissions introduced by the others but also to overexploitation by their (legal or de facto) owners.

Some politicians and academics may disagree with the relatively important role being assigned to state agencies under the proposed quota regime. Even if these groups agree on the proposed change of regime, they would probably prefer less established public policy designs (general quotas) and a more communitarian form of governance. Indeed, it is possible to imaging the quotas working under a common pool resources regime. The seven basic conditions for the successful functioning of such common pool regimes have been clearly demonstrated by Ostrom and her research team (Ostrom 1992). However, we believe that the relatively peaceful common pool regimes only work when the naturally produced goods and services are used in a relatively homogeneous way (e.g. fishery, hunting, irrigation, etc.). In the present European context, at least, the threat of natural resource productivity stems from the heterogeneous rather than homogeneous rival consumption by various social groups with rather conflictual political relationships. Under these conditions, the most powerful users risk calling for politically imposed hierarchical and unfairly exclusive solutions to the detriment of the others. Furthermore, the deterioration of the capacity of nature to produce these goods and services is not only caused by their intentional active consumption but also by emission activities stemming from outside of the producing ecosystems (e.g. fish stocks in overfished water basins which in addition are highly polluted by industrial effluence). Therefore, we cannot reasonably expect the emergence of new self-organised common pool regimes fulfilling the conditions of social sustainability. There is a need for redistribution within integrated regimes (Kissling-Näf, Knoepfel, Varone 2000) to be granted by state interventions attributing the rival uses of one and the same good and service to competing public policy purposes and controlling the key variables of the quota markets.

The proposed system could also provoke fear in those who remain sceptical about the capacity of the market mechanisms to achieve a socially adequate allocation of resources. Although we do not share these criticisms in theory, they could well be founded in practice for those countries, regions, etc., where market mechanisms are, in general, seriously threatened by cultural, political or economic factors, or where "the marketing of nature" could be highly controversial in political or moral terms. In such circumstances, it is possible to imagine other ways of breaking down the general quotas: e.g. by assigning (part of) them in an initial step to national or regional public agencies (against payment or not). The final breakdown to individual users would then be implemented by means of the traditional permit system or (probably more sensibly) by means of royalties ("concessions") limited in time and entirely controlled by state agencies. Moreover, we could conceive the transfer of individual quotas to local governments, collective public or private actors (such as fishery, hunting, etc. corporations or co-operatives) and to non-governmental organisations (such as WWF, Greenpeace etc.).

For all these reasons, in addition to general emission-oriented protection schemes, the selective introduction of general and individual quotas for precisely defined goods and services provided by natural resources will be needed, even if we are aware of their inevitable governance problems. When establishing such regimes on a step-by-step basis, it would certainly pay to study the empirical functioning of similar, perhaps less far-reaching and more homogeneous regimes which operated in the past for specific natural resources, e.g. as the above mentioned old royal prerogative systems governing the extraction of natural raw material such as crude metal, sand, gravel or coal, or hunting and fishing regimes. We must focus our interest in their transitions over the past hundred or so years in modern European countries. As proposed by Kissling-Näf, Knoepfel and Varone (2000), such empirical studies must also include traditional permit systems with *de jure* or *de facto* quotas applied in the field of either state-owned resources or of hitherto formally *res nullius* resources. These studies will probably show that, at least in certain areas, we are already living with natural resource quota regimes but are simply not aware of it.

References

Bobbio, Luigi, Zeppetella, Alberico 1999. Shifting tools and shifting meanings in urban traffic policy: the case of Turin, in: Grant, Wyn, Knoepfel, Peter, Perl, Antony (eds.) 1999, *The Politics of Improving Urban Air Quality*. Aldershot/Hants: Edward Elgar Publishers: 52–72

Breitenmeier, Helmut 1996. *Wie entstehen globale Umweltregime?* Opladen: Leske & Buderich

Bromley, Daniel W. 1991. *Environment and Econiomy. Property Rights and Public Policy*. Oxford UK/Cambridge US: Blackwell

Dalay-Clayton, B. 1996. *Great Expectations? Green Planning in Industrial Countries*. Paper presented at the International Conference "The Environment in the 21st Century", at Abbaye-de-Fontevraud (F), 8–11 September

Devlin, Rose Ann, Grafton, Quentin R. 1998. *Economic Rights and Environmental Wrongs. Property Rights for the Common Good*. Cheltenham: Edward Elgar Publishers

Federal Office for the Environment, the Landscape and the Forests 1997. *The State of the Environment of Switzerland*. Bern: Federal Government-Press

Finger Matthias 1997. *Le New Public Management – reflet et initiateur d'un changement dans la gestion des affaires publiques*, travaux CETEL, no 48 – février 1997. Genève: CETEL: 41–60

Hartje, Volkmar J. 1986. Discussion: Saving Species for the Future, in: Schnaiberg, Allan et al. (eds.) 1986, *Distributional Conflicts in Environmental – Resource Policy*. Aldershot/Berlin: Gower: 405–418

Jäger, J., Liberatore, A., Grundlach, K. (eds.)1995: *Global environmental change and sustainable development in Europe*, European Commission, Directorate General XII Science Research and Development. Luxemburg: Office for Official Publications of the European Communities

Jänicke, Martin, Weidner, Helmut (eds.) 1996: *National Environmental Policies, a Comparative Study of Capacity-Building*, Berlin etc.: Springer

Kissling-Näf I., Knoepfel P., Varone, F. 2000. Broadening Property Rights Theory by Policy Analysis. Framework for Analysing Institutional Resource Regimes, *Swiss Political Science Revue*, 2000 (volume spécial).

Knoepfel, Peter, Descloux, Martin. 1991. Valeurs limites d'immissions: choix politiques ou déterminations scientifiques?, in: Theys, J. (assisté de Liber, V. et Palacios, M.-P., *Environnement, science et politique*, les Cahiers du Germes 13. Paris: Association Germes: 255–280 (texte du Cahier de l'IDHEAP no 48)

Knoepfel, Peter, Achermann, Daniel, Zimmermann, Willi 1996: *Bilanzstudie Bodenpolitik 1990 – 1995*, Cahiers de l'IDHEAP no 162, Chavannes-près-Renens: IDHEAP

Knoepfel, Peter, Grant, Wyn, Perl, Antony 1999. Conclusion: institution building for sustainable urban mobility policies, in Grant ,W., Knoepfel, P., Perl, A. (eds.), *The Politics of Improving Urban Air Quality*. Aldershot/Hants: Edward Elgar Publishers: 144–167

Losada, Antonio 1998. *El impacto de una política sobre la consolidación institucional de la autonomía: Análisis de la Política de Pesca en Galicia.* Universidad de Santiago de Compostela. mimeo (doctorate thesis)

Marlot, Grégoire, Perl, Antony 1999. Lyon's urban transportation policy and the air quality problem: a policy network approach, in: Grant, Wyn, Knoepfel, Peter, Perl, Antony (eds) 1999. *The Politics of Improving Urban Air Quality*. Aldershot/Hants: Edward Elgar Publishers: 107–126

Marty, Frank 1997. *International River Management. The Political Determinants of Success and Failure*. Zurich: Institut für Politikwissenschaft

Mastronardi, Philippe 1997. *Aspects juridiques du nouveau management public*, Travaux CETEL, no 48 – février 1997. Genève: CETEL: 93–104

Mauch, Corine, Buser, Felix 1998. *Hat Biodiversität einen Marktwert? Untersuchung der ökonomischen Bewertung von ökologisch wertvollen Flächen am Beispiel der Bonitierung im Rahmen von drei Meliorationen im Kanton Bern sowie von Handänderungen im Kanton Zürich*, Semesterarbeit, IDHEAP (Cours: "politique de l'environnement et de la durabilité"), 1998

Meadows, Dennis L. 1972. *The Limits of the Growth*. New York: Universe Books

Minsch, J. 1993. *Nachhaltige Entwicklung: Idee und Kernpostulate.* St. Gallen: IVÖ – Diskussionsbeitrag Nr. 14

Minsch, J. 1994. Von der Vision zur Strategie – zu den Prinzipien einer ökologisch nachhaltigen Wirtschaft, *Bulletin der schweizerischen Gesellschaft für Umweltschutz,* Nr. 3: 12–15

Minsch, J., Eberle, A., Meier, B., Schneidewind, U. 1996. *Mut zum ökologischen Umbau, Innovationsstrategien für Unternehmen, Politik und Akteurnetze.* Basel: Birkhäuser

O'Riordan, T. O., Jäger, J. 1995. Global environmental change and sustainable development, in: Jäger, J., Liberatore, A., Grundlach, K. (eds.)1995: *Global environmental change and sustainable development in Europe*, European Commission, Directorate General XII Science Research and Development. Luxemburg: Office for Official Publications of the European Communities: 17–34

Opschoor, J. B., Costanza, R. 1995. Environmental performance indicators, environmental space and the preservation of ecosystem health, in: Jäger, J., Liberatore, A., Grundlach, K. (eds.)1995: *Global environmental change and sustainable development in Europe*, European Commission, Directorate General XII Science Research and Development. Luxemburg: Office for Official Publications of the European Communities: 157–190

Organization for Economic Co-operation and Development (OECD), Group of Economic and Environment Policy Integration / Expert Group on Economic Aspects of Biodiversity 1995. *Making Markets Work for Biological Diversity: The Role of Economic Incentive Measures*, revised draft final report, version 024221, Dokument ENV/EPOC/GEEI/ BIO (95) 1/REV 1. Paris: OECD

Ostrom, Eliane 1990. *Governing the Commons: the Evolution of Insitutions for Collective Action*. Cambridge: CUP

Ostrom, Elinor 1992. *Grafting Institutions for Self-Governing Irrigation Systems*. San Francisco (Cal.): ICS Press Institute for Contemporary Studies

Ostrom, Elinor, Gardner, Roy, Walker, James 1994. *Rules, Games and Common-Pool Resources*. Michigan: The University of Michigan Press

Schnaiberg, A., Watts, N., Zimmermann, K. (eds.) 1986. *Distributional Conflicts in Environmental – Resource Policy*. Aldershot/Berlin: Gower

Schnaiberg, Allan 1986. Future Trajectories of Resource Distributional Conflicts: Trends and Projections, in: Schnaiberg, A., Watts, N., Zimmermann, K. 1986 (eds.), *Distributional Conflicts in Environmental – Resource Policy*. Aldershot/Berlin: Gower: 35–444

Siebert, Horst 1983. *Ökonomische Theorie natürlicher Ressourcen*. Tübingen: Mohr

Underdal, April 1998. *The Politics of International Environmental Management*. Neinhofen: Kluwer

United Nations Conférence on Environment and Development 1992. *Convention on Biological Diversity* (adopted in May 1992). New York: United Nations Organisations (cf. for Switzerland: Botschaft des Bundesrates vom 25. Mai 1994 zum Übereinkommen der Vereinten Nationen über biologische Vielfalt, BBl 1994 III: 182 ff.).

United Nations Environment Programme (UNEP) 1995. *Global Biodiversity Assessment*. Cambridge: University Press

Wachter, D. 1996. Nachhaltige Entwicklung: Folgerungen für die schweizerische Bodenpolitik, *DISP* 124/1996: 10–16

Weidner, H., Knoepfel, P. 1981. Implementationschancen der EG-Richtlinie zur SO_2-Luftreinhaltepolitik. Ein kritischer Beitrag zur Internationalisierung von Umweltpolitik, *Zeitschrift für Umweltpolitik*, 1/81: 27–68

Winter, Gerd 1986. Einführung, in Winter, G. (ed.) 1986, *Grenzwerte. Interdisziplinäre Untersuchungen zu einer Rechtsfigur des Umwelt-, Arbeits– und Lebensmittelschutzes*. Düsseldorf: Werner-Verlag: 1–26

World Commission on Environment and Development 1987. *Our Common Future*. New York: Oxford University Press

Young, Oran 1997. *Global Governments. Drawing in Sites from Environmental Experience*. Cambridge: MIT-Press

15
Institutional Regimes for Natural Resources: An Innovative Theoretical Framework for Sustainability [1] (2007)

Peter Knoepfel, Stéphane Nahrath, Frédéric Varone

There are few terms that are used in such an inflated manner as the word "sustainability". Politicians, businesspeople, scientists and all kinds of advertisers consider themselves, their proposals, their articles and their beliefs to be more sustainable than the ones of their competitors. Listening to them, one gets the impression that our world is the most sustainable one imaginable. Looking at reality, in most situations the exact opposite is true. Globalization accelerates all kinds of industrial, domestic and urban metabolisms and increasingly unbundled market mechanisms are becoming a serious threat for the survival of the reproductive capacities of our common natural resources. With the advancement of globalization and market liberalization, the need for solid institutional mechanisms capable of guaranteeing the survival of normally local and/or regional natural resources has tremendously increased in the last twenty years. Traditional environmental protection policies are incapable of doing this job. Like many other scholars and politicians, we believe that fundamental changes in the way we manage our common natural resources are inevitable if we claim to fight against the "plundering of our common wealth" (Bollier 2002).

This contribution is meant to be an innovative (even still modest) contribution to the theoretical and empirical thoughts towards a solution of the vital problem of overexploitation of natural resources. It firstly explores the traditional response to this question which consists of a wide range of environmental policies, each of which has the objective of providing a solution to the collective problem of protecting the resource against pollution. Taking into account the current situation, which shows increasingly clear indications of a lack of sustainability in the exploitation of these resources, we then undertake to demonstrate that these traditional policies – essentially based on the objectives and instruments of emissions restrictions – have reached their limits (15.1). With the help of more detailed ob-

1 Similar version: Knoepfel, Nahrath: Cahier de l'IDHEAP 226/2005.

servations of empirical processes involving the exploitation of natural resources and of the institutional rules that currently regulate their uses and the rights of users (15.2), we present a new reading of the empirical world based on the concept of institutional regimes for natural resources (IRR), a concept that takes into account the property, disposal and use rights of actors who use these resources along with the public policies that govern their behaviour (15.3). In our view, the analytical capacity of the IRR concept is superior to that of both public policy analysis and institutional resource economics. This belief is backed by a number of empirical research projects – carried out by our team since 1998[2] – and dealing with various resources: soils, forests, wild life, water (in a European comparative context) and landscape. Other applications are actually developed in the field of the national memory and built housing stocks[3]. The section 15.4 presents the IRR concept and its main theoretical bases while also adding some relatively new dimensions to the issues surrounding its empirical and practical application which have not yet been published elsewhere (15.5)[4]. The concluding section 15.6) presents some suggestions concerning the application of the IRR concept to other types of resources besides natural ones.

---

2 This group, which met frequently in the course of numerous working sessions, includes, in addition to the three authors, David Aubin, Kurt Bisang, Jean-David Gerber, Ingrid Kissling-Näf, Corine Mauch, Emmanuel Reynard, Raimund Rodewald, Jérôme Savary and Adèle Thorens. The research carried out by the group was financed through three FNS (Swiss National Science Foundation) projects (Division I and Division IV) and by the Swiss Federal Office for Education and Science. The following researchers also participated in the development of this analytical framework in the context of the European "Euwareness" project (EVK1CT-99-0038): Hans Bressers and Stefan Kuks, Corinne Larrue, Bruno Dente, Joan Subirats and their colleagues. Numerous students from the IDHEAP and the Postgraduate Diploma (DESS) in Urban Studies of the University of Lausanne also contributed by means of studies (seminar papers and diploma theses) involving the application of this concept.

3 All these project have been financed by the Swiss National Science Foundation and by the European Union for the project Euwareness.

4 Other presentations of the analytical framework of INRR can be found in Kissling-Naef and Varone (2000a), (2000b); Knoepfel, Kissling-Naef and Varone (2001:11-48), (2003: 1-58); Varone et al. (2002); Nahrath (2003a: 5-55).

15.1 The limits of traditional environmental policies

Anyone proposing to manage natural resources in a way that is compatible with "the environment" or which is "sustainable" (according to the paradigm and language adopted in the aftermath of the Rio Earth Summit in 1992), will refer to the enormous body of environmental protection legislation that exists in Switzerland, in all European countries and on the level of the European Union as well. The aim of this legislation is to protect human beings, plants and their biocenosis against hazardous or noxious substances by reducing them to a level that can be considered tolerable. These policies are conceived as a means of fighting immissions[5] (i.e. environmental impacts) through the imposition of reductions in emissions. While it cannot be denied that such policies have produced results, in particular with respect to the protection of water bodies, the protection of the air and the treatment of waste (Varone 2004), they have been clearly less successful in the area of the (qualitative and quantitative) protection of soils, nature and landscape (OCDE 1998; Knoepfel and Varone 2000; ARE 2005). Moreover, in Switzerland we are now seeing the re-emergence of problems in areas that had shown positive developments over the last three decades of the 20th century. These include, in particular, new increases in atmospheric pollution, the repeated failure to respect residual flows in certain watercourses, the increase of urban waste produced, not to mention the rises in CO_2 production and energy consumption (OFEN 2005). Even more worryingly, there are few indications that these policies have made any effective contribution to increasing the sustainability of our use of renewable and non-renewable natural resources. On the contrary, the current debates on the (un)sustainability of their use show clear signs of a very limited contribution of these policies to the achievement of truly sustainable development, particularly in urban areas (OFS, OFEFP, ARE 2003).

As a result of these phenomena, observers now have an obligation to review in detail the conception of these environmental policies whose purpose is to protect natural resources. Based on the analysis of these policies, it is relatively easy to demonstrate that one of their main weaknesses is that these policies originate in – and again are often partly based on – rights of adjoining owners (*droit de voisinage*) (Knoepfel 2000b). Thus their essential objective is to protect the environment and natural resources only from the effects of "immissions" or impacts originating from the emission of *pollutants*. It does not address *the effects triggered by other forms of exploitation* such as water withdrawal, clear cutting, or construction. Accord-

[5] *Immissionsschutzpolitiken* in German.

ing to this conception, it only becomes necessary to protect resources when the processes of extraction, exploitation and processing of natural resources give rise to emissions that are likely to affect the resources in question or other resources. Thus, any exploitation of (primary or recycled) natural resources that does not produce emissions that cause harm or discomfort to human beings or other resources will *not* be governed by environmental policy. Therefore, there is a risk that the successful implementation of an environmental policy that aims to fight immissions will pave the way for the "legitimate over-exploitation" of natural resources based on the principle of "the lower the emissions, the greater the admissible level of exploitation" (e.g. improving water quality allows increasing water withdrawal). When considered from the perspective of the management of natural resources, the real paradox of traditional environmental policies is encapsulated in this statement. In effect, very few environmental policies exist today that explicitly forgo this requirement that emissions be present. Indeed, this requirement does not exist in the case of policies for the conservation and protection of nature and the protection of landscape (Knoepfel 2000b: 199) which, according to the available data, are both among the least effective environmental policies – particularly because they encounter difficulties in making threats visible (Larrue and Knoepfel 1998: 192 ff.) and attributable to effectively "hazardous" activities which lends them a legitimacy in the eyes of those who are accustomed to fighting hazards affecting health or ecosystems. Furthermore, the landscape and biomass have very few clearly identifiable owners who are likely to institute proceedings in favour of their protection based on the "vigilant neighbour" model.[6]

These policies are characterized by yet another weakness. Since the 1980s, environmental policy analysts have highlighted the need to take into consideration the spatial and environmental repercussions of policies that have an influence – voluntary or involuntary – on the behaviour of potential producers of emissions. They emphasize the fact that many *non-environmental* policies exist that contribute directly or indirectly to the generation of considerable burdens on the environment. Urban development (soil), transport (air and urban surfaces), agriculture (water, air, soil or landscape), economic promotion (all areas combined) and energy (wa-

[6] The absence in many cases of a "neighbour" likely to intervene in the event of damage caused to their property was historically one of the main arguments in favour of the introduction of the right of appeal by environmental protection and nature conservation organizations into the Swiss Federal Law on the Protection of Nature and the Landscape (Loi fédérale sur la protection de la nature et du paysage) of 1 July 1966 (art. 12), RS 451.

ter, landscape or air) were also analyzed in terms of their spatial and environmental impacts. These analyses clearly show the power and increasing dynamics of these exploitation policies which are capable of challenging the limited successes of environmental protection policies (Benninghoff *et al.* 2004: 697 ff.).

The limits of traditional environmental policies can also be explained by the fact that they generally only concern a single use of a resource, i.e. the absorption, dilution, decomposition or transport of noxious emissions. However, as we know from resource economics, all resources are likely to be subject to a large number of *different*, and often *simultaneous*, uses ("goods and services") that potentially compete with the use of the resource for the absorption of pollutants, which is normally regulated by traditional environmental protection policies (Knoepfel and Savary 2002). It is interesting and important to confirm the existence of a clear causal link between the progressive recognition of the utility and, hence, the status of the goods and/or services among the different services provided by a natural resource, on the one hand, and the creation of very varied policies whose precise objective is the regulation of (homogeneous or heterogeneous) rivalries between the different uses and groups of users, on the other. However, in most cases, these policies concern a specific use, which is generally considered from the perspective of the regulation of activities[7] and the relations between users rather than from the perspective of the resource itself. Moreover, this is the reason why these policies are still rarely coordinated today and why they are often managed by specialized administrative bodies. These administrative bodies seldom have regular contacts with each other, which would enable truly coordinated management of natural resources and the various, already regulated goods and services produced by them (chapter 6 of this book).

As we shall show in the next section, this approach based on resource economics has gained significant relevance since the birth of the concept of sustainable use of natural resources. This concept highlights the integrated character of both a resource system and the goods and services derived from it, even though the latter are regulated by numerous specific policies. Thus, there is no reason to consider, for example, the urban resources located within the territory of a town or city and those located outside of urban centres separately as, in reality, they often belong to the same resource system. Similarly, it no longer makes sense to consider and, hence, regulate the uses of each of the goods and services provided by a

[7] These activities are normally regulated in accordance with a sectoral use logic based on economic promotion within agricultural, fishing, forestry, energy, transport, industrial promotion, tourism and urbanism policies.

single resource separately. In fact, the regulation of one of these goods and services may undermine the regulation of all of the other goods and services during periods when the resource in question is subject to overexploitation. The artificial division resulting from the current organization and structuring of policies between resources exploited for production within multiple industrial or urban metabolisms, on the one hand, and resources used on the basis of their absorption capacity, on the other, is hardly compatible with the objective of their sustainable management, the principles of intergenerational and interregional solidarity, and the integration of the global in the local (Di Giulio 2004: 151 ff.; OFS, OFEFP, ARE 2003: 12 ff.). Thus, the perspectives of the users of all of the goods and services provided by a single resource must be taken into account. Such an "actors-based" approach (Berthelot 1990: 76) should include all institutional rules (Scharpf 1997) that influence the users of a single resource and, hence, their behaviour. This group is almost always significantly larger than that of the producers of emissions who constitute the target group of traditional environmental policies.

15.2. A resource-based approach of sustainability

15.2.1 Alternative views of sustainability

Taking the above into account, the management of natural resources should concern itself with renewable natural resources since non-renewable resources are basically – and in some cases probably incorrectly – considered easier to substitute with the help of technological processes and innovations (O'Connor 2002; Devlin and Grafton 1998; Bromley 1991, 1992). Ensuring the sustainable existence of a renewable resource constitutes the best guarantee that it will be possible to obtain the goods and services derived from it now and, above all, in the future. The sustainable management of these resources must incorporate both the boundaries of the resources and all of the goods and services derived from them. This last and, at an initial glance, innocuous statement has fundamental implications for the way in which the sustainable management of local/regional resources is conceived. Basically, it is possible to identify in the literature – very schematically – three levels of conception of the sustainable management of local/regional resources which vary according to the extent of their requirements (Figure 1).

The "traditional" conception (1) of environmental policies is undoubtedly the least sustainable of the three to the extent that its rationale is limited to the *restriction of pollutant emissions*, i.e. in many cases the restrictions are still applied without consideration of the actual absorption capacity of the resource. This is best illustrated by the often almost uncompelling and, in certain cases even, non-existent relationship between the definition of emission standards and environmental quality standards.

Figure 1: Different conceptions of sustainability

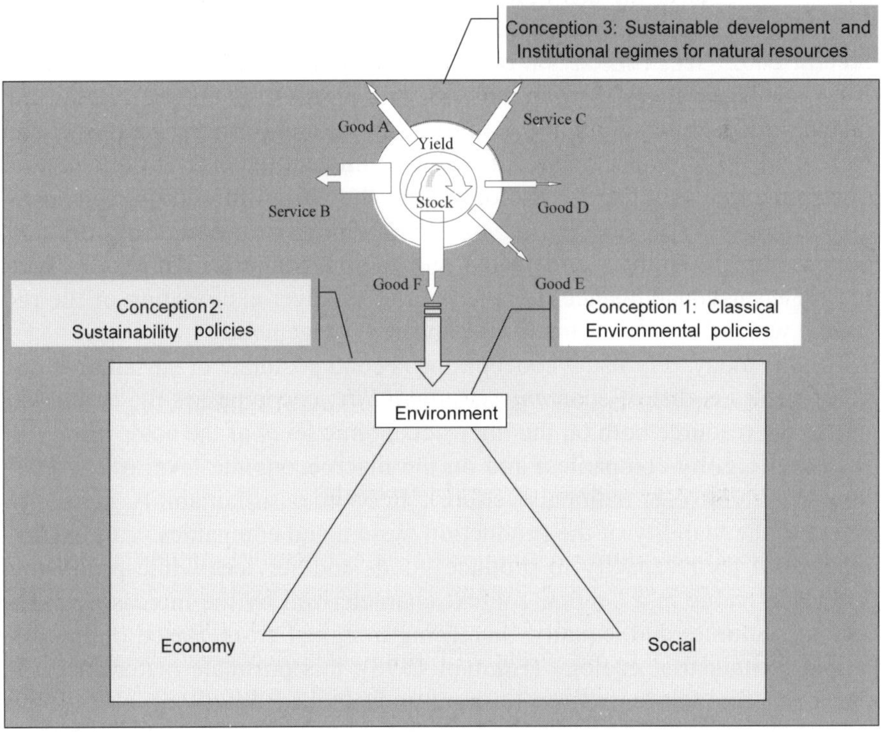

Conception (2) conveyed via the discourse and, when present, implementation of sustainable development policies often consists in a tentative to coordinate environmental requirements (which are still essentially based on the restriction of emissions) with the social and economic requirements or interests affected by these restrictions. The ecological requirement of the famous sustainability triangle incorporates all legal norms defined in the traditional environmental legislation, which are often very detailed and contain quantifiable indicators. However, unlike traditional policies, this ecological requirement actually often targets not only the owners of sources of environmental nuisance for an area, but also those who produce emissions that are harmful to their own resources. Thus, the requirement of damages induced to other resource owners is increasingly given up. What is clearly involved here is the reinforcement of the protective potential of these traditional environmental policies, despite the fact that it remains within the scope of a relatively weak conception of sustainability.[8]

In fact, the empirical processes relating to "sustainable development" basically focus their attention on the *modes of uses* of goods and services provided by natural resources, i.e. the regulation of these uses are supposed to guarantee the ecologically, economically and socially sustainable and equitable *exploitation* of resources (Di Giulio 2004: 49 ff.; Conseil fédéral 2002: 9ff.; United Nations 2002; World Commission 1987). However, this focus takes for granted that it is possible to obtain a sufficient quantity of resource units in the form of goods and services; yet this is far from evident. In reality, there is nothing to prevent the extraction or use of a resource in a way that is "ecological" in the sense of traditional environmental policies (i.e. one that *is not a source of environmental nuisance* to human beings, animals and plants and their biocenosis) but that is ultimately unsustainable since it could lead to the over-exploitation of the resource which is likely to impair its capacity to regenerate.

In this today very usual concept, the second postulate of sustainable development, i.e. that of *economic sustainability*, encompasses the economic use of the resource both on the micro-economic level of the actor using the resource (e.g. by companies) and on the macroeconomic level of the economic system (e.g. national welfare). Economic sustainability generally refers to the viability of the production system and companies or the maintenance of their capacity to produce goods and services while producing additional value in a general context characterized by the increasing scarcity of resources and, equally, increasing marginal costs. Based on the example of industrial ecology (Erkman 1998), this postulate materializes in particular through legislation (current and planned) involving obligations to recycle or reduce the quantity of waste and emissions produced from reusable materials or to prioritize the allocation of raw materials during periods of scarcity. While traditional environmental protection requirements are relatively clear today since they are based *inter alia* on scientific experience, economic sustainability is more politically and scientifically controversial. Even outside of the eternal debate of planned economy ver-

[8] According to the authorities responsible for this issue (OFS, OFEFP, ARE 2003: 14 ff.) the Swiss federal government's strategy starts from a position of so-called "weak sustainability" (Knoepfel, Münster 2004: 80 ff.). The latter "sanctions capital compensation but only if the use of capital is not irreversible and does not represent a threat to the survival of humanity, an area normally governed by limit values" (*ibid* : 80).

sus market economy,[9] now less virulent than it was in the 1960s, the question of the allocative optimization of production factors within what is considered generally a market economy remains controversial. This is why state allocation mechanisms based on authorizations, licenses and quotas have been maintained for certain goods and services whereas the allocation mechanisms set by the market constitute the rule for others. Mixed systems can also be observed that combine a state-imposed global quota on the extraction/withdrawal of resources (in particular for goods and services considered at risk due to the over-exploitation of the resource) and the distribution of these quotas among individual users on the basis of market mechanisms or, in some cases, their allocation by authorities (Kirchgässner 2002; Varone 2002). The fact remains that the concretization of the principle of economic sustainability (and equally of social sustainability) will remain significantly more controversial than that of ecological sustainability and this is clearly evident in the difficulty that exists in formulating concrete postulates and deducing universally accepted indicators for it (Knoepfel 2005a).

A similar, and probably even greater, problem arises when it comes to the concretization of the third postulate, i.e. that of *socially* sustainable development. In reality, the definition of genuine rights of equal access to the goods and services derived from natural resources *in situ* is far from being accepted by all users and is the subject of serious conflict between the holders and non-holders of use rights (Barnes 2001; Behan 2001; Bollier 2002; Radin 1996). The current debate on the limits of the welfare state and the past controversies surrounding the enshrining of social rights in national constitutions and the European Constitutional Charter bear witness to this. The absence of real legal norms and clearly quantifiable indicators concerning the minimum rights of access to the different vital goods and services supplied by natural resources is a clear sign of the highly political issues concealed behind the notion of socially sustainable development.

Finally, as suggested by the distinction between the three sustainability conception levels represented in Figure 1, the sustainability policies emerging over the past twelve years or so undoubtedly constitute an important initial step in the direction of sustainability. However, in their cur-

[9] According to the representatives of the neo-Marxist movement of the 1960s, only collective, state or community planning would be effectively capable of combating the abusive exploitation of our natural resources and manufactured resources, referred to using the generic term of "factors of production". This postulate would affect land/soil in particular and would lead to various propositions of its nationalization.

rent state of development, they are not in a position to guarantee the truly sustainable management of (natural) resources, in the majority of cases. The situation in which we find ourselves today regarding policies for ecological, economic and social sustainability is comparable to the paradox of traditional environmental policies referred to in section 15.1. In fact, there is a significant risk that the pursuit of social, economic and even ecological sustainability at the level of selected goods and services will ultimately lead to the non-sustainable management of the resource. This is the case when the extraction and distribution of the resource's goods and services are carried out on the basis of the simple logic of pollution limitation or internalization of negative externalities ("polluter pays" principle), i.e. independently of the estimated reproduction capacities of the different resource systems.

In our view, *the sustainability postulate (conception 3) requires the clear distinction between the sustainability of the resource (system) and the ecological, economic and social sustainability of its different uses*. In reality, it is only possible to exploit the goods and services of a resource in a sustainable way if its reproduction capacity is not put at risk. Such an objective can only be attained if sustainability policies undergo a fundamental conversion, which they have hitherto mostly failed to do, from the logic of control and restriction of pollutant emissions (management and internalization of negative externalities) to policies focusing on the management of the stocks and reproductive capacities of resource systems. It should be noted that a shift of this kind comes down to recognizing the primacy of the ecological pole over the economic and social poles of the sustainability triangle. It reminds us of a basic principle – at the same time as an obvious fact – of the heuristics of sustainable development, a principle too often forgotten in the context of the development and implementation of contemporary sustainability policies: in other words the ecological sustainability of different resource systems constitutes an necessary (but far from sufficient) condition for the existence of the sustainability of social, economic and ecological uses of the goods and services provided by resources. The latter can only be guaranteed if all of the users jointly ensure that the quantities they extract or withdraw from a resource do not reach the limit of the reproductive capacity of the resource system, a requirement that should in principle give rise to inconveniences for all users (the symmetry of sacrifices made being one of the conditions of social sustainability). Given that all natural resources today are at least in part the product of human activities, this objective often is only attainable under the condition that the appropriators and users limit the quantities of goods and services they extract or even contribute actively to the conservation of the resource through investments coming from human, manufactured or cultural capital.[10]

15.2.2 From global (resource) quota to individual use rights

In this sense, sustainable development strategies likely to guarantee or re-establish the reproductive capacity of the resource in question should govern *all* of its units ("fruits") considered extractable in a given time and space. A glance at the origins of the modes of political regulation shows that this requires three successive operations containing each one important political choices.

Firstly, regulators normally define a maximum *global quota* for the extraction/withdrawal of resource units that incorporates both quantitative and qualitative criteria that are compatible with the ecological requirements for the renewal of the resource system. We suspect that the definition of such a quota constitutes a central political issue to the extent that it expresses the conception of sustainability (i.e. strong, weak, etc.) adopted by a given society. *Secondly,* this global quota will be shared between the different rival uses (goods and services), ideally as a function of the principles of social and economic sustainability. *Thirdly*, the quota of resource units attributed to a specific good or service will again be distributed between the different user groups. All together, such regulations constitute what we refer to as an *institutional regime for natural resources (IRR).*

Theoretically, it is possible to model the emergence of regulation procedures while differentiating between different stages of resource use, moving from a so-called "normal" situation involving the use of goods and services (Phase 1) to a situation of uncontrolled increase in the extraction of certain goods and services likely to exceed the annually acceptable limit (Phase 2), and ending up with a situation in which the reproductive capacity of the resource is called into question (Phase 3). In such situations, we see the initial attempts at quantitative and qualitative regulation of these extractions (i.e. global quota) and this process corresponds to the political "birth" of the resource (Phase 4). The restriction on the allowable extraction ultimately involves the adaptation of the behaviour of individual users who are forced to limit their actual extractions (Phase 5).

In Phase 3, at the latest, we see the emergence of more or less robust and exclusive regulations whose objective is to guarantee users stabilized access in time and space to one of the specific goods or services that they

---

[10] According to the terminology of the World Bank (World Bank 1995). Cf. Knoepfel (2005b).

appropriated or claimed in the preceding phase (Phase 2). These regulations provide a precise definition of the good or service in question, and – in an initial period independent of truly resource-based considerations – the *property and use rights* to this good or service as well as the modes of its attribution to different groups of users. The aim of these regulations consists in controlling the behaviour of users by means of (stabilizing or modifying) intervention in the relationship that previously existed between them and the good or service in question. This is achieved through the attribution of universally recognized use rights. These relationships may be created, quantitatively and qualitatively redefined, eliminated etc. by means of either the modification of formal property rights (Civil Code), the modification of rights of disposal and use, or the redefinition of the obligations imposed to the actors within specific policies (e.g. within water withdrawal concession, planning permission etc.).

This need to formulate rights may emerge prior to the arrival of any threat to the resource in question and generally results from the need to stabilize and render more predictable the relations between an "owner" and his or her "property", in particular for economic reasons. This need is found at the origin of Roman Law, in the phenomenon of "enclosures" and, again, in the generalization of private or exclusive property after the French Revolution (Aubin, Nahrath, Varone 2004).[11] Obviously, this need would become even more evident when the threat of over-exploitation of a resource exists. Under such increasingly frequent conditions, the main concern of legislators is not only guaranteeing this individual right in the interest of maintaining the stability of the conditions necessary for economic activities, but rendering it compatible with the need to maintain the resource and its reproduction capacity as a common good (Bromley 1991; Devlin and Grafton 1998; Holzinger 2002; Ostrom 2002).

For this reason, in the case of a real scarcity of one of the resource's goods and services which risks leading to its over or under-exploitation (excessive or under-use of the global quota), the need arises for a second layer of regulation whose aim is to alter the behaviour of the user based on the public problem of the survival of the resource. Therefore, its objective consists in the regulation of the behaviour of users based on a politically defined global quota. In fact, this second layer of regulation should dominate the first. However, the mere existence of such rights of use is a precondition for any public action that aims to regulate all of these uses: it is only possible to change the behaviour of users institutionally on a global

[11] Such a need even exists among nomad populations who frequently integrate forms of ownership, possession, right of usufruct etc. into their legal order. (Hagmann 2004).

level if their uses of the resources were previously regulated individually through the rights of use. Thus, one of the main conditions for the effectiveness of a policy goes back not only to the instruments the policy has at its disposal, but also and probably primarily to the characteristics of the target groups it is aimed at: it has a greater chance of having a substantial effect on the problem to be resolved if it targets the actors who hold the effective rights of use to the resource. In order to be able to accomplish its aim of regulating the entire resource system and not only one good or service in isolation, this second layer must be capable of relating in a sustainably and globally restrictive manner all those in possession of rights to the resource (Figure 2).

Figure 2: The "two layers" regulation

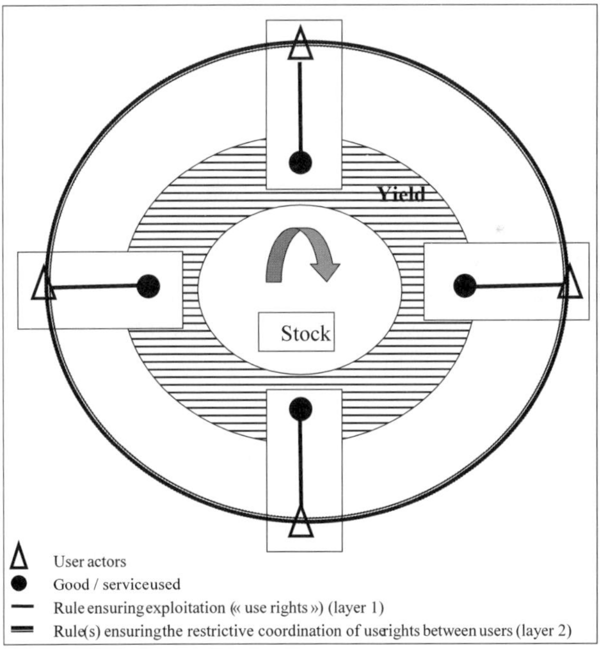

By doing this, these rules construct an institutional mechanism that is capable of defining and redefining the individual use rights corresponding to the quantities – variable in time and space – of resources, the admissible extraction of which is defined politically. In other words, this regulation should provide a restrictive mechanism for the attribution, redistribution, and quantitative and qualitative redefinition of individual quotas on the basis of global quotas (chapter 14 of this book). Figure 3 presents in schematic form two possible opposing mechanisms for the shifts from a global quota to individual quotas.

Figure 3: Models for shifts from a global quota to individual quotas

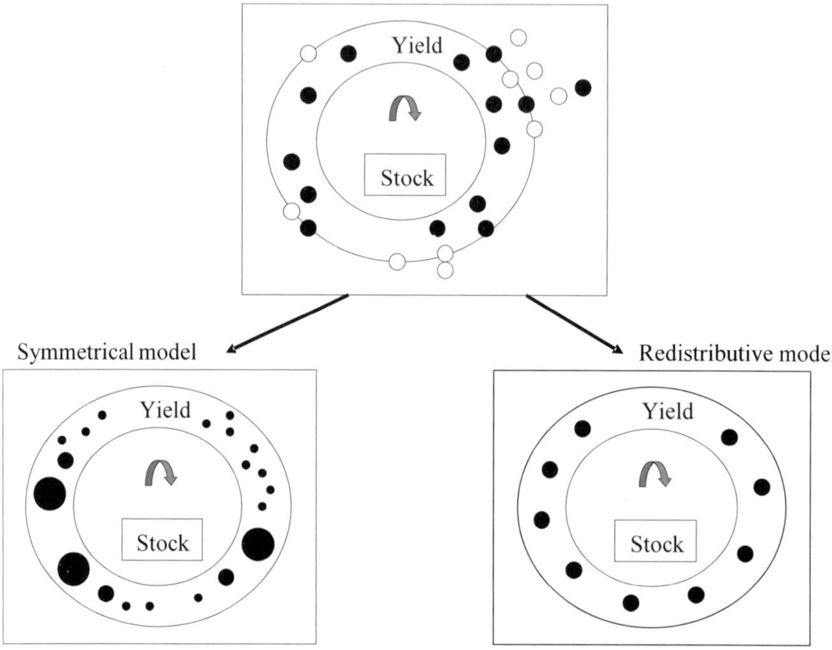

The "symmetrical" (statu quo or grand fathering) model considers all of the goods and services used *a priori* as "legitimate" and contents itself with redefining them *in a restrictively symmetrical* way. In such a model, which is based on the symmetry of sacrifices, the inequalities with respect to access and rights of use of the resource are maintained proportionally in the case of reduction of the global quota and, therefore, of individual quotas. Conversely, the "redistributive" model takes the view that some uses are more crucial or simply more opportune than others and that these should be prioritized in the case of a reduction of the global quota. Therefore, the shift from the global quota to the different individual quotas (through goods and services) involves a choice that is politically more costly and consists in the reallocation – possibly through the expropriation of previous holders of use rights – of the resource units that were previously allocated to secondary uses to the more important uses.[12]

15.3 Institutional regimes for natural resources (IRR): theoretical bases

In searching for a conceptual framework capable of accommodating these sets of regulations – formal and informal, institutionalized or in the process of becoming so, centred on intentional actors whose behaviour is likely to be controlled or influenced by public policy (Berthelot 1990; Knoepfel *et al.* 2001a; Scharpf 1997) – we have found a suitable theory to complement that of public policies: *resource institutional economics* and *property rights theory* (cf. in particular Bromley (1991, 1992); Devlin and Grafton (1998); Endres and Querner (1993); Ostrom (1990, 2000, 2002); Schlager and Ostrom (1992); Siebert (1983)). This approach is appropriately preoccupied with the issue of use rights, their definition, their allocation, and their redistribution. It addresses the rules governing the behaviour of the holders of property rights and considers these rights not only from an economic perspective, but also, and above all, from that of the sustainable management of natural resources. The crucial importance of property rights for the sustainable management of resources can be explained by the fact that the latter constitute *common property* or *common pool resources* (Holzinger 2002; Ostrom 1990), which, unlike public goods, are characterized by use rivalries and by the impossibility of excluding entire social groups from their enjoyment or exploitation; this impossibility is the result of either their physical characteristics or categorical normative imperatives. Therefore, for this branch of research, the critical collective action in terms of the degree of sustainability of our development resides essentially in the (good or bad) definition of these rights (Coase 1960), and in the (more or less adequate) mechanisms adopted for the allocation of these rights.

Thus, the combination of these economic approaches enables us to describe, analyze and explain the problems of the sustainable natural resource management outlined above which, according to the preceding account, are rarely made compatible with the theoretical frameworks familiar to us as political scientists. For reasons explained elsewhere (Knoepfel *et al.* 2003: 31 ff.), we have named this new analytical framework "institutional regimes for natural resources" (IRR). In accordance with the ideas presented in section 15.2, these regimes incorporate all of the formal and

[12] For other modes of allocation see for example Varone (2002).

informal rules that regulate all of the different uses (in terms of goods and services) of a resource system in the context of a given area. The crucial characteristics of these regimes, which determine the sustainable or unsustainable character of resource management and use, can only be successfully identified through theoretical reflection and regular empirical observation.

To do this, the concept of IRR combines the contributions of the theoretical approaches of policy analysis, on the one hand, and institutional resource economics, on the other, with a view to proposing an analytical framework that can overcome their respective individual limitations (Table 1).[13]

[13] Cf. for more details and literature: Kissling-Näf and Varone (2000); Knoepfel *et al.* (2001a); Knoepfel *et al.* (2003); Nahrath (2003a); Gerber (2005); Bressers and Kux (2004).

Table 1: Contributions and limits of policy analysis approach to the IRR framework

Contributions (theoretical and normative)	Limits (theoretical and normative)[14]
• Conceptual instruments suitable for the analysis of the modes of state intervention/regulation (taking into account the interventions of political-administrative actors). • Identification and analysis of the six products of a policy: problem definition, political-administrative programme, action plan, political-administrative arrangement (PAA), outputs and evaluative statements.[15] • Capacity to conceptualize complex use situations, including heterogeneous rivalries (shared uses), in which policies play a central role.[16]	• Sectoral approach to the phenomena arising from the artificial logic of the division of the world produced by the policies themselves. • Difficulty in conceptualizing the coordination between *protection* policies and *use* (exploitation?) policies. • Conception of environmental policies as remedial policies rather than preventive ones. • Focus on the management of pollutant emissions insufficient to guarantee the integrated protection and management of resources. • Failure to take property rights into account making it difficult to understand the main obstacles to policy implementation arising from the social and political resistance exerted by the target groups who hold property rights to the regulated resources.

[14] See section 15.1.
[15] Cf. Knoepfel *et al.* (2001b: 142 ff.).
[16] Cf. Table 3.

The theory of property rights developed, in particular at the "Workshop in political theory and policy analysis" directed by Elinor Ostrom at the University of Indiana at Bloomington (USA), essentially results from the detailed analysis of institutional arrangements (mechanisms for the definition and attribution of property rights and for (self-) control of their implementation). These institutional arrangements are based on a communal definition of property (*common property*), as prevails, for example, in the case of irrigation systems, common pasture lands in the Swiss Alps (*Allmende*), fisheries, hunting leases etc., and have historically demonstrated their capacity to guarantee the sustainable use of a resource by local self-organized groups. The contributions and limits of this theory (i.e. institutional economics) in the case of institutional regimes are listed in Table 2.

Table 2: Contributions and limits of institutional resource economics to the IRR framework

Contributions (theoretical and normative)	Limits (theoretical and normative)
• Resource-based approach founded on the concepts of "resource" and "goods and services". • Capacity to envisage the coordinated and global management of the resource and all of its uses (anticipatory approach). • Clear conceptual definition of (ecological and economic) "sustainability". • Focus on the institutional arrangements and the property rights as form/vector/instrument of the regulation processes. • Establishment of an explanatory link between over-exploitation of resources and the absence of property rights (e.g. air, landscape, genetic resources, global commons etc.).	• The validity of the *common pool resources* (CPR) approach is limited to use situations involving a homogeneous group of users (common use), i.e. difficulty of analyzing situations involving multiple or shared uses.[17] • Ideological bias favouring solutions involving local and self-organized regulations (rejection of the relevance of market or state regulation). • Ignorance of state regulations (policies). • Reductive typology of property regimes due to the absence of legal analysis.

One can hypothesize that there is a link between the type of use situations of a resource and the theoretical approach relevant to their analysis. Tables 1 and 2 demonstrate that the institutional economics approach is particularly well suited to use situations described as "common", in which several users find themselves in the position of rivals competing for one and the same good or service provided by a natural resource. Table 3 suggests that the institutional regimes approach is more relevant for the analysis of use situations described as "joint", in which several users find themselves as rivals with respect to the heterogeneous uses of one and the same

[17] Cf. Table 3.

resource. Contractual arrangements based on civil law are generally used for the regulation of the two other far less complex situations involving "individual" and "multiple" uses.

Table 3: Positioning the IRR framework according to resources uses

<table>
<tr><td colspan="2" rowspan="2">Classification of use situations
(based on the example of water)</td><td colspan="2">Types of use (in terms of goods and services used)</td></tr>
<tr><td>Homogenous Uses</td><td>Heterogeneous Uses</td></tr>
<tr><td rowspan="2">Number of users</td><td>Individual user or group of users</td><td>"Individual use"
Example: (exclusive) use of a stream for the production of drinking water.</td><td>"Multiple uses"
Example: construction of a communal dyke with the dual function of flood protection and constitution of water reserves.</td></tr>
<tr><td>Several users or groups of users</td><td>"Common use"
Example: division of a stream amongst farmers within an irrigation system.
Self organized Common Pool Resources (CPR) regime (Ostrom)</td><td>"Joint uses"
Example: definition of minimum flows to be respected by a hydro-electric power plant so as to protect the biotopes of the fish populations and to guarantee the supply of the irrigation system of farmers located upstream.
Institutional Regimes for natural resources (IRR)</td></tr>
</table>

Source: Knoepfel *et al.* (2001b: 16), based on Young (1992: 103).

15.4 The IRR concept and its operationalization

The IRR concept as applied to the resources soil (Nahrath 2003a, 2003b); water (Reynard and Mauch 2003; Bressers and Kuks 2004; Kissling-Näf and Kuks 2004), forest (Bisang and Schenkel 2003); landscape (Gerber 2005; Rodewald *et al.* 2005); air (Mariéthoz and Savary 2004; Savary and Knoepfel 2005) and built heritage (Knoepfel and Kohler 2005) is composed of two types of regulation (public policies and property rights). Its dynamic depends on two dimensions, *extent* and *coherence*, both of which may vary in time and space according to the type of regime that prevails. In this section, we shall examine the main characteristics of this concept.

15.4.1 Public policies and property rights

The concept of IRR is primarily a framework for the analysis of institutional arrangements concerning the regulation of the collective and individual uses of a resource. This regulation is generally highly complex and composed of legislation and parliamentary, administrative or legal implementation decisions originating from several levels of the state (i.e. municipal, cantonal, federal and sometimes even international). It brings together the public rules, whose objective is the regulation of the behaviour of users of goods and services originating from a resource system, the boundary of which is generally regional. These rules appear either in substantial policies intended to tackle collective problems associated with the protection or exploitation of the resources in question or in the basic property rights order[18] which is far more stable as it is based on (*quasi*) constitutional decisions or civil law (e.g. Swiss Civil Code) and rooted in political convictions that concern fundamental rights, social justice, and the basis of the economic, political and social functioning of our society. An institutional regime represents a combination of these two types of regulation, i.e. (a) policies contained in what we call the *policy design* (PD); and (b) all of the property rights contained in what we call the property rights system (PRS). The contribution of these two types of regulation may vary significantly according to the resources in question and historical periods of the development of a regime, which is why it is possible to identify historically regimes that are essentially based on the regulatory system ("property rights driven regimes") and, conversely those based on policy ("policy driven regimes").[19]

Policy design (PD)
The policies that constitute an IRR contain all of the substantial and institutional elements relative to the programming and implementation of all use and protection policies affecting the management of a resource. The different constitutive elements of this kind of policy design are:[20]

[18] *Eigentumsrechtliche Grundordnung* in German.
[19] Kissling-Näf and Varone (2000b).
[20] Knoepfel *et al.* (2001: 101 ff.).

The *definition of the different collective problem(s) to be resolved* based on the periods being analyzed, and the different objectives sought by the related state intervention. The policy design is often the product of a historical process involving the sometimes uncoordinated accumulation and sedimentation of the successive definitions of the collective problems to be resolved.

The *causal and intervention hypotheses* forming the causal models, which change as a function of the variations in time and space of the definition of the problems to be addressed. The causal model defines the actors (target group) considered responsible for the existence of the problem and the modes of intervention believed capable of producing the desired changes in the behaviour of the target group(s), thus enabling the resolution or attenuation of the problem and hence an improvement of the situation of those who suffer due to the existence of this problem (i.e. end beneficiaries).

The *target groups* and the *beneficiaries* of the various public policies constitutive of the policy design that form, together with the *intervening political-administrative actors*, the "basic triangle" of policy actors.

The *instruments* (regulatory, economic, persuasive, etc.) selected according to the different intervention hypotheses and applied in the implementation of the policies.

The *political-administrative arrangements* involved in the implementation of the concerned policies. These arrangements generally involve one or more municipal, cantonal and/or federal administrative services with a portfolio of various resources and are more or less coordinated (horizontally or vertically) through administrative procedures.

The actual *outputs* of public policies take the form of individual and concrete acts of application in the field of political-administrative legislative programmes.

Table 4 presents the different policy elements and provides a few examples using policies concerning the resource water.

Table 4: Regulation through public policies: elements of the policy design (PD)

***Policy design* (PD) (all of the policies governing the use and protection of a resource)**	**Examples: resource *water* (historical evolution 20th century)**
Policies constitutive of the policy design	Infrastructure policies (flood banks), agricultural policy, policy for the qualitative protection of water, spatial planning, energy policy, nature and landscape protection policy, environmental policy etc.
Definition of collective problems to be resolved and objectives of the state intervention	• Floods • Pollution of water bodies • Strong increase in the quantity of water consumed per capita, reduction of groundwater levels • Drying up of water bodies and destruction of aquatic ecosystems down stream of dams • Diffuse pollution of surface and underground water bodies (run-off and elutriation of agricultural soils)
Causal and intervention hypotheses	• Correction of water courses and draining of wetlands • Mechanical reoxygenation of lakes • Systematic treatment of waste water prior to disposal • Reduction of pollutants contained in industrial and household water • Introduction of the "polluter pays" principle • Fight against non-point source pollution due to intensive agriculture • Obligation to maintain a minimum flow in rivers • Renaturation of water courses
Target groups	Industries (in particular chemical industries), households, dam operators, farmers, owners of waste disposal sites and contaminated sites etc.
Instruments	• Mandatory connection to a treatment plant • Ban on phosphates in detergent products • Measures for the extensification of agriculture, restricted access to fertilizers • Minimum flow rates • Purification taxes • Separation of clean and waste water collection

***Policy design* (PD) (all of the policies governing the use and protection of a resource)**	**Examples: resource *water* (historical evolution 20th century)**
Political-administrative arrangements	• Regional and local implementation • Creation of specialized administrations
Outputs	• Concessions for water withdrawal • Bans on the spreading of manure

Cf. also Varone et al. 2002.

Public actors mobilise existing policies or newly created ones (change in the institutional regime) to modify the behaviour of the users of a resource on the basis of the two following types of intervention or regulation:

Type 1 modes of regulation: This involves the implementation of incentive-based instruments that do not have any impact on the content of the property and use rights of a resource's owners and/or users. This mode of regulation includes such instruments as information campaigns, the payment of subsidies in exchange for the desired behaviour (e.g. specific ecological services), tax relief (e.g. for cars fitted with a catalytic converter or households not introducing water into the general sewage system), etc.

Type 2 modes of regulation: This involves the implementation of instruments with perceptible impacts on the rights of disposal and/or use of actor-users by means of clarifications (often restrictive) of the content of these rights. The most widespread examples of intervention involving rights of disposal are the restrictions on the circulation of property titles in the form of bans on the sale or purchase of these titles by certain categories of buyers (for example, bans on the sale of agricultural land to non-farmers or on the sale of a plot of land to individuals intending to use it for the construction of a holiday property etc.), or restrictions on rights of rental or transfer of a property to people who are not suitably qualified to take care of it. Far more common, however, are the multiple and very varied restrictions on use rights such as, for example, restrictions on construction, on the emission of atmospheric pollutants or liquids, on harvesting (wood, rare plants and game), and on rights of access (to lakeshores, forests and fragile biotopes). These restrictions are aimed at users who hold formal property title or rights of disposal or of derived use (concessions, leases etc.)[21].

Figure 4 is a schematic representation of these two modes of regulation and intervention based on public policies (policy design).

[21] A recent analysis of the federal legislation of Switzerland showes the existence of 94 different limitations of land property stipulated by public policies ("öffentlich-rechtliche Eigentumsbeschränkungen") (Knoepfel, Wey 2006). According to a governmental project on geographical information (Geoinformationsgesetz), the most important of these use right restrictions must figure in a register which should complement the civil law based land register.

Figure 4: Policy-based interventions targeting the resource uses (PD)

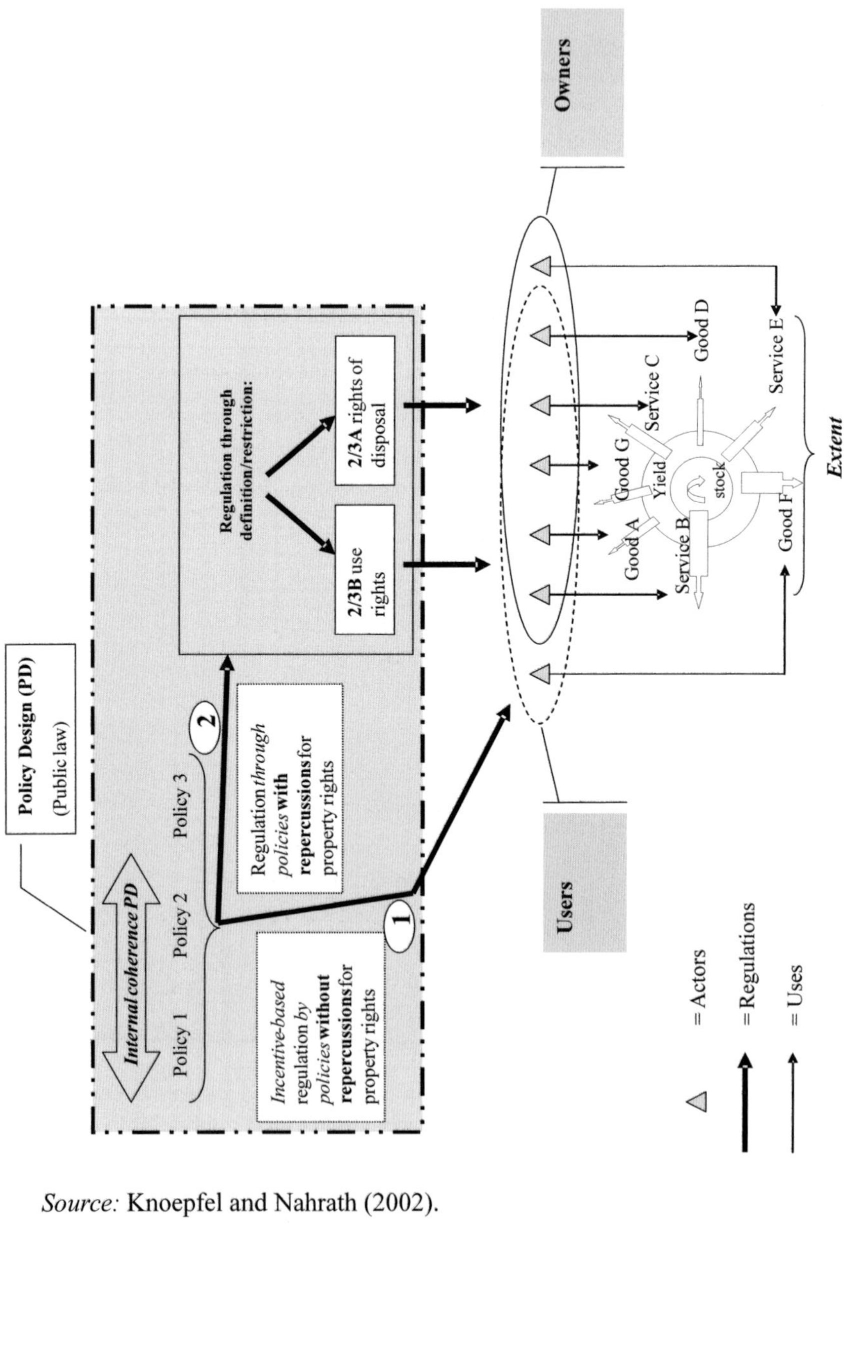

Source: Knoepfel and Nahrath (2002).

Property rights system (PRS)

A property rights system is composed of all of the formal property rights, as well as all of the rights of disposal and use arising from them, that apply to a resource. The content of these disposal and use rights depends on the definition of property used by the society in question (e.g. private, collective/communal) and applicable to this resource. An analysis of the PRS applies just as well to the entire resource system as it does to the individual units used to provide the different goods and services.

Table 5 shows the range of regulations based on property rights (components of the PRS) using the example of the resource ground/soil (ground law).

Table 5 Regulation through property rights: elements of the property rights system (PRS).

Property Rights System	**Examples based on the resource ground/soil**[22]
Formal property rights	Land ownership title
Rights of disposal	Right to: • sale • gift • rental (leasing) • mortgage • inheritance • etc. one's real estate.
Use rights	Right to: • construct on • deposit (waste) on • use (agriculture) • destroy • protect • etc. one's real estate.

[22] According to Nahrath [2003a].

When public or private actors consider these rights as ineffective, too costly in administrative terms, or simply no longer corresponding to the prevailing political values, they may try to resolve the problems associated with the use of goods and services through the modification of these rights. Compared to policy changes, such a strategy is sometimes considered more long-lasting, less fragile, and more likely to improve the predictability of these regulations, as well as the framework conditions that promote a good climate for investment. For this reasons such modifications will find their concretization on the individual level in annotations within the land register, contrary to restrictions stemming from public policies. This kind of more radical and sometimes even revolutionary change modifies the IRR through the restructuring of the PRS on the basis of the following two possible modes of regulation and/or intervention:

Type 3 modes of regulation: This involves different types of possible modifications of the definition of the institution of formal property that have an impact on the scope and content of the disposal and use rights of all holders of such rights. The most important example occurred with the introduction of the Swiss Civil Code in 1907, which created a unified definition of property rights and possible contractually agreeable restrictions at the federal level and abolished in one fell swoop the old use and disposal rights, particularly in the area of ground law. More recent examples include the introduction into the Civil Code of the law of condominium ownership (1965) and the introduction of the new property regime (1969) through the *Bodenrechtsartikel* (*constitutional guarantee of real estate property*), which practically established the principle of the right of compensation in the case of material expropriation in the Swiss Federal Constitution.[23]

Type 4 modes of regulation: This involves different possible forms of intervention of a general nature or, conversely, affecting a particular area whose objective is to redefine the structure of distribution of property titles and/or their restrictions by contractual agreements (e.g. easements, such as path or construction servitudes, etc.). These modes may consist of both an intervention as radical as the privatization or nationalization of land for all kinds of reasons (increased economic efficiency, efforts to counteract speculation or concentration and de-individualization of property ownership, security of supply, etc.) and a more punctual and limited intervention consisting of formal expropriation (e.g. for the implementation of infrastructure projects), targeted public property acquisitions (due to an active public property strategy at municipal level) or contracting specific property

[23] On this point, see Nahrath (2003a, 2005).

limitations regulated by civil law (civil code). Such regulations are normally considered leading to lasting results which in practice therefore figure within land registers.

Figure 5 contains a schematic representation of interventions made through the PRS and affecting the modes of resource exploitation.

Figure 5: Property rights-based interventions regulating the resource uses (PRS)

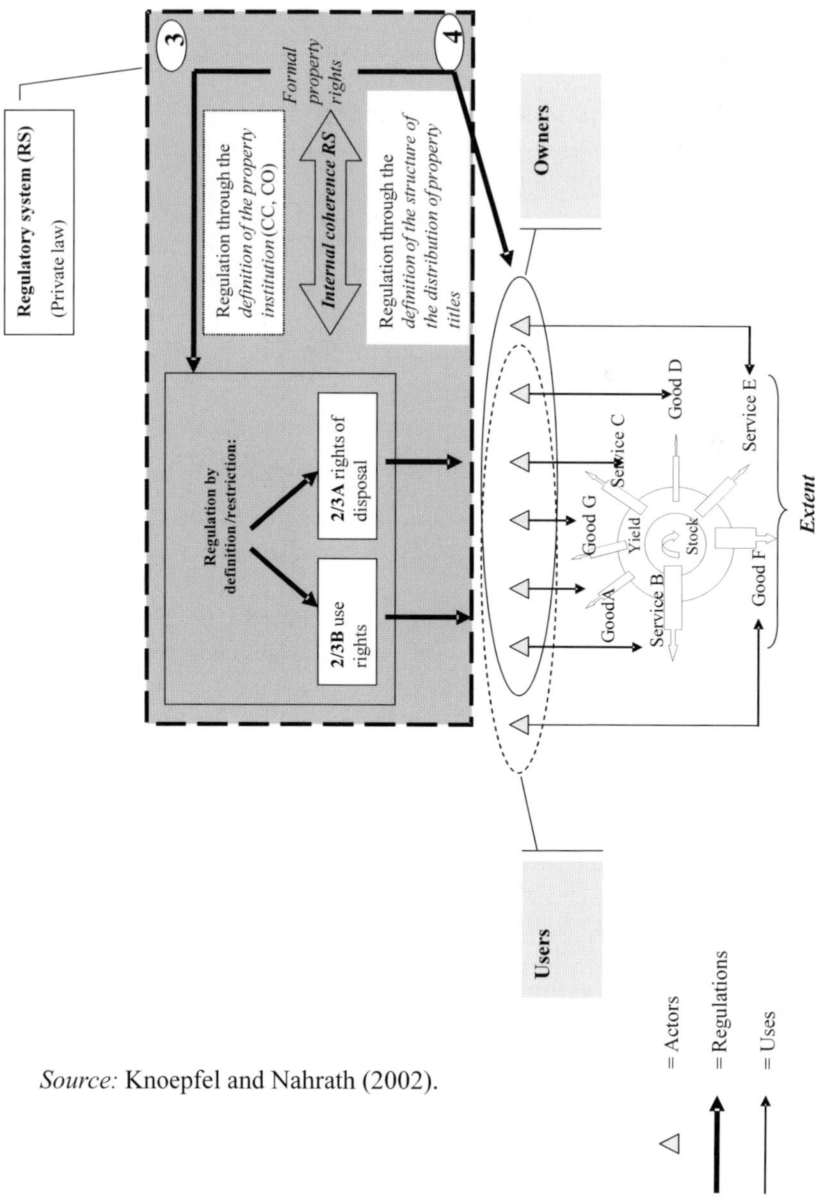

Source: Knoepfel and Nahrath (2002).

As clearly demonstrated in Figure 6, which combines Figures 4 and 5, the scope and content of the modes of regulation affecting rights of disposal (2/3A) and rights of use (2/3B) depend on the link between the two components of the IRR, i.e. the PD and the PRS. These two modes (2/3A and 2/3B), which are by far the most common in practice, constitute the core of the IRR of the principle natural resources.

Figure 6: IRR combining the modes of regulation by PD and PRS

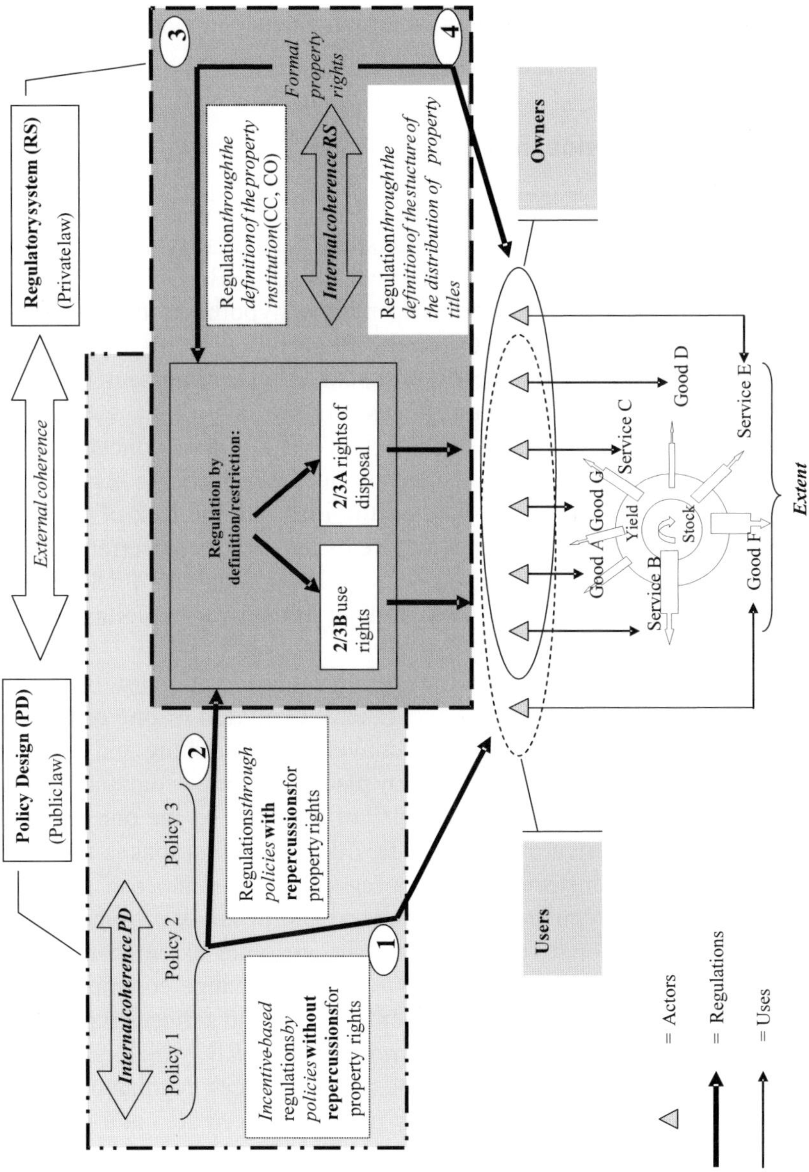

Source: Knoepfel and Nahrath (2002).

The main forms of resource regulation (in terms of frequency, scale and impact) consist of the restriction of disposal and use rights (2/3A and 2/3B). These interventions are located at the junction between policies and property rights, which clearly demonstrates the fundamental interdependence of these two forms of political regulation. Only an approach based on "institutional regimes" can take this interdependence into account.

15.4.2 The coherence and extent of IRR[24]

Institutional regimes for natural resources may be defined and categorized on the basis of their specific characteristics, particularly with the help of the dimensions "extent" and "coherence" of an IRR. On this basis, the concept of the IRR enables one to formulate hypotheses concerning the existence of causal relations between the characteristics of a regime and its contribution to the sustainable or unsustainable development of the resource, to whose regulation it contributes.

The dimension referred to as the "extent" of a regime concerns simply whether or not the different goods and services of a resource actually used are regulated. The analysis is based on the idea that the lack of regulation of the behaviour of users, through a more or less precise description of use rights via public policies and/or property rights defined in a PRS, risks engendering strategic behaviours that can lead to the over-exploitation of the resource during times of scarcity.

Depending on the objectives pursued through the adoption of the IRR concept – whether for describing the general evolution of one or more resource regimes in time and space or, conversely, describing and analyzing a particular regime in action at a given place and time – the analysis will focus on the so-called *"absolute extent"* of the IRR, on the one hand, and on its so-called "relative extent", on the other. The first makes it possible to take into account historical changes regarding the number of goods and services effectively regulated by the federal and possibly cantonal components of an institutional regime. Here, the absolute extent constitutes a good indicator for identifying periods of regime change corresponding to an increase or decreasing of the number of goods and services regulated by a regime. Conversely, the *"relative extent"* of the IRR, which represents a quotient relating the number of goods and services regulated with the number of goods and services actually used in a given area at a given mo-

[24] In accordance with Knoepfel (2003). Cf. also Nahrath (2003a); Bressers and Kuks (2004).

ment in time, tends to be used in the analysis of active empirical regimes. If this quotient is less than 1, the existence of unregulated rivalries may be expected and, therefore, the resource is at risk of over-exploitation. In the opposite case, a situation of over-regulation prevails, which may also pose problems, above all on the level of the economic modes of exploitation of the resource (lack of allocative efficiency). In the empirical world, regimes can often be observed that are characterized by an excessively weak absolute extent. This is due to the fact that in the majority of cases the regulation of use behaviours only emerges as a reactive measure and very often when it is too late, i.e. after a particular use has developed to a point at which it represents a serious threat to the reproductive capacity of the resource (e.g. CO^2 emissions and immissions and climate change).

The criterion of "coherence" is based on the content and connection of the regulations established by the regime. It presupposes, firstly, that the definition of individual use rights, created by the public policies and/or property rights, does not exceed the global quota of the goods and services considered extractable without putting the reproductive capacity of the stock at risk.[25] Incoherencies between these regulations will be more likely to emerge as their number increases (i.e. elevated absolute extent). Such incoherencies may be due to regulations originating in the PRS and/or PD or the connection between the two. Thus, we make a distinction between three types of coherences/incoherencies:

The *internal coherence of the PRS* concerns the degree of clarity of the definition of the property titles or the use rights arising from them. While property rights are generally clearly defined for resources such as water or soil[26], they are far less clearly defined in the case of resources considered "ownerless property" from a legal point of view and in the case of resources that do not belong to the category of "material objects" according to the Civil Code, such as landscape, biodiversity or air. For example, incoherencies in the PRS may originate from the fact that there are more property titles or use rights for a single resource or a single good or service than resource units available or extractable in accordance with the definition of the global maximum quota. Such situations regularly exist in the case of the resource air (e.g. unlimited use rights assigned with vehicle licensing certificates) or the resource water (e.g. the unregulated free pumping of water from the water table for private bore holes).

---

[25] In the sense of layer 2 in Figure 2.

[26] Nevertheless, we can find contradictions between property rights and easements introduced in the land register creating incertainties about the existence of use rights.

The *internal coherence of the PD* concerns the coordination between policies governing the use and protection of natural resources. In the 1990s,[27] this was frequently very weak and sometimes even non-existent (e.g. the contradiction between the renewable energies policy and the liberalization of the electricity market, or between the agriculture policy promoting intensive production and the protection of soil and water quality). Conversely, contradictions within the same policy – for example, between the problem definition, the causal hypotheses adopted, the choice of target groups, the definition of intervention instruments, the capacity for action of the political-administrative arrangement etc. – are more rare. Incoherent policy designs normally produce regulations that are incompatible with each other in relation to the different goods and services.

The external coherence of a IRR concerns the mode of connection between the PRS and the PD. It is expressed particularly through the correspondence between the target groups of the policy design and the holders of rights in accordance with the property rights system. This correspondence is lacking when policies address target groups that do not have use rights and whose eventual changes in behaviour do not have any real effect on the actual uses of the resource. Other external incoherencies consist in the relatively common case whereby policies simply do not have sufficient coercive power to actually restrict the use rights of the users of a resource. An example of this is the incapacity to impose the minimum flow rates stipulated in the legislation on the protection of water (institutional regime for water) on the holders of dam concessions. Another example is the capacity of land owners to resist the implementation of zoning in the context of landuse planning (institutional regime for land and soil). Empirical research shows that a good indicator of the external incoherence of a regime is the "judicialization" of the implementation of the policy design (Rothmayr 2000, Nahrath 2005), and in the central role played by the jurisprudence of the courts in connecting the two components of the regime (e.g. material expropriation).

Thus, the identification of gaps or incoherencies in a regime often helps to explain the empirically observed phenomena associated with overexploitation.

[27] See section 15.1.

15.4.3 Typology of regimes and their repercussions for sustainable development

As stated above, the two key dimensions of extent and coherence enable an initial relatively simple typology of institutional regimes to be defined and hypotheses concerning the existence of possible links between their characteristics, as well as their supposedly variable contribution to the sustainable management of a resource to be established.

Figure 7 identifies the four main types of regime, which we describe as follows:

Figure 7: Four types of IRR

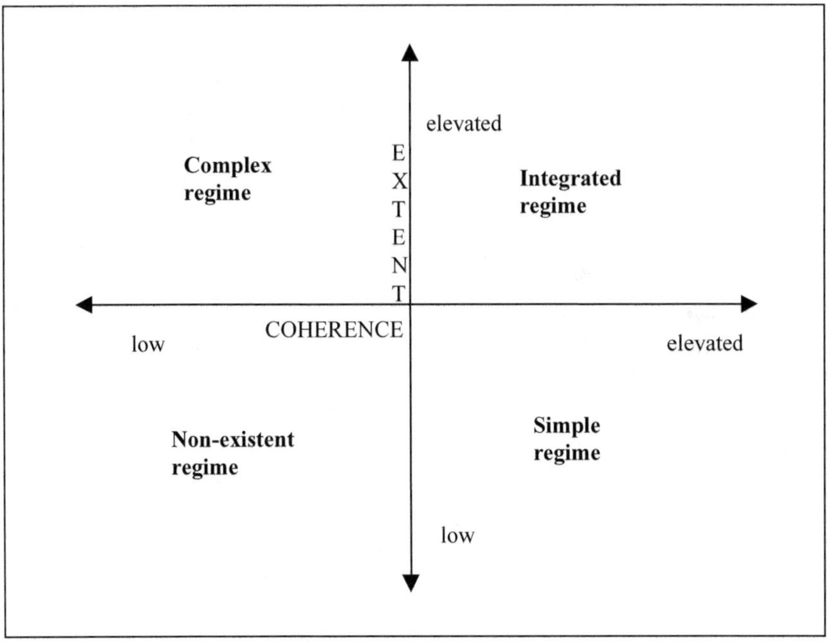

Source: Knoepfel *et al.* (2001:38)

Non-existent regime: Situation whereby the resource does not have any kind of property right associated with it or any kind of regulation of any of its goods and services by any kind of policy. Such a situation prevails, for example, when the need to regulate a resource has not been politically acknowledged despite the fact that the resource is subject to a range of exploitation. The empirical identification of situations involving non-existent regimes is only possible through the screening of institutional regimes for very long periods. Non-existent regimes may be encountered, for example, in the area of the resource air or landscape during the periods preceding the introduction of the legislation to counteract atmospheric pollution or prior to the constitutional recognition in 1962 of the need to protect the landscape.[28]

Simple regime: Situation whereby a limited number of goods and services (lower than the number of goods and services actually used) are regulated in a coherent way; the coherence of the regime results specifically from the low number of regulations in force and, hence, the low risk of contradiction between them. Such a situation can arise, for example, following an initial effort to regulate a resource by attempting to coordinate the uses of the resource that have led to rivalries between users. This initial form of regulation may consist, for example, in the adoption of initial policies creating *de facto* use rights (initial policies for the prevention of air pollution) or, conversely, involve the creation of an initial body of property rights (creation of the Swiss federal civil code in the early 20th century) independent of the existence of any policies. Such a regime may equally be the result of the "disintegration" of a previously complex or integrated regime (in the case of deregulation, e. g. in the field of spatial planning policies). Moreover, in many cases, the *raison d'être* of such regimes is not the protection of the resource, but instead, as shown in section 15.2, that of guaranteeing access to the resource in the long term with a view to its economic exploitation or to the amortization of the operating installations required within a concession regime. Therefore a simple regime normally cannot guaranty sustainability.

Complex regime: Situation whereby the majority of the goods and services actually used is regulated, but in a way that is incoherent in part. This situation corresponds to most of the late 20th century regimes in Switzerland due to the extensive development of sectoral use and protection policies from the 1950s which are largely uncoordinated. Such a regime may also be the result of the disintegration of a previously integrated regime. Unlike simple regimes, according to the empirical data available, complex regimes are essentially the outcome of a political mobilization that aims to deal with problems surrounding resource rivalry and reproduction, the resolution of which, it is assumed, lies in the introduction of more regulations governing the goods and services of the resource in question. These regimes all involve more or less advanced attempts to formulate quotas for specific goods and/or services of a given resource, at least at the level of the goods and/or services regulated. From the perspective of resource economics, these regimes are characterized, however, by flawed mechanisms

[28] Swiss Federal Constitution of 18 April 1999 (RS 101) (preamble and Art. 73).

for the coordination of global quotas with the individual quotas (by use sector).

Integrated regime: Situation whereby all goods and services produced by a resource and actually used are regulated in a coherent way. According to our research findings, such regimes remain very rare in the early 21st century. Examples in Switzerland may be found in the areas of landscape (Rodewald *et al.* 2005: 347 ff.; Gerber 2005: 374 ff.) and water (in the course of being integrated (Reynard and Mauch 2003). Such regimes are found, in particular, where resources are largely in public ownership (e.g. forests) or under the control of a powerful collective actor (e.g. self governing CPR institutions (Ostrom 1990) like a *Bourgeoisie* or an *Allmende association*, or nature conservation organization such as Pro Natura).

The main hypothesis of the INRR concept is based on this typology. It presupposes the existence of a causal relationship between, on the one hand, the regime type (i.e. its extent and coherence) and its regulatory capacity, and, on the other, the sustainability of the uses arising from the regime's resources as well as status of the resulting reproductive capacity of the resource system. More concretely, *the closer a resource situation moves towards an integrated regime, the greater the likelihood of creating sustainable use conditions for the resource*. Conversely, *the less developed the regime is, i.e. the less elevated its coherence and extent (in particular relative), the greater the risks of over-exploitation of the resource*. The validity of this research hypothesis has been largely confirmed by the empirical research carried out in this area up to now.[29]

15.5 Empirical applications

The applications of the IRR concept may be both *scientific* (analytical) and *normative* (prescriptive) in nature. Thus we will present two applications of the concept in the remaining section of this chapter. The first is a guide for carrying out an analysis of the existing empirical regulation of the uses of a resource from the perspective of institutional regimes. The second, more normative one, proposes a usable framework for the development of (new) institutional regimes based on the more sustainable management of natural resources.

[29] For a synthesis of these studies, see Varone et al. (2003).

15.5.1 Applications of the IRR concept to empirical analysis

The model shown in Figure 8 constitutes the chronological process of the implementation of strategic choices in the context of the emergence of a public intervention measure leading to the adoption of an institutional regime which regulates the uses of a given resource.

Figure 8: Guide for the analysis of existing regulations from the perspective of institutional regimes

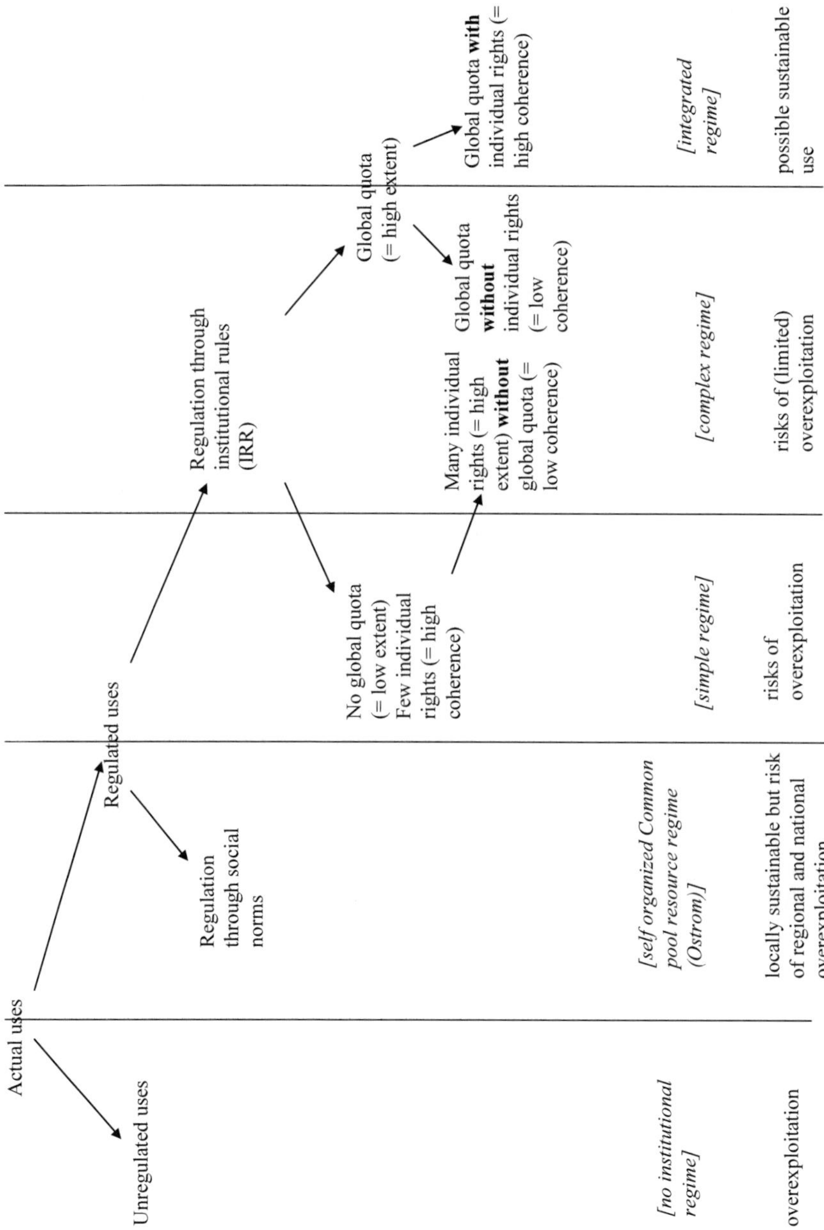

The process proposed here is based on an empirical analysis of the actual uses of the resource, with a distinction being made between the unregulated ("unchecked") and regulated use situations. The researcher then identifies the uses regulated by "social institutions" (customary, associative, self-regulation, etc.) and the uses regulated by "institutional rules" prescribed by a public authority. In the latter case, attention is then focused on the distinction between the regulations that refer to a resource-based framework normally including explicit or implicit global yield quota and those that *do not* refer to such quota.

The latter consists of the regulations relating to property, disposal and use rights stipulated in, for example, the Civil Code, which guarantee to their holders rights of use to the goods and/or services of the resources, which are defined in absolute and inflexible terms as though the resource in question were any kind of material object. It thus corresponds to situations in which individual use rights are defined independently of the definition of a global quota. Such conceptions are based on the (questionable) idea that individual quotas may be regulated in a way that ensures their combined use remains, even in extreme cases, within the limits of sustainability for the use in question. In this case, the risk of resource overexploitation is twofold: (1) because of the low extent of the (simple) regime (few individual rights, but well coordinated whereas many other actual uses remain unregulated) or (2) because of low coherence of the extended (complex) regime. The system of (daily) fishing licenses, and, more worryingly, the system of the unlimited allocation of vehicle licensing certificates, which is independent of any definition of a global quota for registrations that would make it possible to protect the biomass's or the air's self-reproductive capacity, constitute two examples of such situations.

Thus, the researcher distinguishes between these rights, which have no "resource-based connotation", and use regulations that explicitly refer to a resource framework and define the use rights as a function of the reproductive capacity of the resource in question.

Amongst the regulations with a resource-based reference, the research may distinguish, in empirical reality, two different modes of formulation of such regulations. The first mode consists in the definition of quotas for global extractions that are not extended systematically and restrictively to individual use rights. The protection of the resource air illustrates well such a situation in that the global quotas defined by law in the form of ambient air quality standards (Swiss Federal Law on the Protection of the Environment of 7 October 1983) are not systematically translated into legally binding restrictions in the form of individual car emission reductions when the quality standards are exceeded.[30] In such cases, the immission limit value will act as an indicative (target-)value rather than restrictive legal thresholds. The regulatory modes governing such situations tend to be incoherent in nature.

The second mode of regulation, which is also the most able to provide an institutional guarantee of the sustainable management of natural resources, consists both of global quotas and individual use rights and is equipped with a mechanism for creating coherence between the two types of quotas, which is indispensable to the emergence of an integrated regime. It is, however, also possible to find incoherent modes of regulation here when, as seen in the example given in section 15.4, individual quotas are defined, the sum of which exceeds the global quota. Again, such situations may exist in the context of the resource air when the regulations define immission and emission limit values independently of the number of pollutant sources. The coordination of a global quota with all of the individual use rights (which corresponds to a situation of integrated institutional regimes) may only be established if the number of regulated uses corresponds to the number of actual uses in the area in question (e.g. land-use planning fixing global quotas of building zones and dispatching them amongst local communities or forest legislations aiming at maintaining forest surfaces by obligation of compensational reforestations).

It should be noted that the empirical validity of the INRR conceptual framework can be tested via the process described in figure 8. Its application highlights the existent or non-existent relationships between the characteristics of the regime, the reproductive capacity of the resource, and the regime's effects on the actual uses of the resource. Up to now, this test, which has been implemented by our team over the past six years with the help of numerous local case studies, has revealed an explanatory link that fulfils the key dimensions of the institutional resource regimes concept, both in diachronic studies (i.e. historical screening) and synchronic studies (i.e. detailed examinations of the phases of regime change and their concrete impacts on the management of resources).

[30] Despite a provision contained in the Swiss Ordinance on Air Pollution Control of 16 December 1985 which stipulates a systematic adaptation of this kind by the cantons.

15.5.2 The normative use of the concept of institutional regimes as a platform for the development of (new) institutional regulations aimed at the more sustainable management of natural resources

In this final section, we describe in broad and prescriptive terms the four different decisional stages necessary for the creation of an integrated institutional regime for a resource. Each of these stages is based on important political choices, of which the actors involved must be fully aware. This requires awareness-raising and participation processes in the form of, for example, sustainability processes within Agenda 21 projects. Table 6 presents these different stages in a general form.

Table 6: The four stages of the decision-making process in the creation of an integrated institutional regime for a given resource

Stage 1: actual IRR	1) Political construction of the resource Political definition of the problem (scarcity, identification of rivalries between the different uses etc.). Identification of the boundary relevant to the management of the resource. Formulation of a causal hypothesis: identification of users. Inventorization of existing use rights.
Stage 2: Sustainability of the actual regime **Conditions towards integration**	2) Political definition of the (annual) quantity of resource units available Scientific-political consensus on the volume currently available for extraction. Political decision concerning measures enabling the artificial increase/decrease of this quantity (extension of boundary, etc.). Definition of global quota of resource units authorized for withdrawal/extraction.
	3) Evaluation of the legal scope of the definition of the global quota Coercive quota Indicative quota Reference framework in the event of conflicts (to be applied by the courts, for example).

	4) Decision concerning the modes of "translation" of the global quota into individual quotas Through partial global quotas (for example, types of activities or types of territory etc.). Directly from the global quota to individual use rights.
Stage 3: re-design of IRR	5) Decisions concerning the modes of modification of use rights Modification of civil law or ownership rights. Changes in public law (limitation of ownership etc.). Introduction of flexibility clauses (variable use rights according to the definition (variable in time) of the global quota). Weighting of the desirable relationship between social, economic and ecological sustainability (equal weighting, in accordance with the Federal Swiss Constitution). Political choices necessary to modify use rights in accordance with the postulates that concretize these three dimensions of sustainability by area.
	6) Decision concerning the definition of new use rights (in accordance with the principles decided under 5) Definition of eligible users per good and service. Degree of exclusivity of rights. Specific flexibility clauses. Exchangeability, transferability. Spatial or temporal limitation (boundaries). Etc.
Stage 4: monitoring of IRR	7) Institutionalization of monitoring Monitoring of the reproductive capacity of the resource in question over time (so as to evaluate the quality of the global quota definition and possibly adjust it). Monitoring of the actual evolution of the global quota (so as to evaluate or possibly adjust the definition of individual quotas). Monitoring of the actual behaviour of users in light of individual quotas (so as to evaluate the capacity of the individual quotas to actually steer the behaviour of users).

These four stages in the decision-making process are explained in detail in the preceding chapters, thus we will not explain the table in detail here. Every reader, practitioner or academic can easily find examples corresponding to one or other of these stages in their professional experience. Of course, only the future will show us the scope and acuity of the real problems involved in the application of such integrated regulations, problems that reside in the regulatory system founded on the guarantee of private and/or exclusive property which is strongly rooted in political and legal institutions and in Swiss political mores (Aubin, Nahrath and Varone 2004). Indeed, the highlighting of this dimension is one of the main contributions of this analytical framework. It should nevertheless be noted that this concept, which is simple and obvious to all of those concerned with sustainable development, has already met with a certain level of response, albeit still very abstract, for example in the strategy of the Swiss Office for the Environment, Forests and Landscape.[31] Similarly, the European "Euwareness" project has shown that the European Water Framework Directive of 23 October 2000 (2000/60/EU) heads in the same direction as the ideas discussed here on the subject of the necessary integration of institutional regimes (Aubin and Varone 2004). Finally, the processes for the regional planning of forests currently under way in Switzerland also show astonishing similarities with our concept of institutional regimes. As showed elsewhere (Knoepfel 2005b), there is moreover a clear interest in its analogous application to the area of local Agenda 21 sustainable development processes currently under way (winter 2006) in over 140 Swiss municipalities (DuPasquier *et al.* 2003).

15.6 Conclusions

This last application clearly shows that the analytical framework of institutional regimes for natural resources is not solely a conceptual analytical tool with dimensions likely to influence the sustainability of a resource or the social, ecological and economic sustainability relative to its use. The concept is also a potentially relevant political-administrative management tool which will make it possible to improve the efficacy of the regulations and behaviours of the users of natural resources from the perspective of increasing sustainability. When applied to natural resources, it makes it possible to implement some key elements of the constitutional principle of sustainable development in the form of relatively precise public actions. Of

---

[31] See OFEFP (2005).

course, they will need to be described in greater detail in the future in the context of action research within pilot projects. The concept lends itself to areas as wide-ranging as future natural regional parks (Gerber 2004; Gerber 2005; Oppizzi 2003), the battle against atmospheric pollution in urban regions (Mariéthoz and Savary 2004), the regional planning processes for forests in accordance with Article 18 of the Ordinance on Forests of 30 November 1992 (RS 921.01), climate policy (already very advanced implementation of CO_2 regimes at European level in accordance with Directive 2003/87/EU of 13 October 2003), the truly effective battle against urban sprawl and the waste of land (ARE 2005), and the area of water management, in particular in countries prone to increasingly extended periods of drought[32].

Although the action area retained for the development of the concept of institutional resource regimes is primarily that of *common pool resources*, which include natural resources, the concept should also prove its worth in the management of all kinds of non-natural (i.e. artificial and intangible) resources such as cultural, social, human, and institutional resources, which are becoming increasingly important in a society engaged in a process of dematerialization. If, as is the case with natural resources, we exploit these common pool resources (or artificial and intangible resources) in an "unchecked" manner, allowing actor-users uncontrolled appropriation of the goods and services produced by them, we risk finding ourselves faced with situations of over-exploitation as a result of the attribution of use rights to actors who behave primarily as predators rather than as reasonable managers of these resources. Such a situation of over-exploitation of common pool resources risks ultimately leading to violent social struggles, the waste of resources and their associated goods and services and, finally, to a process of destruction of these resources which would probably exceed even the pessimistic prognosis formulated in 1968 by Hardin in his famous article "The tragedy of the commons" (Hardin 1968).

References

ARE 2005., *Rapport 2005 sur le développement territorial*. Berne

Aubin, D., Nahrath, S., Varone, F. 2004. *Paysage et propriété: patrimonialisation, communautarisation ou pluri-domanialisation*, Cahier de l'IDHEAP 219. Chavannes-près-Renens: IDHEAP

[32] Cf. on this point the example of Spain: Costejà *et al.* (2004a); Costejà *et al.* (2004b).

Aubin, D., Varone, F. 2004. The evolution of European water policy, towards integrated resource management at the EU-level, in: Kissling-Näf, I., Kuks, S. (eds.), *The Evolution of National Water Regimes in Europe. Transitions in Water Rights and Water Policies towards Sustainability*. Dordrecht, Boston, London: Kluwer Academic Publishers: 49–86

Barnes, P. 2001. *Who Owns the Sky? Our Common Assets and the Future of Capitalism*. Washington D.C.: Island Press

Behan, R. W. 2001. *Plundered Promise: Capitalism, Politics and the Fate of the Federal Lands*. Washington D.C.: Island Press

Benninghoff, M. et al. 2004. Regional planning, infrastructural policy (transport, energy) and the environment, in: Klöti, U. et al. (dirs.), *Handbook of Swiss Politics*. Zurich: NZZ-Verlag

Berthelot, J.-M. 1990. *L'intelligence du social. Le pluralisme explicatif en sociologie*. Paris: Presses universitaires de France

Bisang, K., Schenkel, W. 2003. Effekte des regionalen Regimewandels der Ressource Wald, in: Knoepfel, P. et al. (dirs.), *Institutionelle Ressourcenregime in Aktion – Régimes institutionnels de ressources naturelles en action*. Bâle: Helbing & Lichtenhahn (série Ecologie & Société, vol. 19): 141–204

Bollier, D. 2002. *Silent Theft. The Private Plunder of our Common Wealth*. New York, London: Routledge

Bressers, H., Kuks, S. (eds.) 2004. *Integrated Governance and Water Basin Management. Conditions for Regime Change and Sustainability*. Dordrecht, Boston, London: Kluwer Academic Publishers

Bromley, D. W. 1991. *Environment and Economy. Property Rights and Public Policy*. Oxford UK, Cambridge USA: Blackwell

Bromley, D. W. 1992. The commons, common property, and environmental policy, *Environmental and Resource Economics*, 2: 1–17

Coase, R. H. 1960. The problem of social cost, *Journal of law and economics*, 3: 1–44

Conseil fédéral 2002. *Stratégies 2002 pour le développement durable*, rapport du Conseil fédéral suisse du 27 mars 2002. Berne

Costejà, M. et al. 2004a. The evolution of the water regime in Spain, in: Kissling-Näf, I., Kuks, S. (eds.), *The Evolution of National Water Regimes in Europe. Transitions in water rights and water policies*. Dordrecht, Boston, London: Kluwer Academic Publishers: 235–264

Costejà, M. et al. 2004b. Redistributing water uses and living with scarcity, in: Bressers, H., Kuks, S. (eds.), *Integrated Governance and Water Basin Management. Conditions for Regime Change and Sustainability*. Dordrecht, Boston, London: Kluwer Academic Publishers: 165–188

Devlin, R. A., Grafton, Q. R. 1998. *Economic Rights and Environmental Wrongs. Property Rights for the Common Good.*. Cheltenham (UK) Northampton MA (USA): Edward Elgar

Di Giulio, A. 2004. *Die Idee der Nachhaltigkeit im Verständnis der Vereinten Nationen – Anspruch, Bedeutung und Schwierigkeiten.* Münster: Lit. Verlag

DuPasquier, A. et al. 2003. *Etat des lieux des démarches de développement durable dans les communes / Bestandesaufnahme der Prozesse zur nachhaltigen Entwicklung auf Gemeindeebene.* Berne: Office fédéral du développement du territoire (ARE), Département fédéral de l'environnement, des transports, de l'énergie et de la communication (DETEC), Chavannes-près-Renens: IDHEAP, www.are.ch/www.idheap.ch

Endres, A., Querner, I. 1993. *Die Ökonomie natürlicher Ressourcen – eine Einführung.* Darmstadt: Wissenschaftliche Buchgesellschaft

Erkman, S. 1998. *L'écologie industrielle. Comment mettre en pratique le développement durable dans une société hyper-industrielle.* Paris: Editions Charles Léopold Mayer

Gerber, J.-D. 2004. *La nouvelle gouvernance comme moyen d'arbitrage entre les intérêts de développement et de protection du paysage. Le cas du parc naturel régional de Chartreuse*, Cahier de l'IDHEAP 215. Chavannes-près-Renens: IDHEAP

Gerber, J.-D. 2006. *Structures de gestion des rivalités d'usage du paysage. Une analyse comparée de trois cas alpins.* Zürich/Chur: Editions Ruegger

Hagmann, T. 2004. *Beyond clannishness and colonialism: understanding political disorder in Ethiopia's Somali region, 1999–2004.* Bern: Swiss Peace

Hardin, G. 1968. The tragedy of the commons, *Science,* 162: 1243–1248

Holzinger, K. 2002. *Transnational Common Goods: Strategic Constellations, Collective Action Problems, and Multi-level Provisions*, unveröffentlichte Habilitationsschrift an der Universität Bamberg, April 2002

Kirchgässner, G. 2002. Kontingentlösungen als Instrumente der Wirtschafts- und insbesondere der Umweltpolitik: Die ökonomische Perspektive, in: Knoepfel, P. (Hrsg.), Begrenzen um mehr zu erreichen, wissenschaftliche Kommentare. Berne: EDMZ: 25–41

Kissling-Näf, I., Varone, F. (Hrsg.) 2000a. *Institutionen für eine nachhaltige Ressourcennutzung. Innovative Steuerungssätze am Beispiel der Ressourcen Luft und Boden.* Zurich: Rüegger Verlag

Kissling-Näf, I., Varone, F. 2000b. *Historical analyses of institutional resource regimes in Switzerland. A Comparison of the Case of Forest, Water, Soil, Air and Landscape*, paper presented at the 8th Biennial Conference of the International Association for the Study of Common

Property (IASCP. Bloomington, Indiana, May 31 – June 4, 2000. Working paper de l'IDHEAP 12/2000. Chavannes-près-Renens: IDHEAP

Kissling-Näf, I., Kuks, S. (eds.) 2004. *The Evolution of National Water Regimes in Europe. Transitions in water rights and water policies.* Dordrecht, Boston, London: Kluwer Academic Publishers

Knoepfel, P. 1995. New Institutional Arrangements for a New Generation of Environmental Policy Instruments: Intra- and Interpolicy Cooperation, in: Dente, B. (éd.), *Environmental Policy in Search of New Instruments*, European Science Foundation. Dordrecht, Boston, London: Kluwer Academic Publishers: 197–233 (= chapter 6 of this book)

Knoepfel, Peter, Varone, Frédéric 2000. Boden- und Luftreinhaltepolitiken – Rationalitätenwechsel in historischer Perspektive, in: Kissling-Näf, I., Varone, F., unter Mitarbeit von Giger, M., Kläy, A., Mauch, C. (éds), *Institutionen für eine nachhaltige Ressourcennutzung. Innovative Steuerungssätze am Beispiel der Ressourcen Luft und Boden.* Zurich: Rüegger: 153-181

Knoepfel, P. 2000a. L'influence de l'écologie s'affirme dans les institutions, in: Programme prioritaire, Demain la Suisse (éd.), *Rapport social 2000*, sous la direction de Suter, Christian, Pahud, Claude, collection Analyses sociales. Zurich: Seismo: 268-296

Knoepfel P. 2000b. Von der Politik zur Politik der institutionellen , in: Kissling-Näf, I., Varone, F., unter Mitarbeit von Giger, M., Kläy, A., Mauch, C. (éds), *Institutionen für eine nachhaltige Ressourcennutzung. Innovative Steuerungssätze am Beispiel der Ressourcen Luft und Boden.* Zurich: Rüegger: 195-210

Knoepfel P. 2000c. *Natural Resource Quotas and Contracts - A New Institutional Regime for our Common Resources*, Cahiers de l'IDHEAP no 178, Chavannes-près-Renens: IDHEAP (= chapter 14 of this book)

Knoepfel, P., Kissling-Näf, I., Varone, F. (éds) 2001a. *Institutionelle Regime für natürliche Ressourcen: Boden, Wasser und Wald im Vergleich – Régimes institutionnels de ressources naturelles: analyse comparée du sol, de l'eau et de la forêt.* Bâle: Helbing & Lichtenhahn (série Ecologie & Société, vol. 17)

Knoepfel, P., Larrue, C., Varone, F. 2001b. *Analyse et pilotage des politiques publiques.* Bâle: Helbing & Lichtenhahn (série Analyse des politiques publiques / Politikanalyse no 2)

Knoepfel P., Nahrath S. 2002. Les régimes institutionnels pour la durabilité des ressources urbaines, *Vues sur la ville,* n°2 avril-mai 2002, Lausanne, Observatoire universitaire de la ville et du développement durable: 3-5

Knoepfel P., Savary J. 2002. Politiques de l'environnement classiques en Suisse, Matériel de cours de l'IDHEAP 5. Chavannes-près-Renens: IDHEAP

Knoepfel, P., Kissling-Näf, I., Varone, F. (éds) 2003. *Institutionelle Ressourcenregime in Aktion – Régimes institutionnels de ressources naturelles en action*. Bâle: Helbing & Lichtenhahn (série Ecologie & Société, vol. 19)

Knoepfel, P., Münster, M. 2004. *Guide des outils d'évaluation de projets selon le développement durable.* Berne: DETEC

Knoepfel P., Kohler, N. 2005. *Institutional regimes for sustainable building stocks (collective housing)*, requête FNRS du 28 février 2005 (acceptée par le FNRS en mai 2005)

Knoepfel, Peter, Nahrath, Stéphane, 2005. *The sustainable management of natural resources: from traditional environmental protection policies towards institutional natural resource regimes (INRR)*, Cahier de l'IDHEAP 226/2005. Chavannes-près-Renens: IDHEAP

Knoepfel, P. 2005a. Quand les indicateurs deviennent la norme – trois variations sur un thème avec finale sin brio, in: Bovay, Benoît, Ngyuen, Minh Son (éds), *Mélanges Pierre Moor. Théorie du droit – Droit administratif – Organisation du territoire*. Berne: Stämpfli: 115-128

Knoepfel, P. 2005b. Régimes institutionnels des ressources naturelles et démarches de développement durable au niveau local (Agenda 21), in: Da Cunha, Antonio, Knoepfel, Peter, Leresche, Jean-Philippe, Nahrath, Stéphane (éds) 2005. *Enjeux du développement urbain durable. Transformations urbaines, gestion des ressources et gouvernance.* Lausanne: Presses polytechniques et universitaires romandes: 329-354

Knoepfel Peter, Wey, Benjamin 2006. *Öffentlich-rechtliche Eigentumsbeschränkungen (ÖREB): Bestand nach Bundesgesetzgebung und ausgewählten Detailuntersuchungen,* Working paper de l'IDHEAP 7/2006. Chavannes-près-Renens: IDHEAP

Larrue, C., Knoepfel,P. 1998. *Environmental Protection, Subsidiarity Principle and Spatial Related Policies*, rapport final à l'intention de la DG XII de l'Union Européenne (Cahier no ENV4–CT96–0238), Observatoire de l'économie et des institutions locales IUP (OEIL), Créteil

Mariéthoz, M., Savary, J. 2004. *Des droits sur l'air? Une analyse de la gestion de l'air en Suisse sous l'angle de l'approche des régimes institutionnels de ressources naturelles,* Cahier de l'IDHEAP 213/2004. Chavannes-près-Renens: IDHEAP

Nahrath, S. 2003a. *La mise en place du régime institutionnel de l'aménagement du territoire en Suisse entre 1960 et 1990*, thèse de doctorat, IDHEAP/Université de Lausanne, http://www2.unil.ch/cyberdocuments/

Nahrath, S. 2003b. Les effets de la mise en place du régime de l'aménagement du territoire sur les rives du lac de Bienne. la colline du "Hueb" et dans l'agglomération lausannoise entre 1960 et 1990, in: Knoepfel, P. et al. (dirs.), *Institutionelle Ressourcenregime in Aktion – Régimes institutionnels de ressources naturelles en action*. Bâle: Helbing & Lichtenhahn (série Ecologie & Société, vol. 19): 59–140

Nahrath, S. 2005. Le rôle de la propriété foncière dans la genèse et la mise en oeuvre de la politique d'aménagement du territoire: quels enseignements pour la durabilité des aménagements urbains?, in: Da Cunha, A., Knoepfel, P., Leresche, J.-Ph., Nahrath, S. (éds) 2005, *Enjeux du développement urbain durable. Transformations urbaines, gestion des ressources et gouvernance*. Lausanne: Presses Polytechniques et Universitaires Romandes: 299–328

O'Connor, M. 2002. Social costs and sustainability, in: Bromley, D., Paavola, J. (eds.), *Economics, Ethics and Environmental Policy. Contested Choices*. Oxford: Blackwell: 181–201

OCDE 1998. *Examen des performances environnementales suisses*. Paris: OCDE

OFEN 2005. *Rapport CO_2*. Berne

OFS, OFEFP, ARE 2003. *Développement durable en Suisse, indicateurs et commentaires*. Neuchâtel

Oppizzi, D. 2003. *Parcs, gouvernance et durabilité. Analyse fine de trois parcs naturels et paysagers européens*, Diplôme MPA de l'IDHEAP. Chavannes-près-Renens

Ostrom, E. 1990. *Governing the Commons. The Evolution of Institutions for Collective Action*. Cambridge (UK): Cambridge University Press

Ostrom, E. 2000. Reformulating the Commons, *Swiss Political Science Review*, 6, 1: 29–52

Ostrom, E. 2002. Property Rights Regimes and Common Goods: A Complex Link, in: Héritier, A. (ed), *Common Goods: Reinventing European and International Governance*. Lanham: Rowman and Littlefield: 29–58

Radin, M. J. 1996. *Contested Commodities*. Cambridge MA.: Harvard University Press

Reynard, E., Mauch, C. 2003. Les régimes institutionnels de l'eau en Suisse: les cas du Seetal, de la Dranse de Bagnes, de la Maggia et de la Thur", in: Knoepfel, P., et al. (dirs.), I*nstitutionelle Ressourcenregime in Aktion – Régimes institutionnels de ressources naturelles en action*. Bâle: Helbing & Lichtenhahn (série Ecologie & Société, vol. 19): 205–296

Rodewald, R., Knoepfel, P. (Hrsg.) 2005. *Institutionelle Regime für nachhaltige Landschaftsentwicklung. Régimes institutionnels pour le développement durable du paysage*. Zurich, Coire: Rüegger Verlag

Rothmayr, C. 2000. *Politik vor Gericht: Implementation und Wirkung von Entscheiden des Schweizerischen Bundesgerichts in den Bereichen Fortpflanzungsmedizin, Lohngleichheit von Frau und Mann und Sonntagsarbeit*, Berne: Paul Haupt

Savary, J., Knoepfe,l P. 2005. *Revisiting Urban Mobility Policies,* conference paper at the 5th Swiss Transport Research Conference, Ascona, March 9–11, Zürich: ETHZ/EPFZ, www.strc.ch

Scharpf, F. W. 1997. *Games real actors play. Actor-centred institutionalism in policy research.* Boulder: Westviewpress

Schlager E., Ostrom, E. 1992. Property-rights regimes and natural resources: a conceptual analysis, *Land Economics*, n°3, vol.68: 249–262

Siebert, H. 1983. *Oekonomische Theorie natürlicher Ressoucen*, Tübbingen: Mohr

United Nations 2002. *Report of the World Summit on Sustainable Development, World Summit on Sustainable Development* (a/conf. 199/20), Johannesburg, South Africa, 26 August to 4 September 2002

Varone, Frédéric (2001). Comparaison des régimes institutionnels du sol, de l'eau et de la forêt, in: Knoepfel, Peter, Kissling-Näf, Ingrid, Varone, Frédéric (eds), *Régimes institutionnels de ressources naturelles: analyse comparée du sol, de l'eau et de la forêt / Institutionelle Regime für natürliche Ressourcen: Boden, Wasser und Wald im Vergleich.* Genève/Bâle/Munich: Helbing & Lichtenhahn: 183–219

Varone, F. 2002. Qu'est-ce qu'un quota juste? Les enjeux redistributifs des contingents environnementaux, in: Knoepfel, P. (Hrsg.), *Begrenzen um mehr zu erreichen, wissenschaftliche Kommentare*. Berne: EDMZ: 43–63

Varone, Frédéric, Reynard, Emmanuel, Kissling-Näf, Ingrid, Mauch, Corine 2002. Institutional Resources Regimes. The Case of Water in Switzerland, Integrated Assessment. *An International Journal*, 3(1): 78–94

Varone, Frédéric, Bisang, Kurt, Kissling-Näf, Ingrid 2003. Vergleich der regionalen Ressourcenregime: Lehren für die Nachhaltigkeit, in: Knoepfel, Peter, Kissling-Näf, Ingrid, Varone, Frédéric (eds) 2003. *Institutionelle Regime natürlicher Ressourcen in Aktion / Régimes institutionnels de ressources naturelles en action*. Genève/Bâle/Munich: Helbing & Lichtenhahn: 305–341

Varone, F. 2004. Auf dem Weg zu einer nachhaltigen Nutzung natürlicher Ressourcen?, in: Suter, C., Renschler, I., Joye, D. (Hrsg.), *Sozialbericht 2004*, Zurich: Seismo: 247–272

World Bank 1995. *Monitoring environmental progress. A report on work in progress*. Washington D.C.: World Bank

World Commission on Environment and Development 1987. *Our Common Future*, New York: Oxford

Young, M. D. 1992. *Sustainable investment and resource use*. Paris: U-NESCO

Printing: Krips bv, Meppel
Binding: Stürtz, Würzburg